Fluorinated Coatings and Finishes Handbook

The Definitive User's Guide and Databook

Laurence W. McKeen

William Andrew
publishing

Plastics Design Library and its logo are trademarks of William Andrew, Inc.

ISBN: 978-0-8155-1522-7

Library of Congress Cataloging-in-Publication Data

McKeen, Laurence W.
  Fluorinated coatings and finishes handbook : the definitive user's guide and
databook / Laurence W. McKeen.
    p. cm.
  Includes bibliographical references and index.
  ISBN-13: 978-0-8155-1522-7 (978-0-8155)
  ISBN-10: 0-8155-1522-7 (0-8155)
  1. Plastic coatings—Handbooks, manuals, etc. 2.
Fluoropolymers—Handbooks, manuals, etc. I. Title.

  TP1175.S6M45 2006
  668.4'22—dc22
                            2005033634

Printed in the United States of America

This book is printed on acid-free paper.

10 9 8 7 6 5 4 3 2 1

Published by:
William Andrew Publishing
13 Eaton Avenue
Norwich, NY 13815
1-800-932-7045
www.williamandrew.com

Dedicated to my mother (Helene) and father (Veikko)

William Andrew Publishing

Sina Ebnesajjad, Editor in Chief (External Scientific Advisor)

Contents

7 Additives

8 Substrates and Substrate Preparation

9 Liquid Formulations

10 Application of Liquid Coatings

11 Powder Coating Fluoropolymers

14 Recognizing, Understanding, and Dealing with Coating Defects

15 Commercial Applications and Uses

16 Health and Safety

Appendix I: Chemical Resistance of Fluoropolymers

Appendix II: Permeability of Fluoropolymers

Appendix III: Permeation of Automotive Fuels Through Fluoroplastics

Appendix IV: Permeation of Chemicals Through Fluoroplastics

PDL Fluorocarbon Series Editor's Preface

The original idea for the *Fluorocarbon Series* was conceived in the mid 1990s. Two important rationales required the development of the collection. First, there were no definitive sources for the study of fluorinated polymers, particularly the commercial products. A researcher seeking the properties and characteristics of fluorinated plastics did not have a single source to use as a reference. Information available from commercial manufacturers of polymers had long been the source of choice. Second, waves of the post-war generation (a.k.a., Baby Boomers) were beginning to retire, thus eroding the available knowledge base in the industry and academia.

The scope of the series has been expanded over time to incorporate other important fluorinated materials. Selection of the topics of the books has been based on the importance of practical applications. Inevitably, a number of fluorinated compounds, important in their own right, have been left out of the series. In each case, the size of its audience has been found simply too small to meet the economic hurdles of publishing.

The first two books of the series cover commercial fluoropolymers (ethylenic); the third book is focused on their applications in the chemical processing industries. The fourth book covers fluoroelastomers, the fifth fluorinated coatings and finishes, and the sixth book is about fluorinated ionomers, such as Nafion®. The seventh handbook represents an extension of the scope of the series to non-polymeric materials. It addresses the preparation, properties, and uses of fluorinated chemicals as refrigerants, fire extinguishers, blowing agents, and cleaning gases. All of the titles in the *PDL Fluorocarbon Series* appear in the back of this book in the PDL Library list.

The authors of the handbooks are leaders in their fields who have devoted their professional careers to acquiring expertise. Each book is a product of decades of each author's experience and research into the available body of knowledge. Our hope is that these efforts will meet the needs of the people who work with fluorinated polymers and chemicals. Future revisions are planned to keep this series abreast of progress in the field.

Sina Ebnesajjad   April 2006

Paints and coatings are much more complex than initially meets the eye. My first introduction to paint chemistry came as I was job hunting in my final year of graduate school at the University of Wisconsin. I was trained there as a theoretical and experimental analytical chemist, at the time specializing in high-resolution mass spectroscopy. DuPont called and offered an interview at their technical center of paint technology, Marshall Laboratory in Philadelphia, Pennsylvania. They told me it was their premier paint and coating development laboratory. I remember thinking, "paint…how boring," and I almost did not accept the offer to interview. Fortunately, I gave it a chance and learned how truly complex these paint mixtures were. Interactions among all the components complicate the systems beyond theoretical modeling and analysis in most cases. I also learned that there were uses or functions of paints and coatings beyond appearance and rust protection, and that there were dozens of ways to apply the coatings.

While interviewing for a job in the Marshall Lab's analytical chemistry section, I learned how complex the analytical chemistry of paints was due to the complex mixtures formed in them. Part of my initial assignment was focused on developing ways of analyzing unknown liquid and solid paint samples, including competitive product analysis. This work was enjoyable and interesting. One would be surprised at what can be learned from a *paint chip*. Within a couple years, I expressed an interest in the formulation of paints. DuPont reassigned me to the Teflon® finishes group, where I spent the next twenty-five years learning what I am communicating in this work.

In contrast to the earlier works in the PDL Fluorocarbon Series, this book is not only a reference book but also more tutorial in nature. It teaches the practice and theory of fluorinated coatings at every step. There are literally thousands of people using thousands of different fluoropolymer coatings globally. Much of my time with the DuPont customers has been devoted to teaching the basics of paint formulation, how it impacts their coating processes, and why things did not always work as expected. The contents of this book should help them understand some of the problems encountered, interpret their own observations, and arrive at possible solutions.

A better understanding will also lead to better communication between the coaters and coatings suppliers, which can shorten product and process development time.

Most of the chapters have very few references because what I have included is what I have learned during twenty-five years developing, formulating, and studying fluoropolymer coatings around the world. There are lots of potential references in the public literature, but I did not use them. In sections where I discuss some specific products or product lines, the information reflects what was considered up to date at the time of publication. Product lines are continually modified, added, and eliminated.

The first chapter introduces the fluoropolymers that have been used or could be used in coatings. The structures are given and the basic properties are summarized. Chapter 2 provides background on how fluoropolymers are made and finished, and I have attempted to present the chemistry as simply and clearly as possible.

Chapter 3 is an introduction to paint or coating formulation. It introduces the next four chapters of this text, which discuss the components or ingredients of paints and coatings. Chapter 4 discusses binders, the non-fluoropolymers that are often used in fluorocoatings. It discusses the structures and properties of these materials. The question, "How do you get a non-stick coating to stick to the substrate?" is also answered. Chapter 5 deals with pigments and fillers. The dispersion processes used to prepare them for use in coatings is summarized. Chapter 6 completes the components of paint chapters with an in-depth discussion about solvents. Chapter 7 is a summary of additives, those minor paint components that often make or break the commercial success of a fluorocoating.

Chapter 8 delves into preparing the substrate, the item being coated. This is a critical chapter as substrate preparation impacts adhesion and eventually the performance of the coated item.

Chapter 9 focuses on liquid fluorocoatings. Included in this chapter are some guidelines on making coatings, both aqueous and non-aqueous, and preparing them for use. There are also tables of commercial products with technology and property information when it is publicly available. Chapter 10 summarizes the different ways liquid coatings are applied and what formulation factors are important in each application method.

Chapter 11 focuses on powder coatings. It discusses the powder formulations, commercial powder products, and the application of powders, including the equipment and how it works.

Chapter 12 is called "Fluoropolymer Coating Processing Technology," but it essentially is about baking or curing the fluorocoatings. Included in this chapter are details about what happens during the bake, along with the different technologies used for baking and monitoring that process.

Chapter 13 moves on to coating performance. Many of the performance tests used by coating manufacturers and end-users are discussed. Some of these are specification tests. Others are important in helping a potential customer select a coating. Included are references to standard test protocols. Chapter 14 is related to performance, but it deals specifically with coating defects. Many photographic examples are included along with an explanation of what causes the defects and possible ways to fix them.

Chapter 15 contains a brief history of fluorocoating development and summaries of many commercial uses of fluorocoatings. A history of coating development starts the chapter. Summaries of many uses of fluorocoatings are also included.

Chapter 16 is a discussion of many of the health and safety aspects of using and applying fluorocoatings.

This work closes with a series of appendices that contain useful details about the resistance of several fluoropolymers to many chemicals and solvents. There is also a collection of select permeation data. Together these appendices are useful in determining the fitness for use of coatings based on the various fluoropolymers in different exposure environments.

I am a teacher at heart. I hope you will find this work instructive and useful.

Larry McKeen   December, 2005
Sewell, New Jersey

Acknowledgments

This book is a summary of what I have learned over more than twenty-five years at DuPont. Many colleagues' and coworkers' direct and indirect help has lead to my understanding of the fluorinated coating technologies. Some of the them include, in no particular order, Mr. Michael Witsch, Mr. William McHale, Mr. W. Douglas Obal, Dr. Paul Noyes, Dr. Hank Jakubauskas, Dr. Michael Fryd, Mr. Craig Hennessey, Mr. June Uemura, Mr. Fumio Inomae, Mr. Thomas Concannon (deceased), Dr. Milt Misogianes, Dr. Peter Huesmann, Dr. Seymour Hochberg, Mr. Luk D'Haenens, Mr. Philippe Thomas, Dr. Cliff Strolle (deceased), and Dr. Kenneth Leavell. Dr. Charles DeBoer persuaded me to come to DuPont and was a mentor to me for years. If I have overlooked anyone, please forgive me and drop the publisher a note for correction in future editions.

A number of teachers have greatly influenced my education and deserve special thanks. Ms. Anna Kruse, my high school chemistry teacher (Lyman Hall High School, Wallingford, CT), not only was a great educator but also motivated this particular young student to study chemistry beyond the classroom. That included writing articles for an educational chemistry magazine while in high school, and after-school projects, one that eventually led to the 1969 International Science Fair. The many outstanding chemistry teachers at Rensselaer Polytechnic Institute provided the best and most thorough undergraduate chemistry education in the country. Finally, my major professor at the University of Wisconsin, Professor James W. Taylor, was a great educator and ultimately developed my teaching abilities.

I have taught much to our customers, but they probably do not realize how much more I have learned from them. Many of the sections in this book have resulted from lectures and discussions on those subjects with customers. Their questions have helped focus and crystallize this work and my understanding of the technology. The anticipation and interest of this work has been a constant motivation for me to complete it and I thank them all for that.

The author is especially appreciative of the confidence, support, and patience of my friend and editor Sina Ebnesajjad. Without his support and encouragement, my work on this book would have been further delayed or perhaps abandoned. He was also the primary proofreader of the manuscript.

The support for this book from my direct management, Dr. William Raiford and Dr. George H. Senkler, is sincerely appreciated.

I am sincerely thankful for the support and patience from my editor at William Andrew Publishing, Ms. Millicent Treloar. I thank Ms. Jeanne Roussel and her staff at Write One for not only composing the manuscript into a book, but for the final proof reading. They noticed numerous passages that were not clearly explained. These rewritten passages enhanced the clarity of the book.

My family has been particularly supportive through the long hours of writing and research from my home office. My wife, Linda, has been behind this work 100 percent. My children Lindsey (a junior Biomedical Engineering/Premed major at Rensselaer Polytechnic Institute), Michael (a freshman business major at James Madison University), and Steven (a high school junior at St. Augustine Prep) were directly supportive through their interest and by constantly inquiring about the book's status.

1.1 Introduction

Fluoropolymer coatings are widely used in many industries, though the consumer and many engineers and scientists are only aware of their use as nonstick coatings for cookware. There are hundreds of applications, some of which are discussed later in this book. This work will be useful to students, engineers, paint applicators, and the end-users that buy and specify fluoropolymer-coated parts.

This book aims to:

1. Provide information on coating formulation including what is in a fluoropolymer paint and why.

2. Provide guidelines on the performance of the various types of coatings to aid in selection or provide an understanding on why some coatings work or do not work in an application.

3. Provide collections of data on raw materials for potential formulators.

4. Provide application and curing information.

The first two volumes of this series cover in detail the technology of the fluorinated polymers that are used as raw materials in fluorinated coatings. It is important to understand what these important raw materials are and what their properties are because they are imparted to the coatings in which they are used. One may also want to understand the shortcomings or limits of the finishes. The first two volumes of this series cover in great detail the chemistry and use of the raw fluoropolymers.[1][2] Those texts provide extensive property comparisons. This book on coatings could not possibly contain performance properties of all fluorinated coatings because they number in the thousands.

While there are literally thousands of fluorine-containing polymers described in the technical and academic literature and those known only in the laboratory, this work will focus on the commercially available materials.

1.2 The Discovery of Fluoropolymers

The story of the discovery of fluoropolymers by DuPont has been related by many. It began not by a well-designed purposeful experiment based on polymerization or organic chemistry. The first fluoropolymer was discovered by accident. Dr. Roy Plunkett (earned his Ph.D. at Ohio State University in 1936) of the DuPont Company was the image of a classical chemist: ingenious, but more important, observant.[3] In 1938, he had been at DuPont for only two years and was doing research on the development of fluorinated refrigerants in a joint venture of DuPont and General Motors called Kinetic Chemicals.

This work took place in what is now known as DuPont's Jackson Laboratory on the site of the Chambers Works in New Jersey across the Delaware River from Wilmington. It is now in the shadow of the Delaware Memorial Bridge. He was experimenting with tetrafluoroethylene (TFE) looking for a synthetic route to a useful refrigerant ($CClF_2-CHF_2$).[4] The effort was spurred by the desire to create safe, nonflammable, nontoxic, colorless, and odorless refrigerants.

On the morning of April 6, 1938, when Plunkett checked the pressure on a cylinder of TFE that he was certain was full, he found no pressure. Many chemists might have thought the cylinder had leaked its contents, but the observant Plunkett noted that the cylinder had not lost weight. Charles Pederson reflected on this day in his Nobel Laureate lecture of December 8, 1987, "…. I noticed commotion in the laboratory of Roy Plunkett, which was across the hall from my own. I investigated and witnessed the sawing open of a cylinder from which was obtained the first sample of Teflon® fluoropolymer." After opening, shaking the cylinder upside down yielded a few grams of a waxy looking white powder—the first polymer of *tetrafluoroethylene*. (See Fig. 1.1.)

Plunkett analyzed the white powder and it was conclusively proven to be *polytetrafluoroethylene*

(PTFE). The slippery PTFE could not be dissolved in any solvent, acid, or base, and upon melting formed a stiff clear gel without flow. Later, research led to the discovery of processing techniques similar to those used with metal powders (sintering). At the time, the Manhattan Project was seeking new corrosion-resistant materials for gaskets, packings, and liners for UF_6 handling. PTFE provided the answer and was used in production. The US government maintained a veil of secrecy over the PTFE project until well after the end of World War II. For security reasons, it was called by a code name, K 416. The US government took all of the output from the small, heavily guarded, production plant located in Arlington, New Jersey. Commercial production was therefore delayed until 1947. Plunkett patented it in US Patent 2,230,654 on February 4, 1941. Because of the secrecy, no other patents issued until Brubaker's patent 2,393,967 in 1947.

Large-scale monomer synthesis and controlled polymerization were technical problems to be resolved. Intensive studies solved these problems and small-scale production of Teflon® (trademarked in 1944, but not publicly revealed until 1946) began in 1947. In 1950, DuPont scaled up the commercial production of Teflon® in the USA with the construction of a new plant in Parkersburg, West Virginia. In 1947, Imperial Chemical Industries built the first PTFE plant outside the US, in Western Europe. Since then, many more plants have been built around the globe. Over the last six decades, many forms of PTFE and copolymers of other monomers and TFE have been developed and commercialized.

The words of Plunkett himself best summarize the discovery of PTFE. He recounted the story of Teflon® in a talk to the American Chemical Society at its April 1986 meeting in New York. "The discovery of polytetrafluoroethylene (PTFE) has been variously described as (i) an example of serendipity, (ii) a lucky accident, and (iii) a flash of genius. Perhaps all three were involved. There is complete agreement, however, on the results of that discovery. It revolutionized the plastics industry and led to vigorous applications not otherwise possible." The discovery created a new industry, with annual sales of fluoropolymers over $2 billion and value-added to tens of billions dollars. Fluoropolymers have entered every aspect of human life, often unseen yet functional, and the extension of fluorine chemistry has revolutionized many industries from pharmaceutical to prosthetics to crop protection to space craft.

1.3   What are Fluoropolymers?

The following sections briefly explain the structures and properties of the various fluoropolymers. It is important to keep in mind there are variations of most of these polymers. The most common variation is the molecular weight, which will affect the melting point somewhat, and the viscosity of the polymer above its melt point, properties that are very important in determining processing conditions and use.

1.3.1   Perfluorinated Polymers

Traditionally, a fluoropolymer or fluoroplastic is defined as a polymer consisting of carbon (C) and fluorine (F). Sometimes these are referred to as *perfluoropolymers* to distinguish them from partially fluorinated polymers, fluoroelastomers, and other polymers that contain fluorine in their chemical structure. For example, fluorosilicone and fluoroacrylate polymers are not referred to as fluoropolymers.

1.3.1.1   *Polytetrafluoroethylene (PTFE)*

An example of a linear fluoropolymer is the tetrafluoroethylene polymer (PTFE) discovered by Plunkett. Its structure in simplistic form is:

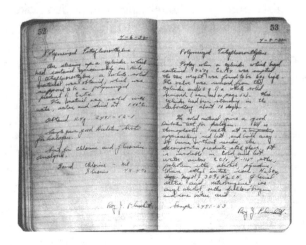

Figure 1.1 Photograph of two pages from the research notebook of Dr. Roy Plunkett recording the discovery of polytetrafluoroethylene on April 6, 1938. *(Courtesy of DuPont.)*

Formed by the polymerization of tetrafluoroethylene (TFE), the (CF_2–CF_2–) groups repeat many thousands of times. To understand some of the properties of PTFE, the structure is better viewed in a ball and stick model or a space filling three-dimensional model, which are shown in Fig. 1.2. It is important to understand the basic properties of fluoropolymers when learning about coating properties. The polymer properties are retained to varying degrees in the formulated finishes. The fundamental properties of fluoropolymers evolve from the atomic structure of fluorine and carbon and their covalent bonding in specific chemical structures. Because PTFE has a linear structure, it is a good subject for discussion of extreme properties. The backbone is formed of carbon-carbon bonds and carbon-fluorine bonds. Both are extremely strong bonds (C–C = 607 kJ/mole and C–F = 552 kJ/mole).

The basic properties of PTFE stem from these two very strong chemical bonds. PTFE as pictured in Fig. 1.2 shows a rodlike shape even though it shows only a small section of the molecule. The size of the fluorine atom allows the formation of a uniform and continuous covering around the carbon-carbon bonds and protects them from attack, thus imparting chemical resistance and stability to the molecule. PTFE is rated for use up to 500°F. PTFE does not dissolve in any common solvent. The fluorine sheath is also responsible for the low surface energy (18 dynes/cm) and low coefficient of friction (0.05–0.8, static) of PTFE. Another attribute of the uniform fluorine sheath is the electrical inertness (or non-polarity) of the PTFE molecule. Electrical fields impart only slight polarization in this molecule, so volume and surface resistivity are high.

The PTFE molecule is simple and is quite ordered, so it can align itself with other molecules or other portions of the same molecule. These three dimension representations are particularly useful in showing this. Disordered regions are called *amorphous* regions. This is important because polymers with high crystallinity require more energy to melt. In other words, they have higher melting points.

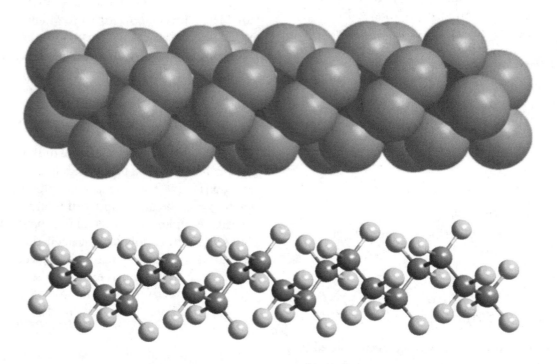

Figure 1.2 Three-dimensional representations of polytetrafluoroethylene (PTFE).

When this happens, it forms what is called a *crystalline* region. Crystalline polymers have a substantial fraction of their mass in the form of parallel, closely packed molecules. High molecular weight PTFE resins have high crystallinity and therefore very high melting points, typically as high as 320°C–342°C (608°F–648°F). The crystallinity in PTFE is typically 92%–98%.[5] Further, the viscosity in the molten state (called *melt viscosity*) is so high that high molecular weight PTFE particles do not melt and flow out. They sinter much like powdered metals; they stick to each other at the contact points. Reducing the molecular weight can significantly change this and these materials are be discussed in Ch.2, Sec. 2.4.4.

The polymerization process must have a start and an end (*initiation* and *termination*). Therefore, there are endgroups that may not be much like the rest of the polymer. Endgroups are present in small concentrations when the molecular weight of the polymer is high; however, their presence and type can affect the degradation behavior.

PTFE is called a *homopolymer*, a polymer made from a single monomer. Recently many PTFE manufacturers have been adding very small amounts of other monomers to their PTFE polymerizations to produce alternate grades of PTFE designed for specific applications. Generally, polymers made from two monomers are called *copolymers*, but fluoropolymer manufacturers still call these grades homopolymer. DuPont grades of this type are called Teflon® NXT Resins. These modified granular PTFE materials retain the exceptional chemical, thermal, anti-stick, and low-friction properties of conventional PTFE resin, but offer some improvements:

1. Improved permeation resistance
2. Less creep
3. Smoother, less porous surfaces
4. Better high-voltage insulation

The copolymers described in the next sections contain significantly more of the non-TFE monomers.

1.3.1.2 *Fluorinated Ethylene Propylene (FEP) Copolymer*

If one of the fluorine atoms on tetrafluoroethylene is replaced with a trifluoromethyl group ($-CF_3$)

then the new monomer is called *hexafluoropropylene* (HFP). Polymerization of monomers (HFP) and TFE yield a different fluoropolymer called *fluorinated ethylene propylene* (FEP). The number of HFP groups is typically five percent or less.

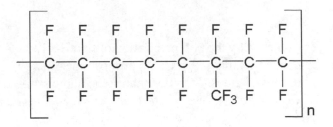

To understand some of the properties of FEP, the structure is better viewed in a ball and stick model or a space filling three-dimensional model as shown in Fig. 1.3.

The effect of this small change is to put a "bump" along the polymer chain. This disrupts the crystallization of the FEP, which has a typical as-polymerized crystallinity of 70% versus 98% for PTFE. It also lowers its melting point. The reduction of the melting point depends on the amount of trifluoromethyl groups added and secondarily on the molecular weight, but most FEP resins melt around 274°C (525°F), though lower melting points are possible. Even high molecular weight FEP will melt and flow. The high chemical resistance, low surface energy, and good electrical insulation properties of PTFE are retained.

1.3.1.3 *Perfluoroalkoxy (PFA) Polymers*

Making a more dramatic change in the side-group, chemists put a perfluoroalkoxy group on the polymer chain. This group is signified as $-O-R_f$, where R_f can be any number of totally fluorinated carbons. A typical one is perfluoropropyl ($-O-CF_2-CF_2-CF_3$). These polymers are called PFA and the perfluoroalkylvinylether group is typically added at only a couple mole percent. Another common a perfluoroalkoxy group is perfluoromethylvinylether ($-O-CF_3$) making a polymer called MFA.

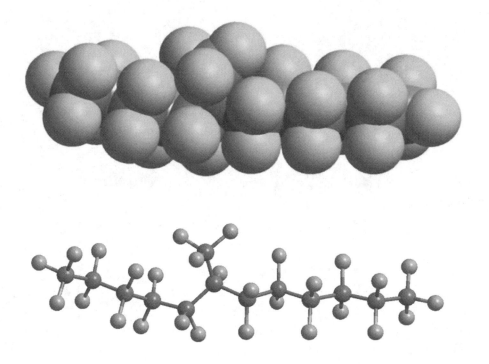

Figure 1.3 Three-dimensional representations of fluorinated ethylene propylene (FEP).

is still perfluorinated as with FEP the high chemical resistance, low surface energy and good electrical insulation properties are retained.

1.3.1.4 Teflon AF®

A perfluorinated polymer made by DuPont called Teflon AF® breaks down the crystallinity completely, hence its designation *amorphous fluoropolymer* (AF). It is a copolymer (Fig. 1.5) made from 2,2-bistrifluoromethyl-4,5-difluoro-1,3-dioxole (PPD) and TFE.

To understand some of the properties of PFA, the structure is better viewed in a ball and stick model or a space filling three-dimensional model as shown in Fig. 1.4.

The large side group reduces the crystallinity drastically. The melting point is generally between 305°C–310°C (581°F–590°F) depending on the molecular weight. The melt viscosity is also dramatically dependent on the molecular weight. Since PFA

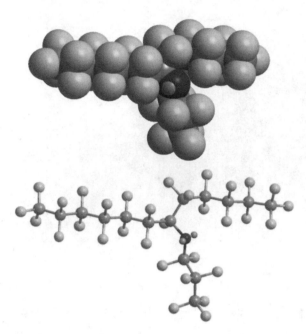

Figure 1.4 Three-dimensional representations of perfluoroalkoxy (PFA).

The Teflon AF® family of amorphous fluoropolymers is similar to Teflon® PTFE and PFA in many of the usual properties but is unique in the following ways. It is:

1. A true amorphous fluoropolymer.

2. Somewhat higher coefficient of friction than Teflon® PTFE and PFA

3. Excellent mechanical and physical properties at end-use temperature up to 300°C (572°F).

4. Excellent light transmission from ultraviolet (UV) through a good portion of infrared (IR).

5. Very low refractive index.

6. Lowest dielectric constant of any plastic even at gigahertz frequencies.

7. Solubility to a limited extent in selected perfluorinated solvents.

Teflon® AF can be designed to have some solubility in selected perfluorinated solvents but remains chemically resistant to all other solvents and process chemicals. The solubility is typically only 3% to 15% by weight, but this allows you to solution-cast extremely thin coatings in the submicron thickness range.

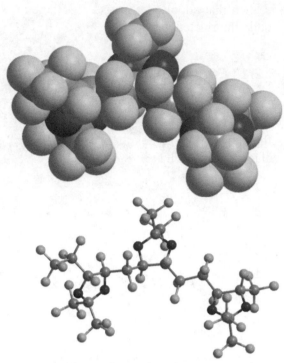

Figure 1.5 Three dimensional representations of Teflon® AF.

1.3.1.5 Other Fully Fluorinated Polymers

Inevitably, polymerization of more than two monomers would be made by the polymer chemists. Terpolymer is the term generally applied to these kinds of polymers. There are a few commercial examples. One such terpolymer is a polymer of tetrafluoroethylene, hexafluoropropylene, and vinylidene fluoride. It provides a combination of performance advantages different than any other melt processable fluorothermoplastic. It is called Dyneon® THV® fluorothermoplastic. Another terpolymer is made from hexafluoropropylene, tetrafluoroethylene, and ethylene and is called Dyneon® THE®. Another example is ETFE, which usually contains a small amount of PFBE to enhance stress crack resistance and mechanical properties.

1.3.2 Partially Fluorinated Polymers

If monomers have other atoms such as hydrogen or chlorine in place of any fluorine atoms then homopolymers made from that monomer are considered partially fluorinated polymers. Similarly, if a copolymer has a monomer that contains other atoms, then that copolymer is also considered to be partially fluorinated.

1.3.2.1 Ethylene-Tetrafluoroethylene (ETFE) Copolymers

ETFE is a copolymer of ethylene and tetrafluoroethylene (Fig. 1.6). The following is the basic molecular structure of ETFE:

$$\left[\begin{array}{cccccccc} F & F & H & H & F & F & H & H \\ | & | & | & | & | & | & | & | \\ C-C-C-C-C-C-C-C \\ | & | & | & | & | & | & | & | \\ F & F & H & H & F & F & H & H \end{array}\right]_n$$

This simplistic structure shows alternating units of TFE and ethylene. While this can be readily made, many grades of ETFE vary the ratio of the two monomers to optimize properties for specific end uses.

ETFE is a fluoroplastic with excellent electrical and chemical properties. It also has excellent mechanical properties. ETFE is especially suited for uses requiring high mechanical strength, chemical, thermal, and/or electrical properties. The mechanical properties of ETFE are superior to those of PTFE and FEP. ETFE has:

1. Excellent resistance to extremes of temperature, ETFE has a working temperature range of –200°C to 150°C.

2. Excellent chemical resistance.

3. Mechanical strength ETFE is good with excellent tensile strength and elongation and has superior physical properties compared to most fluoropolymers.

4. With low smoke and flame characteristics, ETFE is rated 94V–0 by the Underwriters Laboratories Inc. It is odorless and non-toxic.

5. Outstanding resistance to weather and aging.

6. Excellent dielectric properties.

7. Non-stick characteristics.

1.3.2.2 Polyvinylidene Fluoride (PVDF)

The polymers made from 1,1-di-fluoro-ethene (or vinylidene fluoride) are known as polyvinylidene fluoride (PVDF). They are resistant to oils and fats, water and steam, and gas and odors, making them of particular value for the food industry. PVDF is known for its exceptional chemical stability and excellent resistance to ultraviolet radiation. It is used chiefly in the production and coating of equipment used in aggressive environments, and where high levels of mechanical and thermal resistance are required. It has also been used in architectural applications as a coating on metal siding where it provides exceptional resistance to environmental exposure. The chemical structure of PVDF is:

$$\left[\begin{array}{cccccccc} F & H & F & H & F & H & F & H \\ | & | & | & | & | & | & | & | \\ C-C-C-C-C-C-C-C \\ | & | & | & | & | & | & | & | \\ F & H & F & H & F & H & F & H \end{array}\right]_n$$

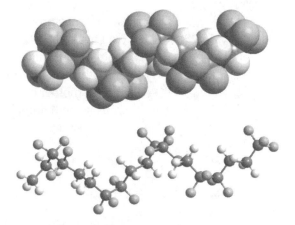

Figure 1.6 Three-dimensional representations of ETFE.

One of the tradenames of PVDF is KYNAR®. The alternating CH_2 and CF_2 groups along the polymer chain (see Fig. 1.7) provide a unique polarity that influences its solubility and electric properties. At elevated temperatures PVDF can be dissolved in polar solvents such as organic esters and amines.

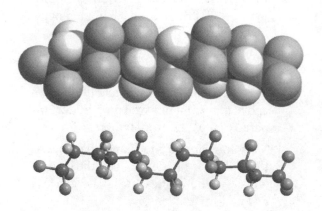

Figure 1.7 Three-dimensional representations of PVDF.

This selective solubility offers a way to prepare corrosion resistant coatings for chemical process equipment and long-life architectural finishes on building panels.

Key attributes of PVDF include:

1. Mechanical strength and toughness
2. High abrasion resistance
3. High thermal stability
4. High dielectric strength
5. High purity
6. Readily melt processable
7. Resistant to most chemicals and solvents
8. Resistant to ultraviolet and nuclear radiation
9. Resistant to weathering
10. Resistant to fungi
11. Low permeability to most gases and liquids
12. Low flame and smoke characteristics

Architectural coatings contain a minimum of 70% by weight of PVDF resin. The coating is usually applied in a factory by coil coating where it can be accurately baked. The coated flat panels are formed by bending or stamping before use. The coating provides the following properties in this application:

1. Color retention
2. Chalk resistance
3. Corrosion resistance
4. Flexibility

5. Stain resistance
6. Overall exterior durability

1.3.2.3 Polyvinyl Fluoride (PVF)

PVF is a homopolymer vnyl fluoride. The following is the molecular structure of PVF and the three-dimensional representation is shown in Fig. 1.8:

$$\left[\begin{array}{cccccccc} F & H & H & H & F & H & H & H \\ | & | & | & | & | & | & | & | \\ -C & -C & -C & -C & -C & -C & -C & -C- \\ | & | & | & | & | & | & | & | \\ H & H & F & H & H & H & F & H \end{array}\right]_n$$

DuPont is the only known manufacturer of this polymer they call Tedlar®. The structure above shows a head-to-tail configuration of the CF monomer; there are no fluorines on adjacent carbons. But in reality, vinyl fluoride polymerizes in both head-to-head and head-to-tail configurations. DuPont's commercial PVF contains 10%–12% of head-to-head and tail-to-tail units, also called *inversions*.[6]

PVF has excellent resistance to weathering, staining, and chemical attack (except ketones and esters). It exhibits very slow burning and low permeability to vapor. It's most visible use in on the interiors of the passenger compartments of commercial aircraft.

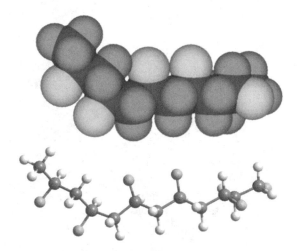

Figure 1.8 Three-dimensional representations of PVF.

1.3.2.4 Ethylene-Chlorotrifluoro-ethylene (E-CTFE) Copolymer

E-CTFE is a copolymer of ethylene and chlorotrifluoroethylene. The following is the molecular structure of E-CTFE (see also Fig. 1.9):

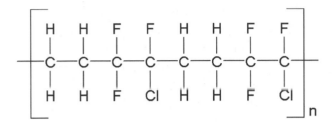

This simplified structure shows the ratio of the monomers being 1–1 and strictly alternating, but this is not required. Commonly known by the tradename, Halar®, E-CTFE is an expensive, melt processable, semicrystalline, whitish semi-opaque thermoplastic with good chemical resistance and barrier properties. It also has good tensile and creep properties and good high frequency electrical characteristics. Applications include chemically resistant linings, valve and pump components, barrier films, and release/vacuum bagging films.

1.3.2.5 Chlorotrifluoroethylene (CTFE) Polymers

CTFE is a homopolymer of chlorotrifluoroethylene, characterized by the following chemical formula (Fig. 1.10 shows a 3-D view):

The addition of the one chlorine bond contributes to lower the melt viscosity to permit extrusion and injection molding. It also contributes to the transparency, the exceptional flow, and the rigidity characteristics of the polymer. Fluorine is responsible for its chemical inertness and zero moisture absorption. Therefore, CTFE has unique properties. Its resis-

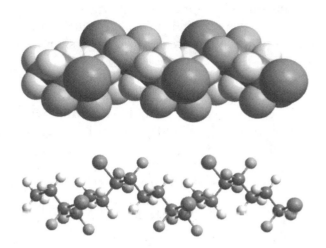

Figure 1.9 Three-dimensional representations of E-CTFE.

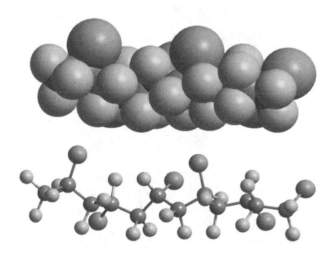

Figure 1.10 Three-dimensional representations of CTFE.

tance to cold flow, dimensional stability, rigidity, low gas permeability, and low moisture absorption is superior to any other fluoropolymer. It can be used at very low temperatures.

1.3.2.6 Fluoroalkyl Modified Polymers

There are many polymers reported in the academic literature and by industry that take standard polymers like polyesters, acrylics, and polyimides and include monomers containing perfluorinated side groups.[7] These perfluorinated side groups frequently impart property improvements even when the amount of fluorine added is small. The groups tend to concentrate themselves on coating surfaces. Partially fluorinated fluoropolymers are significantly different from the perfluoropolymers with respect to properties and processing characteristics. For example, perfluoropolymers are more thermally stable but physically less hard than partially fluorinated polymers. The former has much higher "hardness" than the latter.

While there are literally thousands of fluorine containing polymers described in technical literature, and those known only in the laboratory, this section focuses only on commercially available materials. There are a series of fluoropolymer coatings aimed at architectural applications. Some of these are coatings based on mixtures of non-fluoropolymers with a fluoropolymer such as Kynar®, but a couple are tailored special fluoropolymers.

1.   PPG: Coraflon®

2.   Asahi Glass: Lumiflon®

One polymer deserves special mention, Lumiflon®.

1.3.2.7 Lumiflon®, Coraflon®, ADS (Air-Dried System), FEVE

Lumiflon® is the tradename applied to a series of optimized molecules generally called polyfluoroethylene/vinyl ether, or FEVE.[8] The general structure is shown below.

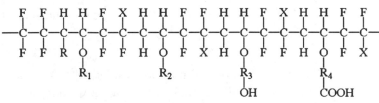

It's solubility, properties, and performance in use can be optimized for individual applications by varying the R-groups and X-groups in this structure. The fact that it is soluble and crosslinkable is significant, and it also can be formulated into air-dry or low-bake coatings.

When properly designed the molecule has these characteristics:

1. Excellent weatherability. It retains a lot of the fluoropolymer properties such as good chemical stability and has excellent weatherability as expected from many common hydrocarbon based coating resins.

2. The polymer can be dissolved in several organic solvents.

3. It can be designed to give you a choice of curing conditions from ambient temperatures to high temperatures.

4. It has good appearance, it is a transparent fluororesin, so both clear film and light solid color are possible including high gloss.

1.4   Comparison of Fluoropolymer Properties

The following tables show comparisons of the mechanical properties (Tables 1.1–1.2), electrical and thermal properties (Tables 1.3–1.4), and chemical properties (Tables 1.5–1.6) of some fluoropolymers used in coatings.

Table 1.1 Mechanical Properties of Homofluoropolymers Used in Coatings

Property	Test	PTFE	PVDF	PVF	CTFE
Specific Gravity (g/cm³)	ASTM D792	2.14-2.22	1.78	1.37-1.39	2.1-2.18
Tensile Strength (MPa)	ASTM D638	20-35	31-52	55-110	31-41
Break Elongation (%)	ASTM D638	300-550	500-250	90-250	80-250
Tensile Modulus (MPa)	ASTM D638	550	1040-2070	2100-2600	1300-1800
Flexural Strength (MPa)	ASTM D790	No Break	45-74		
Flexural Modulus (Mpa at 23°C)	ASTM D790	340-620	1140-2240	1400	1600
Static Coefficient of Friction	ASTM D621	0.1	0.2-0.4		

Table 1.2 Mechanical Properties of Cofluoropolymers Used in Coatings

Property	Test	FEP	PFA	AF	ETFE	E-CTFE	THV
Specific Gravity (g/cm³)	ASTM D792	2.15	2.15		1.71	1.68	1.95-1.98
Tensile Strength (MPa)	ASTM D638	20-28	20-26	24.6-27	45	48	23-24
Break Elongation (%)	ASTM D638	300	300	3-40	150-300	200	500-600
Tensile Modulus (MPa)	ASTM D638	345	276	950-2150	827	1400-1600	
Flexural Strength (MPa)	ASTM D790	No Break			38		
Flexural Modulus (Mpa at 23°C)	ASTM D790	655	551		1034-1171	2000	83-207
Static Coefficient of Friction	ASTM D621	0.2	0.2		0.4		0.8

Table 1.3 Electrical and Thermal Properties of Homofluoropolymers Used in Coatings

Property	Test	PTFE	PVDF	PVF	CTFE
Heat Distortion (°C @ 0.45 MPa)	ASTM D648	122	140-174	120	126
Coefficient of Thermal Expansion (cm/cm/°C×10^5)	ASTM D696	12.6-18	7-15	5-10	7
Continuous Use Temperature (°C)	UL-Sub 94	260	120	120	120
Volume Resistivity (ohm-cm)	ASTM D257	>10^{18}	>10^{14}	10^{13}	>10^{18}
Dielectric Strength (kV/mm)	ASTM D149	19.7	63-67	20	48
Melting Point (°C)	ASTM D4591	320-340	155-192	190	210-215
Melt Viscosity (Pa sec)		10^{10}-10^{12}	0.2-17×10^3		1-10

Table 1.4 Electrical and Thermal Properties of Cofluoropolymers Used in Coatings

Property	Test	FEP	PFA	AF	ETFE	E-CTFE	THV
Heat Distortion (°C @ 0.45 Mpa)	ASTM D648	70	74		81	115	
Coefficient of Thermal Expansion (cm/cm/°C×10^5)	ASTM D696	8.3-10.4	13.7-20.7	8-10	13.1-25.7	8-14	
Continuous Use Temperature (°C)	UL-Sub 94	204	260	290	150		
Volume Resistivity (ohm-cm)	ASTM D257	>10^{18}	>10^{18}		>10^{17}		>10^{15}
Dielectric Strength (kV/mm)	ASTM D149	19.7	19.7		14.6		
Melting Point (°C)	ASTM D4591	260-282	302-310		225-280	240	115-180
Melt Viscosity (Pa sec)			10^4-10^5	10^3-10^4	0.7-3×10^3		
Melt Flow Rate (g/10 min)	ASTM D1238	0.8-27	1-38		2.3-45	1-50	10-20

Table 1.5 Chemical Properties of Homofluoropolymers Used in Coatings*

Property	PTFE	PVDF	PVF	CTFE
Water Absorption (24 hr, weight change %)	0	0.04	0.05	0.01-0.10
Aromatic Hydrocarbon Resistance	Excellent	Excellent		Excellent
Aliphatic Hydrocarbon Resistance	Excellent	Excellent		Excellent
Chlorinated Solvent Resistance	Excellent	Excellent		Good
Ester & Ketone Resistance	Excellent	Good		Excellent
Refractive Index	1.38	1.42	1.46	1.44

*Detailed chemical resistance data are available in Appendix I.

Table 1.6 Chemical Properties of Fluoropolymers Used in Coatings*

Property	FEP	PFA	AF	ETFE	E-CTFE
Water Absorption (24 hr, weight change %)	<0.01	<0.03	0	<0.03	<0.02
Aromatic Hydrocarbon Resistance	Excellent	Excellent		Excellent	
Aliphatic Hydrocarbon Resistance	Excellent	Excellent		Excellent	
Chlorinated Solvent Resistance	Excellent	Excellent		Excellent	
Ester & Ketone Resistance	Excellent	Excellent		Excellent	
Refractive Index	1.344	1.34	1.29-1.31	1.403	

*Detailed chemical resistance data are available in Appendix I.

REFERENCES

1. Ebnesajjad, S., *Fluoroplastics,* Vol. 1: *Non-Melt Processible Fluoroplastics, The Definitive User's Guide and Databook,* William Andrew, Inc., Norwich, NY (2000)

2. Ebnesajjad, S., *Fluoroplastics,* Vol. 2: *Melt Processible Fluoroplastics, The Definitive User's Guide and Databook,* William Andrew, Inc., Norwich, NY (2003)

3. Plunkett, R. J., US Patent 2,230,654, assigned to DuPont Co. (Feb. 4, 1941)

4. Plunkett, R. J., The History of Polytetrafluoroethylene: Discovery and Development, in: High Performance Polymers: Their Origin and Development, *Proc. Symp. Hist. High Perf. Polymers,* at the ACS Meeting in New York, April 1986, (R. B. Seymour and G. S. Kirshenbaum, eds.), Elsevier, New York (1987)

5. Gangal, S. V., Polytetrafluoroethylene, Homopolymers of Tetrafluoroethylene, in: *Encyclopedia of Polymer Science and Engineering,* 3rd ed., 11:25-35, John Wiley & Sons, New York (1980)

6. Lin, F. M. C., Chain Microstructure Studies of Poly(vinyl fluoride) by High Resolution NMR Spectroscopy, Ph.D. dissertation, University of Akron (1981)

7. Anton, D., Surface-Fluorinated Coatings, *Advanced Materials,* 10(15):1197–1205 (1998)

8. Asakawa, A., Performance of Durable Fluoropolymer Coatings, in: *Paint & Coatings Industry*, 19(9) (Sep 2003)

2 Producing Monomers, Polymers, and Fluoropolymer Finishing

2.1 Introduction

This chapter will summarize some of the polymerization processes used to make fluoropolymers. The processing steps used to take the fluoropolymer from the reactor to a form that customers can buy and use will also be covered. This includes dispersions and powders. Tables of commercial dispersions and powders are also included.

2.2 Monomers

In this section, synthesis of PTFE and properties of major monomers for polymerization of melt-processible fluoropolymers used in the majority of coatings are discussed. Tetrafluoroethylene is the primary monomer. As described earlier in Ch. 1, small amounts of other monomers are incorporated in the TFE polymer structure to modify its properties and processing characteristics. These monomers include hexafluoropropylene (HFP) and perfluoroalkyl-vinylethers (PAVE). A number of specialty monomers, though less common, are also used to modify the PTFE structure but are discussed in previous volumes of this series.[1]–[5]

2.2.1 Synthesis of Tetrafluoroethylene

Tetrafluoroethylene ($CF_2=CF_2$, also known as R1114) is the main building block of all perfluorinated polymers. Commercially important techniques for TFE preparation use a mined mineral fluorspar (CaF_2) and sulfuric acid to make HF, and chloroform as the starting ingredients.[6]–[13] The reaction scheme is shown below:

HF preparation:

Eq. (2.1) $CaF_2 + H_2SO_4 \rightarrow 2HF + CaSO_4$

Chloroform preparation:

Eq. (2.2) $CH_4 + 3Cl_2 \rightarrow CHCl_3 + 3HCl$

Chlorodifluoromethane preparation:

Eq. (2.3) $CHCl_3 + 2HF \rightarrow CHClF_2 + 2HCl$
(SbF_3 catalyst)

TFE synthesis:

Eq. (2.4) $2CHClF_2 \rightarrow CF_2=CF_2 + 2HCl$
(pyrolysis)

2.2.2 Synthesis of Hexafluoropropylene

Hexafluoropropylene ($CF_3CF=CF_2$, also known as R1216) is used as a comonomer in a number of fluoropolymers such as fluorinated ethylene-propylene copolymer. It is also used to "modify" the properties of homofluoropolymers. HFP has been prepared in a number of ways. Excellent hexafluoropropylene yields from the thermal degradation of sodium heptafluorobutyrate ($CF_3CF_2CF_2COONa$) have been reported.[14] Cracking tetrafluoroethylene in a stainless steel tube at –700°C–800°C under vacuum also produces HFP. Thermal decomposition of PTFE under 20 torr vacuum at 860°C yields 58% hexafluoropropylene.[15]

A more recently developed technique is the pyrolysis of a mixture of tetrafluoroethylene and carbon dioxide at atmospheric pressure at 700°C–900°C. Conversions of 20%–80% and HFP yields of better than 80% were obtained.[16]

2.2.3   Synthesis of Perfluoroalkylvinylethers

Perfluoroalkylvinylethers are synthesized according to the steps shown below:

1. Hexafluoropropylene (HFP) is converted to hexafluoropropylene epoxy (HFPO) by reaction Eq. (2.5) with hydrogen peroxide or an other oxidizer in a basic pH solution.[17]

Eq. (2.5)

$$CF_3-CF{=}CF_2 + H_2O_2 \rightarrow CF_2-CF-CF_2 + H_2O$$
$$\diagdown \diagup$$
$$O$$

(HFP)   (HFPO)

2. The HFPO is reacted with a perfluorinated acyl fluoride to produce perfluoro-2-alkoxy-propionyl fluoride according to Eq. (2.6):

Eq. (2.6)

$$CF_2-CF-CF_2 + R_f-CF{=}O \rightarrow R_fCF_2OCF-C{=}O$$
$$\diagdown \diagup \qquad\qquad\qquad | \qquad |$$
$$O \qquad\qquad\qquad\qquad\qquad CF_3 \quad F$$

3. Perfluoro-2-alkoxy-propionyl fluoride is reacted with sodium carbonate at high temperature:[18]

Eq. (2.7)

$$R_fCF_2OCF-C{=}O + Na_2CO_3$$
$$| \qquad |$$
$$CF_3 \quad F$$

$$\rightarrow R_fCF_2OCF{=}CF_2 + 2\ CO_2 + 2\ NaF$$

2.2.4   Properties of Monomers

Tetrafluoroethylene is a colorless, odorless, tasteless, non-toxic gas which boils at −76.3°C and freezes at −142.5°C. Critical temperature and pressure of tetrafluoroethylene are 33.3°C and 39.2 MPa. TFE is stored as a liquid; vapor pressure at −20°C is 1 MPa. Its heat of formation is reported to be −151.9 kcal/mole. Polymerization of tetrafluoroethylene is highly exothermic and generates 41.12 kcal/mole heat. The extent of exothermic reaction of TFE polymerization can be seen when it is compared with the polymerization of vinyl chloride and styrene which have heats of polymerization of 23–26 kcal/mole and 16.7 kcal/mole, respectively.[22]

Polymerization of tetrafluoroethylene to high molecular weight requires extremely high purity of the monomer. The removal of all traces of telogenic hydrogen or chlorine-bearing impurities is critically important. The products of the pyrolysis reaction are cooled, scrubbed with a dilute basic solution to remove HCl, and dried. The remaining gas is compressed and distilled to recover the unreacted $CHClF_2$ and to recover high purity TFE.[19] Tetrafluoroethylene can autopolymerize if it is not inhibited (which is how Plunkett discovered PTFE). Common TFE autopolymerization inhibitors include a variety of terpenes, such as *a*-pinene, Terpene B, and *d*-limonene[20] which appear to act as scavengers of oxygen, a polymerization initiator.

Tetrafluoroethylene is highly flammable and can undergo violent deflagration in the absence of air:

Eq. (2.8) $C_2F_4 \rightarrow C + CF_4$

Heat of reaction values between 57–62 kcal/mole (at 25°C and 1 atm) has been reported for TFE deflagration.[21] This is similar to the amount of heat released by the explosion of black gunpowder. Explosion is always a concern to manufacturers. To eliminate transportation accident concerns, TFE preparation and polymerization are usually carried out at the same site. Mixtures of TFE with carbon dioxide are known to be fairly safe.

Hexafluoropropylene ($CF_3CF{=}CF_2$) is used as a comonomer in a number of fluoropolymers including fluorinated ethylene-propylene (FEP) copolymer. It is also used to "modify" the properties of homo-fluoropolymers.

Hexafluoropropylene is a colorless, odorless, tasteless, and relatively low toxicity gas, which boils at -29.4°C and freezes at -156.2°C. Critical temperature and pressure of hexafluoropropylene are 85°C and 3,254 MPa. Unlike tetrafluoroethylene, HFP is extremely stable with respect to autopolymerization and may be stored in liquid state without the addition of telogen.[23]

Perfluoropropylvinylether (PPVE) is a commercially significant example of PAVEs. PPVE is an odorless, colorless liquid at room temperature. It is extremely flammable and burns with a colorless flame. It is less toxic than hexafluoropropylene and copolymerizes with tetrafluoroethylene.

2.3   Polymerization

A general understanding of the polymerization process is useful and can be applied to coatings technology. Impurities and endgroups are affected by the polymerization chemistry and they can impact the performance of coatings. Some of this technology is included at this time. Previous volumes of this series describe the polymerization processes and history in great detail for most of the commercial fluoropolymers. This text will discuss the polymerization process in general, and the details of a couple polymerizations that demonstrate the general concepts, leaving the details to the polymer chemists and engineers.

All the monomers used to make homopolymers and copolymers are *unsaturated* (they have a carbon-carbon double bond). The polymerization is generally free radical as described by the following scheme:

Eq. (2.9)

This reaction is called a polyaddition reaction and generally proceeds by a free radical addition mechanism or ionic polymerization mechanism.[24]

This is a chain reaction, with each monomer being added to the chain end, which lengthens the chain, but keeps the end reactive. Most fluoropolymer producing reactions proceed by the free radical mechanism.

A general description of free radical polymerization makes the understanding of the fluoropolymer variants of it easier to understand. Basically there are four types of chemical reactions occurring.

First, there is a process that starts the polymerization, called *chain initiation*. It is like the spark of the process and it generates a highly reactive molecule called a *radical*. These are generated usually by heating up a reactive molecule called an *initiator*. The following equation describes initiation:

Eq. (2.10) $R–R' \rightarrow R\cdot + R'\cdot$

The *initiator* (R-R') breaks into two *radicals* (R· and R'·). The radicals are highly reactive and generally can not be isolated; they are transient. The double bonds in the monomer molecules (M) are attacked by the radical and opened up transferring the radical to the end of the chain. The process that continues is what is called *chain propagation*.

Eq. (2.11) $R\cdot + M \rightarrow RM\cdot$

The monomer continues to add, building a long linear polymer:

Eq. (2.12) $RM\cdot + M \rightarrow RMM\cdot$

Eq. (2.13) $RMM_n\cdot + M \rightarrow RMM_{n+1}\cdot$

This goes on while there is plenty of monomer present. Eventually, it does stop. The radical can be transferred to a molecule (YZ) in a step called *chain transfer*. The molecule YZ is usually called a *chain transfer agent*.

Eq. (2.14) $RMM_n\cdot + YZ \rightarrow RMM_nY + Z\cdot$

The polymer molecule stops growing, but a new radical is formed that could start a new polymer molecule. Eventually, two radicals meet and react to cancel out the two reactive radicals, halting the chain reaction.

Eq. (2.15) $RMM_n \cdot + \cdot M_mR \rightarrow RM_nM_mR$

A simple understanding of this scheme should make the following discussion of PTFE polymerization easier to understand.

2.3.1   Polymerization of Homofluoropolymer PTFE

Polymerization of TFE to make PTFE is the only polymerization that discussed in detail in this book. It indicates some of the complexity and engineering difficulties found in fluoropolymer manufacturing. Further discussions can be found in earlier volumes of this series.[1][25]

Tetrafluoroethylene (TFE) is the monomer used for making polytetrafluoroethylene (PTFE). TFE is polymerized in water in the presence of an initiator, a surfactant, and other additives. Two different methods of free radical polymerization are common for production of different types of PTFE. *Suspension* polymerization is used to produce *granular* PTFE resins. Granular PTFE is extremely high molecular weight and is used in molding applications. TFE is polymerized in water in the presence of a very small amount of dispersant, or no surfactant accompanied by vigorous agitation. The dispersant, which acts like a soap that surrounds the particles in the water and is rapidly consumed during the initial phase of polymerization reaction, forming small particles that seed the aqueous medium. Further polymerization occurs in gas phase because of the absence of a surfactant, thus leading to the precipitation of the polymer.

Emulsion or *dispersion* polymerization is the method used to manufacture dispersion and fine powder PTFE products. These are the PTFE materials generally used in coatings. Fine powder resins are also called *coagulated dispersion,* which is descriptive of their production method, described later in this chapter (Sec. 2.4.4). Mild agitation, plentiful dispersant, and a wax additive set the dispersion polymerization apart from the suspension method. Dispersion and fine powder products are polymerized by the same method. The finishing steps convert the polymerization product, which is dispersion, to the two different product forms.

Familiarity with the important types of polytetrafluoroethylene and the commercially signifi-cant technologies of producing them helps with better understanding coatings. Although the exact polymerization technologies being practiced by resin manufacturers are closely guarded secrets, the descriptions and discussions of the important public disclosures in patents and other publications should provide an adequate understanding of the subject.

Polymerization of tetrafluoroethylene proceeds by a free radical mechanism. An initiator (sometimes called a catalyst) starts the reaction. The choice of the initiator is based on the desired reaction temperature. If polymerization is carried out at low temperatures (<30°C), a redox catalyst, such as potassium permanganate is used. (These compounds ionize into charged fragments such as $KMnO_4 \xrightarrow{water} K^+ + MNO_4^-$). A bisulfite or persulfate is the typical initiator for higher temperature TFE polymerization. The initiator affects the endgroups of the PTFE at a different stage of the polymerization. The key point is that there is no sulfur when persulfate is the initiator. Bisulfite initiators form sulfonic acid endgroups, which can influent thermal stability and color.

The free radical mechanism discussed here is exactly analogous to the general discussion of free radical polymerization discussed earlier in this chapter (Sec. 2.3). The reaction is initiated by a catalyst or by an initiator. A persulfate is the typical initiator for higher temperature TFE polymerization. The reaction scheme for persulfate initiation is shown below.

Initiator fragments, or *free radicals,* are formed by degradation of persulfate under heat:

Eq. (2.16) $K_2S_2O_8 + Heat \rightarrow 2SO_4^{\cdot -} + 2K^+$

Initiation takes place by formation of new free radicals by reaction of persulfate fragments with tetrafluoroethylene dissolved in the aqueous phase:

Eq. (2.17) $SO_4^{\cdot -} + CF_2{=}CF_2 \rightarrow {}^-SO_4CF_2{-}CF_2\cdot$

Propagation is the growth of the free radicals of the initiation step by further addition of tetrafluoroethylene:

Eq. (2.18) ${}^-SO_4(CF_2{-}CF_2)\cdot + nCF_2{=}CF_2$

$$\rightarrow {}^-SO_4(CF_2{-}CF_2)_n{-}(CF_2{-}CF_2)\cdot$$

Free radicals undergo hydrolysis where a hydroxyl endgroup replaces the sulfate eventually forming an acid endgroup:

Eq.(2.19) $^{-}SO_4(CF_2-CF_2)_n-(CF_2-CF_2)\cdot + H_2O$

$$\rightarrow HO(CF_2-CF_2)_n-(CF_2-CF_2)\cdot + H_2SO_4$$

Eq.(2.20) $HO(CF_2-CF_2)_n-(CF_2-CF_2)\cdot + H_2O$

$$\rightarrow COOHCF_2-(CF_2-CF_2)_n\cdot + 2HF$$

Termination is the last step before the growth in molecular weight of the free radicals halts:

Eq. (2.21) $COOH-CF_2-(CF_2-CF_2)_n\cdot$

$$+ COOH-CF_2-(CF_2-CF_2)_m\cdot$$

$$\rightarrow COOH-(CF_2-CF_2)_{m+n+1}-COOH$$

Alternative courses of hydrolysis can affect the endgroups at a different stage of the polymerization.

Tetrafluoroethylene polymerizes completely linearly without branching. This gives rise to a virtually perfect linear chain structure even at high molecular weights. The chains have minimal interactions and crystallize to form a nearly 100% crystalline structure. Controlling the crystallinity of the polymer is important in the development of good mechanical properties. The only means of controlling the extent of recrystallization after melting in homopolymers of TFE (no other comonomer) is by driving up the molecular weight of the polymer. The extremely long chains of PTFE have a much better probability of chain entanglement in the molten phase and little chance to crystallize to the premelt extent (>90%—95%). This is precisely the reason that it is essential to polymerize TFE to 10^6-10^7 for commercial applications. It is speculated that molecular weight may be as high as fifty million.[26] Molecular weight of PTFE can be controlled by means of certain polymerization parameters such as initiator content, telogens, and chain transfer agents.

Because of the very high molecular weight of PTFE, its melt viscosity is extremely high. The melt creep viscosity of PTFE is 10 GPa (10^{11} poise) at 380°C.[27] This is more than a million times too viscous for melt processing in extrusion or injection molding. PTFE may be a thermoplastic, but it develops no flow upon melting. The closure of voids in articles made from this polymer does not take place with the ease and completeness of the other thermoplastics such as polyolefins. A small fraction of void volume remains in parts made from homopolymers of PTFE due to the difficulty and slow rate of void closure in this polymer. Voids affect permeation and mechanical properties such as flex life and stress crack resistance.

The residual voids must be eliminated to improve mechanical properties and resistance to permeation. A reduction in the viscosity of PTFE without extensive recrystallization is required. Many manufacturers have polymerized a small amount of a comonomer with tetrafluoroethylene to disrupt the crystalline structure of PTFE.

All forms of PTFE are produced by batch polymerization under elevated pressure in specially designed reactors. Polymerization media is high purity water, which is virtually devoid of inorganic and organic impurities that impact the reaction by inhibition and retardation of the free radical polymerization. The surfactant of choice in these reactions is anionic, and often a perfluorinated carboxylic ammonium salt. In general, the important characteristics of the polymerization processes include little or no dispersing agent, and vigorous agitation at elevated temperature and pressure.

Polymerization of tetrafluoroethylene is done under constant pressure conditions to control the molecular weight and its distribution. It also affects the kinetics of polymerization. Pressure ranges from 0.03–3.5 MPa[28] and is held constant by feeding monomer into the reactor. The initiator and temperature are interrelated. Ionic, inorganic initiators such as ammonium persulfate or alkali metal persulfates such as potassium and lithium persulfates would be effective in the range of 40°C–90°C. Organic peroxides such as bis (β-carboxypropionyl) peroxide also called *disuccinic acid peroxide* can also initiate the polymerization.[29] As the temperature is lowered, at some point the effectiveness of the persulfates is diminished due to insufficient decomposition rate. Redox initiators such as potassium permanganate must replace them.

Water is primarily the heat transfer medium and does not interfere with the reaction but, even in low concentrations, most organic chemicals do. Metal ions such as iron may impart unwanted color.

A small quantity of an anionic dispersing agent that is predominantly non-telogenic is added to seed the polymerization. The most common dispersants are the ammonium salts of perfluorocarboxylic acids, containing 7–20 carbon atoms. Typical concentration of dispersants is 5–500 ppm, which is insufficient to cause formation of colloidal polymer particles. In the early stages of polymerization, a dispersion forms which becomes unstable as soon as the dispersing agent is consumed. The instability occurs at fairly low solids content, about 0.2% by weight. From there on, most of the polymerization occurs directly onto the larger granular particles, which are, porous water repellent, therefore, float on the water. The reaction tends to continue for some time, even after agitation is stopped, supporting the direct polymerization hypothesis.

Tetrafluoroethylene easily polymerizes at moderate pressures and temperatures. It is necessary to control the rate and to transfer the significant heat generated by the exothermic polymerization reaction. This is accomplished by circulating a cold fluid through the polymerization reactor jacket and cooling the aqueous phase, which is the heat transfer media. An important concern in the suspension process is the build-up of PTFE on the inner wall of the reactor, which reduces heat transfer. Development of hot spots on the reactor wall can result in deflagration (exothermic and explosive), if it goes unchecked.

A typical batch begins with the charging of highly purified water (18 MΩ) to a reactor which is equipped with a stirrer, followed by evacuation, and pressurization with tetrafluoroethylene. The feed rate of TFE is controlled to maintain a constant pressure in the reactor throughout the polymerization. The content of the reactor is vigorously agitated at 0.0004–0.002 kg·m/sec/ml. Temperature is controlled by adjusting the temperature of the coolant medium in the jacket. Stopping the monomer after a certain feed weight has been reached ends the polymerization. The reaction is allowed to continue in order to consume the majority of the remaining TFE.

A number of different techniques are used for the polymerization of various monomers. Distinctions from conventional methods include variables such as aqueous or nonaqueous medium, batch or continuous production, and suspension or emulsion regimes.

2.3.2 Copolymer and Terpolymer Polymerization

A key concept to remember is reactivity which characterizes the different rates that different monomers react with each other and themselves. This means that if you start a polymerization reactor with monomers in a given ratio, you probably will not get a polymer containing the same ratio of monomer units, or they might not be statistically distributed along the polymer chain. The details of these reactions can be found in Vol. 2 of this series.[25]

2.3.3 Core-Shell Polymerization

Most fluoropolymer polymerizations involve feeding monomer(s) into the reactor (also known as an autoclave). One does not have to feed all the monomers at the same time nor in the same ratio. At the start of a reaction, all TFE could be forming PTFE. One could then feed in TFE and a comonomer or a modifier, making the outside of the fluoropolymer particle a different polymer than the inside. This regime of polymerization allows the production of particles that constitute different polymers at different depths inside the particle. This is descriptively called a "core-shell" polymer; there are many examples in the patent literature and several commercial products are made this way.[46]

The modifier may be introduced at any time during the polymerization. For example, if it is introduced after 70% of the TFE monomer to be polymerized has been consumed, each PTFE particle will contain a core of high molecular weight PTFE and a shell containing the low molecular weight modified polytetrafluoroethylene. In this example 30% of the outer shell of the particle, by weight, has been modified. The total modifier content of the polymer can be extremely small, but the impact on the properties can be profound. In one example, the melt creep viscosity of the core-shell polymer was 3–6×10^{10} poise. Compare that to the polymer made under identical conditions without the modifier which had a melt creep viscosity of 10×10^{10} poise.[30]

2.3.4 Polymerization in Supercritical Carbon Dioxide

Polymerization of fluorinated monomers in carbon dioxide above its critical temperature and pressure is now becoming a commercial process. Supercritical CO_2 has been used as an environmentally friendly solvent to replace organic solvents. The critical state for a pure substance is the state of temperature and pressure at which the gas and liquid phases are so similar that they can not be present separately. The critical temperature is the maximum temperature at which the gas and liquid phases can exist as separate phases; critical pressure is the corresponding pressure. Past its critical pressure (7.4 MPa) and temperature (31°C), carbon dioxide becomes dense, similar to a liquid, yet it maintains its gas-like ability to flow without significant viscosity or surface tension.[31] Polymerization in this medium is one of its more recent applications.[32]

Publications by a number of authors[33]–[37] have shown that carbon dioxide is a favorable medium for free radical polymerization. Other than environmental advantages, there are a number of important advantages[38] in the use of CO_2 for polymerizing fluoropolymers.

1. The possibility of the removal of initiator residues and degradation products by supercritical carbon dioxide extraction.

2. Possibility of creating new morphologies due to the solubility of supercritical CO_2 in fluoropolymer.

3. Mixtures of tetrafluoroethylene and carbon dioxide are much safer and can be operated at higher polymerization pressure than TFE by itself.

2.3.5 Endgroups

An important issue is the stability of the polymer endgroups, which depends on their chemical structure. Chemistry of the polymerization and the polymerization conditions determine the nature of the endgroups. Unstable endgroups degrade and usually produce gases during polymer storage, part fabrication, or when thick coatings are being processed.

Melt processible fluoroplastics such as perfluoroalkoxy polymer (PFA), fluorinated ethylene propylene polymer (FEP), and polyvinylidene fluoride (PVDF) can be polymerized in aqueous mediums or in chlorofluorocarbon solvents. More recently, they have been polymerized in an aqueous medium due to the detrimental effect of the chlorofluorocarbons on the ozone layer. Fluoropolymers made in an aqueous medium contain carboxylic and acyl fluoride endgroups, both of which are unstable at processing temperatures. Thermal degradation of the two endgroups leads to hydrofluoric acid production. Additional processing is required to stabilize/eliminate the unstable endgroups to prevent equipment damage and accommodate the applications that require low extractable fluoride content in the polymer. That extra processing can make a significant contribution to increasing the cost of the fluoropolymers.

Polymers made in the chlorofluorocarbon medium contain significantly fewer unstable endgroups compared to those produced in water. A great deal of work has been done to find alternative solvents for fluoroolefin polymerization. Selection of a solvent is further complicated by the highly electrophilic nature of fluorinated free radicals. They readily abstract hydrogen atoms from almost any hydrocarbon, thus rendering these solvents useless as a polymerization reaction medium for high molecular weight fluoropolymers. Perfluorocarbon solvents such as perfluorohexane would work well, but they are extremely expensive, thus, not feasible for commercial scale operations. These factors have spurred research into carbon dioxide as a safe, functional, and economical reaction medium.

Figure 2.1 shows models of three possible fluoropolymer endgroups. The first one is a fully fluorinated endgroup of $-CF_3$. A process for producing this endgroup has been patented by DuPont. The last two models show common fluoropolymer endgroups that are reactive and will also decompose at lower temperatures.

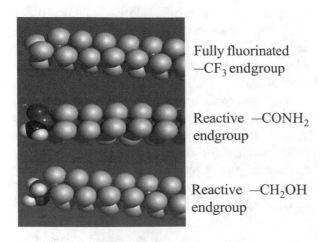

Fully fluorinated
—CF₃ endgroup

Reactive —CONH₂
endgroup

Reactive —CH₂OH
endgroup

Figure 2.1 Models of fluoropolymer endgroups.

2.4 Finishing

As-produced dispersions are called *raw* because PTFE dispersion out of the reactor is of little use. Its solids concentration is too low and it is typically not stable enough for shipping and storage. These are rarely used as-is in coatings products. The processing involved in converting the raw dispersion to final usable forms is called finishing. Normally the raw PTFE dispersion is processed into one of two forms:

1. Stabilized concentrated dispersion
2. Fine powder

2.4.1 Dispersion Concentration

Raw dispersions are produced by dispersion polymerization in the range of <10% to 45% by weight of PTFE or a TFE copolymer in water. To insure a commercially viable product, the dispersion must be sufficiently stable for transportation, storage, and handling. This means that it should not form precipitated PTFE particles commonly called coagulum or grit, which cannot be reincorporated in the liquid phase by simple mixing. It should also have the minimum possible water content to decrease the transportation cost and add to the formulation flexibility.

Historically, there have been many ingenious ways to concentrate and stabilize PTFE dispersions. These are discussed elsewhere [39] and only current commercial approaches will be discussed here.

Marks and Whipple[40] proposed a process for concentrating emulsions containing 30%–45% colloidal particles of polytetrafluoroethylene. The process was accomplished by the addition of 0.01%–1.0% by weight of the dispersion of sodium hydroxide, ammonium hydroxide, or ammonium carbonate. Next, 6%–12% by weight of dispersion solids of a nonionic surfactant was added. The structure of the surfactant being [R → C₆H₄–(OCH₂CH₂)ₙOH] where R is a monovalent hydrocarbon with 8–10 carbon atoms and $n = R + 1$ or R $+ 2$ (R = 8–10), or [(tertiary octyl) → C₆H₄ → (OCH₂CH₂)₉₋₁₀OH]. After stirring, the mixture was heated to a temperature of 50°C–80°C. A cloudy appearance indicated that the nonionic surfactant had begun to become insoluble, a point commonly known as the *cloud point*. PTFE particles settled after a period of time and formed a layer at the bottom of the container. The upper layer, which was relatively clear, was decanted and the lower layer was recovered. The solid content of this lower layer of dispersion was colloidal PTFE particles at 55%–75% solids content by weight free of coagulated polymer particles.

In another procedure for concentration[41] an acrylic polymer containing a large amount of acid groups or its salts were added to PTFE dispersions. The acid content of the acrylic polymer was at least 20% and they had a weight-average molecular weight of 50M–500M; polyacrylic acid was the preferred polymer. Addition of a small amount (0.01%–0.5%) of this type of acrylic polymer to the dispersion caused a phase separation to occur. The lower phase contained 50%–70% by weight of PTFE and was recovered by decantation. To reduce viscosity and increase stability of the concentrated dispersion, either an ionic or nonionic surfactant could be added before the addition of the acrylic polymer. Other fluoropolymer dispersions are finished similarly.

Stability is a key requirement of the final product. Any dispersion must have a reasonable shelf life of a few weeks to as long as one year. It should also be able to withstand transportation and handling during processing. The shear rate inherent in these activities must not lead to coagulation of the polymer particles. For a given dispersion, stability is a function of solids content, pH and viscosity. These properties can be adjusted by systematic study or by trial and error to improve the stability.

2.4.2 Commercial Dispersions and Properties

There are specific ASTM (American Society for Testing and Materials) test protocols for some fluoropolymer dispersions. For example, D4441-04 is the "Standard Specification for Aqueous Dispersions of Polytetrafluoroethylene." However, individual tests are used for each chatacteristic measurement. Table 2.1 lists the test methods used to measure the characteristics of fluoropolymer dispersions given in Tables 2.2 through 2.7.

Almost any dispersion can be formulated into a coating. There are dozens of PTFE dispersions because there are so many uses for them. Properties of commercial dispersions for possible use in coatings are shown in Tables 2.2–2.6.

There are fewer commercial dispersions available of other fluoropolymers. Several are listed in Table 2.7. Generally there are other dispersions available, but they may not appear on lists of those freely available.

Table 2.1 Definitions of Basic Properties of Fluoropolymer Dispersions

Property	Definition	Reference Test Methods
Weight Solids Content	The amount of fluoropolymer in the dispersion as weight %	ASTM D4441
Surfactant Content	The amount of surfactant in the dispersion as weight %	ASTM D4441
Dispersion Particle Size	The average particle size of the fluoropolymer particles in the dispersion (microns)	ASTM D4464 or ISO 13321
Viscosity	The viscosity of the dispersion in centipoise (cps)	ASTM D2196
pH	Acidity/alkalinity of the dispersion	ASTM E70
Melting Point	Peak melting point of dried polymer from dispersion by differential scanning calorimetry (DSC) (°C)	ASTM D4591
Melt Flow Rate (MFR) or Melt Flow Index (MFI)	An indirect measure of the melt viscosity by extrusion plastometer (grams of polymer flow in 10 min)*	ASTM D1238

* The melt flow rate or melt flow index is run at a temperature that depends on the polymer being tested:

- 265°C for THV
- 297°C for ETFE, 372°C for FEP, PFA, and PTFE micropowders
- 237°C for PVDF

Table 2.2 DuPont Aqueous PTFE Dispersions

Product Code	PTFE Solids, Weight %	Surfactant Solids, Weight %	PTFE Particle Size, μm	Viscosity, cps	pH	PTFE Melt Point ($1^{st}/2^{nd}$), °C	Comments
30	60	3.6	0.22	20	>9.5	337/327	
30B	60	4.8	0.22	20	>9.5	337/327	
35	32.5	2.5	0.05–0.5		4	337/327	
B	60		0.22		<8.5	337/327	
304A	45	2.7	0.22	1700	>9.5	337/327	
305A	60	4.8	0.22	20	>9.5	337/327	
307A	60	3.6	0.16	20	>9.5	337/327	
313A	60	4.2	0.22	20	>9.5	337/327	
FPD3584	60	3.6	0.20	20	>9.5	337/327	
K-20	33	1.2	0.22	15	>9.5	337/327	
TE-3667N	60		0.22	20	>9.5		MFR 4-30
TE-5070AN	56		<0.11		>9.5		MFR 1-13
TE-3823	60	6	0.27	20	10	344/327	

Table 2.3 Dyneon Aqueous PTFE Dispersions

Product Code	PTFE Solids, Weight %	Surfactant Solids, Weight %	PTFE Particle Size, μm	Viscosity, cps	pH
TF 5032	60	3	0.16	9	10
TF 5033	35	1.4	0.16	3	10
TF 5035	62	2	0.225	12	9
TF 5039	55	5.5	0.225		8.5
TF 5041	60	4.8	0.225	50	9.5
TF 5050	58	20.9	0.22	8	10
TF 5060	60	3	0.22	9	9
TF 5065	59	2	0.22	11	9
PA 5958	60	5	0.16	9.5	≥9
PA 5959	58	5	0.22	9.5	≥9.5

Table 2.4 Solvay Solexis Aqueous PTFE Dispersions

Product Code	PTFE Solids, Weight %	Surfactant Solids, Weight %	PTFE Particle Size, μm	Viscosity, cps	pH
D3300	59	3.5	0.22	25	>9
D3000	59	3	0.22	15	>9
D60/A	60	3	0.24	20	>9
D60/G	60	4	0.24	25	>9
D1100	60	3.5	0.24	25	>9
D1000	60	3	0.24	20	>9

Table 2.5 Daiken Aqueous PTFE Dispersions

Product Code	PTFE Solids, Weight %	SurfactantSolids, Weight %	PTFE Particle Size, μm	Viscosity, cps	pH	PTFE Melt Point °C
D-2	59-61	3.7-4.5	0.2-0.4	15-30	9-11	335
D-2C	59-61	3.7-4.5	0.2-0.4	15-30	10-11	335
D-3A	59-61	3.6-4.8		15-25	8.5-10	
D-3B	59-61	3.7-4.8		15-25	8.5-10	
D-46	58-60	3.5-4.5		18-28	9-11	
D-6A	59-61	3.7-4.5		15-30	9-11	
D-6B	59-61	3.7-4.5		15-30	9-11	
LDW-40	40	2.4	.18	8	8-9	330

Table 2.6 Asahi Glass Aqueous PTFE Dispersions

Product Code	PTFE Solids, Weight %	Surfactant Solids, Weight %	PTFE Particle Size, μm	Viscosity, cps	pH
XAD911	60	3	0.25	25	9.5
XAD912	55	5.2	0.25	25	9.5
XAD938	60	3	0.3	25	9.5
AD1	60	3	0.25	30	9.5
AD639	57	6.0	0.25	22	9.5
AD936	60	3	0.3	19	9.5

Table 2.7 Other Fluoropolymer Dispersions

Product Code	Fluoropolymer Type	Fluoropolymer Solids, Weight %	Particle Size, microns	pH	Melt Point, °C	MFR/MFI, grams/10 min
DuPont 335A	PFA	60	0.20	9.5	305	1–3
Dyneon 6900N	PFA	50	0.235	<7	310	15
Dyneon 6910N	PFA	22	0.09	9	306	15
DuPont 121A	FEP	54	0.18	9.5	260	
Dyneon X6300	FEP	50	0.15	9.5	255	10
Dyneon THV 220D	THV	31	0.095	2.5	120	20 (265°C)
Dyneon THV 340C	THV	50	0.09	9.5	145	40 (265°C)
Dyneon THV 510D	THV	34	0.09	7	165	10
Hylar 301F	PVDF				160	29–33
Dyneon ET6425	ETFE	20	0.07	7	270	10

2.4.3 Fine Powder Production

Most fine PTFE powders are used in molding applications rather than coating applications, though the use of these powders in coatings is not out of the realm of possibility. To produce fine powder from the polymerization dispersion, three processing steps have to take place.

1. Coagulation of the colloidal particles.
2. Separation of the agglomerates from the aqueous phase.
3. Drying the agglomerates.

Diluting the raw dispersion to a polymer concentration of 10%–20% by weight, agitation, and adjusting the pH to neutral or basic[37] results in coagulation. A coagulating agent such as a water-soluble organic compound or inorganic salt or acid can be added to the dispersion. Examples of the organic compound include methanol and acetone. Inorganic salts such as potassium nitrate and ammonium carbonate, and inorganic acids like nitric acid and hydrochloric acid can aid coagulation. The diluted dispersion is then agitated vigorously. The primary PTFE particles form agglomerates which are isolated by skimming or filtration.

Drying of the polytetrafluoroethylene agglomerates is carried out by vacuum, high frequency, or heated air such that the wet powder is not excessively fluidized.[37] Friction or contact between the particles, especially at a high temperature, adversely affects the fine powder because of easy fibrillation and loss of its particulate structure leading to poor properties of parts made from this resin. Fibrillated PTFE is shown in Fig. 2.2.

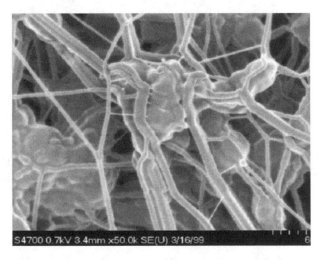

Figure 2.2 Fibrillated PTFE.

Fine powder resins must be protected from fibrillation after drying. PTFE does not fibrillate below its transition point (19°C for TFE homopolymers) during normal handling and transportation. Storage and transportation of the resin after refrigeration below its transition point is the normal commercial practice for handling fine powder polytetrafluoroethylene resins.

2.4.4 PTFE Micropowder Production

Fluoroadditives or micropowders are finely divided low molecular weight polytetrafluoroethylene powders, which are nearly entirely consumed by industries outside fluoropolymers. Micropowders are added to a great number of other products and compounded to enhance the properties of those products. The addition of a fluoroadditive imparts some of the properties of fluoropolymers to a host system. There are few applications, such as dry film lubricants, where a micropowder is used by itself.

In general, fluoroadditives have small particle size of the order of a few microns, hence, the word *micropowders*. These powders are either granular (suspension polymerized) or fine powder based (dispersion polymerized), which have different particle morphologies, therefore, different properties and incorporation manners in the host material. Their molecular weight is in the range of a few tens of thousands to a few hundreds of thousands, compared to several million for the molding (granular and fine powder) resins. This section describes the properties, methods of production, and some of the applications of micropowders.

Fluoroadditives are primarily produced by reducing the particle size of granular and fine powder polytetrafluoroethylene particles. Some companies recycle molded PTFE parts into micropowders. A few micropowders are produced by direct polymerization of tetrafluoroethylene, but these are limited to dispersion-polymerized material.

The extremely high molecular weight of granular PTFE prevents grinding the particles below 20–25 μm. To grind them to fine particles, the molecular weight must be reduced. That makes the resin brittle and easier to grind, and reduces fibrillation. When molecular weight is decreased sufficiently, one can reduce the size of the particles to a few microns, which is required in most micropowder applications.

Fine powder particles are 500 μm agglomerates of small (<0.25 μm) primary particles. It is impossible to directly deagglomerate these particles because of fibrillation and subsequent entanglement of the particles. The molecular weight has to be reduced to prevent fibrillation during the shearing to which the particles have to be subjected during deagglomeration or grinding. Sufficiently low molecular weight fine powder resins do not fibrillate.

The feedstock for micropowder production can come from a variety of sources. Micropowders were first developed as an outlet for the disposal of scrap resin. The highest quality micropowders are obtained by irradiation of first quality, or virgin, high molecular weight PTFE. Second quality resins could make high quality feedstock depending on the reason they are second quality. Contamination in the second quality feedstock would preclude production of high quality micropowder.

Thermal and radiation degradation of polytetrafluoroethylene are the two ways to reduce PTFE molecular weight. Heat and radiation are capable of degrading polytetrafluoroethylene if they are supplied in sufficient quantities to reach the temperature or radiation dose at which the polymer chain degrades.

For those PTFEs that are thermally decomposed, the atmospheric environment has a big effect on the properties exhibited by the resulting micropowders. Thermal decomposition of PTFE under vacuum or in the presence of inert gases is different than decomposition in an atmosphere of oxygen or air. Various studies have reported somewhat different degradation temperatures, which are attributable to the differences in the experimental conditions and polymer type. One common conclusion of all the studies is that PTFE degrades more at a lower temperature and more rapidly in the presence of oxygen or air. Another difference is in the type of products of degradation. Under vacuum or inert gases, the product of decomposition is mostly TFE and other small molecules; and for PTFE degradation this does not start until 500°C.[42] Under oxygen or air, smaller polymer chains are the decomposition products.

Table 2.8 shows a comparison of degradation temperatures for PTFE and other polymers. In all cases, oxygen appears to promote the polymer decomposition. Polymer structure and monomer type influence thermal degradation.

Table 2.8 Degradation and Weight Loss of Fluoropolymers in Various Atmospheres at the Given Temperatures[43]

Atmosphere	Vacuum	Oxygen	Nitrogen
Weight loss, after 2 hrs	25%	25%	3%
PTFE	494°C	482°C	460°C
FEP	481°C	417°C	-
PVF	403°C	354°C	-

PTFE radicals react with oxygen the same way regardless of whether they have been produced by thermal decomposition or irradiation. X-ray photoelectron spectroscopy analysis of the surfaces of irradiated and un-irradiated PTFE indicates significant oxygen content as a result of irradiation in air. The following reaction scheme has been widely accepted:

Eq. (2.22)

$$-CF_2-CF_2- + \text{Heat or Irradiation}$$
$$\rightarrow -CF_2-CF_2\cdot + -CF_2-CF_2-CF_2\cdot-$$

$$-CF_2-CF_2\cdot + O_2 \rightarrow -CF_2CFO$$

$$-CF_2-CF_2-CF_2- + O_2 \rightarrow -CF_2-CF_2-CF_2-$$
$$O\cdot$$
$$\rightarrow -CF_2-CF_2\cdot + -CF_2CFO$$

The endgroup of degraded polytetrafluoroethylene is acyl fluoride (–CFO). This group reacts with water and forms a carboxylic acid group (–COOH) and evolves into hydrofluoric acid (HF). Endgroups can usually be identified by infrared spectroscopy. The number of endgroups is too low, in most cases, to have a significant effect on the final properties of micropowders. In some applications, the endgroups can have an effect. For example, the endgroups can promote adhesion of micropowders to metals.

An alternate approach to reducing the molecular weight of polytetrafluoroethylene is exposure to high energy radiation such as x-ray, gamma ray, and electron beam. The high-energy radiation breaks down carbon-carbon bonds in the molecule's chain. When irradiated in vacuum or inert atmosphere, the cleavage of the bonds produces highly stable radicals. The recombination of those stable radicals prevents rapid degradation of PTFE, as the molecular weight rebuilds. When irradiation is conducted in the air, the radicals react with oxygen leading to smaller molecular weight PTFE chains fairly quickly.

2.4.4.1 Production of Fluoroadditives by Electron Beam Irradiation

Electron beam is the most common commercial method of converting high molecular weight PTFE to a grindable form. Electron beam irradiation is relatively a simple process. In practice, a continuous process is used to improve the economics of the process. The PTFE resin is spread on a conveyor belt at a specified thickness and is passed under the electron beam. The speed of the conveyor belt is used to control the dose that the PTFE is subjected to (the common unit is megarad, abbreviated Mrad, or 2.30 calories of energy absorbed per gram or material). Normally, multiple passes are made to expose the PTFE to higher doses. After the total dose has been received, the irradiated material is removed for grinding.

The conveyor belt can be shaped circularly to carry the resin under the electron beam several times. Multiple pass irradiation allows the polymer to cool after each pass, since dissipation of electron beam irradiation in polytetrafluoroethylene heats up the resin. Without removal of heat, the PTFE will get very hot. It may even melt, which will lead to sticking of the individual particles to each other and that complicates the grinding process. The dose delivered in each pass is additive, such that ten passes of 1 Mrad dose is equal to 1 to 10 Mrad exposure in terms of molecular weight reduction. Irradiation causes cleavage of bonds and generates off-gases such as hydrofluoric acid, which must be removed by means of adequate ventilation from the processing areas. The stack effluents might have to be treated to remove the entrained particles and the evolved species prior to venting to the atmosphere. The nature of the stack gas treatment depends on its contents and the governing emission rules. Details of the electron beam equipment and operation are found in the literature.[44][45]

As the electron beam irradiation dosage increases, smaller particle sizes can be produced by grinding as shown in Table 2.9. This data compares the effect of dose in Mrads to the particle size of the PTFE micropowder obtained when ground under identical conditions. Temperature of the resin is held below 121°C during the irradiation. Particle size decreases rapidly as irradiation doses increase from 5 to 25 Mrad. The melt flow rate goes up, as the molecular weight of the powder goes down with the increasing amounts of radiation applied.

2.4.4.2 Grinding Irradiated PTFE

The irradiated PTFE resin is ready for sized reduction. This is accomplished by milling. Two methods of milling are generally used for fluoropolymers. A *jet mill* is one milling approach, and the second is a *hammer mill*.

A jet mill, in the simplest view, shoots the particles at each other at high velocity with compressed air or other gas through nozzles called jets. It is sometimes called a *fluid energy mill*. The particles strike each other often, causing them to fracture into smaller particles. Such mills are designed to separate or classify particles below a specific size and remove them from the mill. Those particles that remain large are recycled through, or remain in, the mill, until they have been reduced sufficiently in size. Sometime the grinding is done cryogenically with fluids such as liquid nitrogen for particularly difficult to grind materials.

A schematic diagram of the basic elements of a fluid energy jet mill is shown in Fig. 2.3. This is a diagram of a laboratory mill. On the left hand side is a sideview drawing of the mill, often called a donut. A diagram of the interior structure is on the right. Compressed air is forced into an outer ring within the mill. The powder to be ground is added at the raw material inlet. Compressed air shoots through pulverizing nozzles or jets on an inner ring at nearly supersonic speeds. The particles to be ground flow at very high speed in a circular path near the inner ring. Centrifugal force keeps large particles to the outer area. These particles collide with each other and the walls on the inner ring in the mill. As the particles are made smaller by the collisions, they move towards the product outlet. When they are small enough, they fall out of the mill and are collected. Large particles tend to stay to the outside of the chamber since centrifugal forces are greater than small particles, which tend to move towards the center of the chamber.

2.4.4.3 Regulatory Compliance

A majority of the fluoroadditives is produced by irradiation of high molecular weight polytetrafluoroethylene to facilitate their grinding into small particles. Food and Drug Administration (FDA)

Table 2.9 Effect of Irradiation Dose on PTFE Micropowder Particle Size[44]

Test Case	Irradiation Time, sec	Dose, Mrad	Average Particle Size, μm
1	2.5	5	11.1
2	5.0	10	5.3
3	7.5	15	2.5
4	10	20	1.5
5	12.5	25	0.9

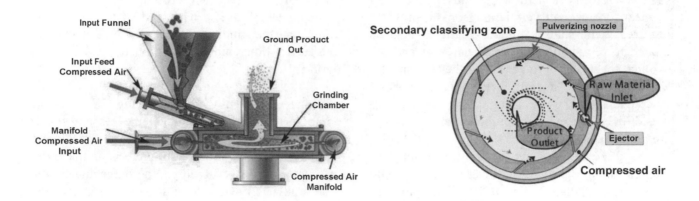

Figure 2.3 Jet mill diagrams. (*left*) A sideview drawing of a mill, often called a donut. (*right*) A topview diagram of the interior structure of the donut.

rule 21CFR177.1550, paragraph (c) specifies maximum allowable doses of radiation and maximum particle size processed by irradiation. This rule restricts the application of components containing irradiated fluoroadditives intended for repeated use in contact with food. Anyone planning to produce articles, which come in contact with food should be sure of FDA compliance beforehand. Fluoroadditive manufacturers can usually supply FDA compliance information.

There are a number of other regulatory agencies including the FDA, the U.S. Department of Agriculture, and the U.S. Pharmacopia. There are similar agencies in other parts of the world. It is important to investigate compliance issues when planning to formulate fluoroadditives into articles for use in applications where food, produce, and pharmaceutical contact may occur. Ultimately it is the responsibility of those selling produced parts to certify that their product is compliant under the appropriate regulations.

2.4.4.4 Commercial Micropowder Products

Polytetrafluoroethylene fluoroadditives (micropowders) are produced by irradiation of high molecular weight PTFE or by direct polymerization (dispersion). They have finely divided particles that are smaller than the particles of other PTFE types. Micropowders are mainly intended for use as a minor constituent of mixtures with other solids or liquids. They can impart some of the properties of fluoropolymers to the host systems. Fluoropolymer manufacturers offer a variety of virgin micropowders. Other companies supply fluoroadditives made from the irradiation of scrap and second grade PTFE.

Like for dispersions, there is a specific ASTM test protocol for fluoropolymer micropowders. ASTM D5675 is the "Standard Specification for Fluoropolymer Micropowders." However, individual tests are used for each characteristic measurement. Table 2.10 lists the test methods used to measure the characteristics for the fluoropolymer micropowders given in Tables 2.11 through 2.15.

Table 2.10 Definitions of Basic Properties of Fluoropolymer Micropowders

Property	Definition	Reference Test Methods
Particle Size	The average particle size of the fluoropolymer particles in measured by laser light scattering (microns)	ASTM D4464
Bulk Density or Apparent Density	Mass of fluoropolymer powder per liter of powder measured under specific conditions (g/1000 ml)	ASTM D895
Specific Surface Area	The surface area of the particles per unit weight (m^2/g)	ASTM D4567 or DIN 66131**
Melting Point	Peak melting point of dried polymer from dispersion by differential scanning calorimetry (DSC) (°C)	ASTM D4591
Melt Flow Rate (MFR) or Melt Flow Index (MFI)*	An indirect measure of the melt viscosity by extrusion plastometer (grams of polymer flow in 10 min)*	ASTM D1238

* The melt flow rate or melt flow index is run at a temperature that depends on the polymer being tested:

- 265°C for THV
- 297°C for ETFE, 372°C for FEP, PFA, and PTFE micropowders
- 237°C for PVDF

** DIN (German Industrial Standard)

Table 2.11 Commercial Dyneon PTFE Micropowders and Properties

Product Code	Particle Size, μ	Bulk Density, g/1000 ml	MFI, g/10 cm³	Specific Surface Area, m²/g	Melting Point, °C
Dyneon PA 5952	15	450			
Dyneon PA 5951	6	350			
Dyneon PA 5953	8	400			
Dyneon PA 5954	4	280			
Dyneon PA 5955	4	280	4		
Dyneon PA 5956	6	250			
J24	20	400			327
J14	6	250			322
TF9201	6	350	<2	10	330
TF9205	8	400	12	12	325
TF9207	4	280	4	17	329

Table 2.12 Commercial Dupont PTFE Micropowders and Properties

Product Code	Melting Peak Temperature, °C	Melt Flow Rate, g/10min	Average Particle Size, μm	Bulk Density, g/1000ml	Specific Surface Area, m²/g
MP1000	325		12	500	5-10
MP1100	325	>1.0	4	300	5-10
MP1150	325		10	450	5-10
MP1200	325		3	450	1.5-3
MP1300	325	>0.1	12	425	1.5-3
MP1400	325	>0.1	10	425	1.5-3
MP1500J	330		20	425	8-12
MP1600N	325	4-30	12	350	8-12
TE-5069		1-17		450	10-30
TE-3807A		5-20	5-20	450	4.6-15

Table 2.13 Commercial Asahi PTFE Micropowders and Properties

Product Code	Average Particle Size, μm	Bulk Density, g/1000ml	Specific Surface Area, m²/g
Fluon 1680	13	450	0.8
Fluon 1690	21	480	1.0
Fluon 1700		530	3.1
Fluon 1710	9	400	2.3

Table 2.14 Commercial Daikin PTFE Micropowders and Properties

Product Code	Melting Peak Temperature °C	Average Particle Size, μm	Bulk Density g/1000ml	Specific Surface Area, m²/g
Polyflon L-5F	327	5	400	11
Polyflon L-2	330	4	400	9

Table 2.15 Commercial Solvay Solexis PTFE Micropowders and Properties

Product Code	Melting Peak Temperature, °C	Melt Flow Rate, g/10min	Average Particle Size, μm	Bulk Density, g/1000ml	Specific Surface Area, m²/g
Polymist® F-5	320-325	20	<10	380	3
Polymist® F-5A	320-325	20	<6	400	3
Polymist® F-5A EX	325-330	0.25	<7	400	3
Polymist® F-510	325-330	0.25	<25	475	3
Algoflon® L203	329-332	0.2	<6	310	9
Algoflon® L206	329-332	0.1	<7	330	9
Algoflon® L101X	330		<7		14

2.4.5 Dispersion Coagulation

FEP as recovered from the coagulation of the reactor dispersion is called *fluff* or *powder*. It is treated to eliminate the reactive ends before extrusion and pelletization into a granular form. The granules were further hardened by heat treatment. The hardening phenomenon occurs when the powder is heated to about 25°C below its melting point.

2.4.6 Spray Drying

Spray drying, as the name implies, is a process for drying, utilizing a spray. A spray dryer mixes a heated gas with a sprayed (atomized) liquid stream, such as a fluoropolymer dispersion, within a drying chamber to accomplish evaporation of the liquid from the small atomized droplets forming a dry fine powder. This process is commonly used to make powdered milk and some pharmaceuticals.

A spray dryer typically includes the following key components as shown in Fig. 2.4.

1. A mixer to keep the dispersion uniformly dispersed during the process.

2. A pump to deliver the liquid to an atomizer.

3. The atomizer, typically a rotating disk or nozzle, produces a fine spray into a large chamber.

4. An air/gas heater that supplies the hot dry air into the chamber with the atomized dispersion.

5. The *chamber* with adequate residence time to dry the droplets without hitting and sticking to the chamber walls.

6. A means for *recovering* the powder from the chamber.

7. A bag house to remove fines from the air/gas removed from the chamber and expelled to the atmosphere.

Operational considerations include:

- Feed rate of dispersion
- Dispersion stability
- Atomizing efficiency (droplet size)
- Atomizer temperature
- Chamber temperature – hot enough to dry, but not to melt or decompose

2.4.7 Spray Sintering

Spray sintering is essentially spray drying at temperatures high enough to melt the dry powder particles, but cool them, before they are collected. The particles made this way are quite spherical and relatively hard.

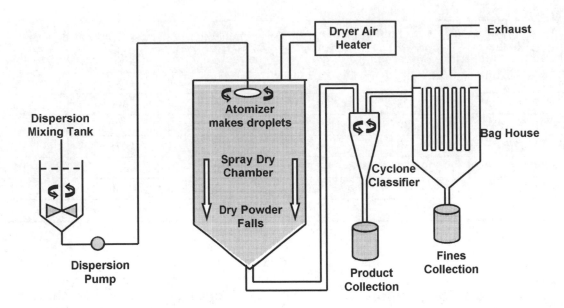

Figure 2.4 Schematic of a spray drying facility.

REFERENCES

1. Ebnesajjad, S., *Fluoroplastics,* Vol. 1: *Non-Melt Processible Fluoroplastics, The Definitive User's Guide and Databook*, William Andrew, Inc., Norwich, NY (2000)

2. Cardinal, A. J., Edens, W. L., and Van Dyk, J. W., US Patent 3,142,665, assigned to DuPont (Jul 28, 1964)

3. Holmes, D. A. and Fasig, E. W., US Patent 3,819,594, assigned to DuPont (Jun 25, 1974)

4. Mueller, M. B. Salatiello, P. P., and Kaufman, H. S., US Patent 3,655,611 assigned to Allied (Apr 11, 1972)

5. Doughty, T. R., Sperati, C. A. and Un, H. W., US Patent 3,855,191 assigned to DuPont (Dec 17, 1974)

6. Park, J. D., et al., *Ind. Eng. Chem.,* 39:354 (1947)

7. Hamilton, J. M., in: *Advances in Fluorine Chemistry* (M. Stacey, J. C. Tatlow, and A. G. Sharpe, eds.), 3:117, Butterworth & Co., Ltd., Kent, U. K. (1963)

8. Edwards, J. W., and Small, P. A., *Nature*, 202:1329 (1964)

9. Gozzo, F., and Patrick, C. R., *Nature*, 202:80 (1964)

10. Hisazumi, M., and Shingu, H., Japanese Patent 60 15,353

11. Scherer, O., et al., US Patent 2,994,723, assigned to Farbewerke Hoechst (Aug 1, 1961)

12. Edwards, J. W., Sherratt, S., and Small, P. A., British Patent 960,309, assigned to ICI (Jun 10, 1964)

13. Ukahashi, H., and Hisasne, M., US Patent 3,459,818, assigned to Asahi Glass Co. (Aug 5, 1969)

14. Locke, E. G., Brode, W. R., and Henne, A. L., *J. Am. Chem. Soc.*, 56:1726–1728 (1934)

15. Lewis, E. E., and Naylor, M. A., *J. Am. Chem. Soc.*, 69:1968–1970 (1947)

16. West, N. E., US Patent 3,873,630, assigned to DuPont (Mar 25, 1975)

17. Carlson, D. P., US Patent 3,536,733 assigned to DuPont (Oct 27, 1970)

18. Fritz, G. G., and Selman, S., US Patent 3,291,843, assigned to DuPont (Dec 13, 1966)

19. Sherratt, S., in: *Kirk-Othmer Encyclopedia of Chemical Technology*, 2nd ed., (A. Standen, ed.), 9:805–831, Interscience Publishers, Div. of John Wiley & Sons, New York (1966)

20. Dietrich, M. A., and Joyce, R. M., US Patent 2,407,405, assigned to DuPont (Sep 10, 1946)

21. Duus, H. C., Thermochemical Studies on Fluorocarbons, *Indus. Eng. & Chem.*, 47:1445–1449 (1955)

22. Refrew, M. M. and Lewis, E. E., *Ind. Eng. Chem.*, 38:870-877 (1946)

23. Gangal, S. V., Fluorine Compounds, Organic (Polymers), Perfluorinated Ethylene-Propylene Co-polymers, in: *Kirk-Othmer Encyclopedia of Chemical Technology*, 4th ed., 11:644-656, John Wiley & Sons, New York (1994)

24. Allcock, H. R. and Lampe, F. W., *Contemporary Polymer Chemistry*, pp. 48–66, Prentice-Hall, Inc., Englewoods Cliffs, NJ (1981)

25. Ebnesajjad, S., *Fluoroplastics*, Vol. 2: *Melt Processible Fluoroplastics, The Definitive User's Guide and Databook*, William Andrew, Inc., Norwich, NY (2003)

26. Fluorocarbon Resins from the Original PTFE to the Latest Melt Processible Copolymers, Technical Paper, Reg. Technical Conference SPE, Mid Ohio Valley Bicentennial Conference On Plastics (Nov 30–Dec 1, 1976)

27. Gangal, S. V., Polytetrafluoroethylene, Homopolymers of Tetrafluoroethylene, in: *Encyclopedia of Polymer Science and Engineering*, 2nd ed., 17:577–600, John Wiley & Sons, New York (1989)

28. Sherratt, S., in: *Kirk-Othmer Encyclopedia of Chemical Technology*, 2nd ed., (A. Standen, ed.), 9:805–831, Interscience Publishers, Div. of John Wiley and Sons, New York (1966)

29. Renfrew, M. M., US Patent 2,534,058, assigned to DuPont (Dec 12, 1950)

30. Cardinal, A. J, Edens W. L., and Van Dyk, J. W., US Patent 3,142,665, assigned to DuPont (Jul 28, 1964)

31. Wilson, E. K., Materials Made under Pressure, *Chemical and Engineering News* (Dec 18, 2000)

32. DeSimone, J. M., Maury, E. E., Combes, J. R., and Menceloglu, Y. Z., US Patent 5,312,882, assigned to the University of North Carolina at Chapel Hill, NC (May 17, 1994)

33. DeSimone, J. M., Guan, Z., and Elsbernd, C. S., *Science*, 257:945 (1992)

34. Guan, Z., Combes, J. R., Menceloglu, Y. Z., and DeSimone, J. M., *Macromolecules*, 26:2663 (1993)

35. DeSimone, J. M., Maury, E. E., Menceloglu, Y. Z., McClain, J. B., Romack, T. J., and Combes, J. R., *Science*, pp. 256–356 (1994)

36. Dada, E. A., et al., US Patent 5,328,972, assigned to Rohm and Haas Co. (Jul 12, 1994)

37. Fukuia, K., et al., US Patent 3,522,228, assigned to Sumitomo Chemical Co. (Jul 28, 1970)

38. DeSimone, J. M., Romack, T. J., and Treat, T. A., *Macromolecules*, 28(24):8429 (1995)

39. Ebnesajjad, S., *Fluoroplastics*, Vol. 2: *Melt Processible Fluoroplastics, The Definitive User's Guide and Databook*, William Andrew, Inc., Norwich, NY (2003)

40. Marks, B. M., and Whipple, G. H., US Patent 2,037,953, assigned to DuPont (Jun 5, 1962)

41. Jones, C. W., US Patent 5,272,186, assigned to DuPont (Dec 21, 1993)

42. Lewis, R. F., and Naylor, A., *J. Am. Chem. Society*, 69:1968 (1947)

43. Critchley, J. P., Knight, G. J., and Wright, W. W., *Heat Resistant Polymers*, Plenum Press, New York (1983)

44. US Patent 3,766,031, Dillon, J. A., assigned to Garlock, Inc. (Oct 16, 1973)

45. US Patent 4,220,511, Darbyshire, R. L., assigned to Radiation Dynamics, Inc. (Sep 2, 1980)

46. Morgan, R. A., and Stewart, C. W., US Patent 4,904,726, assigned to DuPont (Feb 27, 1990)

3 Introductory Fluoropolymer Coating Formulations

3.1 Introduction

While there are relatively few commercial fluoropolymers, there are thousands of fluoropolymer paint formulations. Previous volumes of this series have dealt with the details of each of the polymers (see Ch. 2, Refs. 1 and 25). The same approach is not practical in this work. Instead, overall technology is discussed and the differences between various formulations must be discussed with the respective manufacturers.

Nearly all fluoropolymer coatings come in two forms. They are either dry powders or liquids. This book does not cover laminates (where a film is glued to a surface) or exotic approaches like gas or plasma phase reactions, and depositions. A basic understanding of paint formulation technology would facilitate understanding fluoropolymers coatings. This chapter provides a general overview and details follow in later chapters.

3.2 Components of Paint

The components of paint generally include the following:

- Binder
- Solvents (except for dry powder coatings)
- Pigments and fillers
- Additives

Coatings and paints always have what formulators call a *binder*. A binder is generally a polymeric material that is solid, or becomes solid, and forms the paint film. The polymeric material generally is classified as *thermoset* or *thermoplastic*. Thermoplastic coatings typically melt when reheated, whereas thermosets undergo a chemical reaction during curing that prevents remelting. Unless the coating is a dry powder, the binder is in a liquid called the *solvent* or *carrier.* Common binders found in household products include materials like acrylics, alkyds, epoxies, or urethanes. Fluoropolymers are usually binders, though they can also be thought of as fillers or additives in some applications.

The binder is *dissolved, dispersed,* or *suspended* in the solvent. The solvent is usually a mixture. It liquefies the other paint components allowing them to spread out over the substrate being coated. Water is considered an important solvent.

Several terms are common in the paint industry. These include medium, vehicle, and carrier. The generally accepted definitions are as follows:

- *Vehicle* is the liquid portion of paint. The vehicle is composed mainly of solvents, resins, and oils.
- *Carrier* usually refers to the solvent.
- *Medium* is the continuous phase in which the pigment is dispersed; it is synonymous with vehicle.

Pigments and *fillers* are small particles added to paints to impart color, affect physical properties such as hardness or abrasion resistance or affect corrosion resistance. Pigments can also be used to influence viscosity, cost, adhesion, moisture permeability, gloss, abrasion resistance, electrical and thermal conductivity, and other properties.

Additives are chemicals added to paints, usually in small amounts to achieve specific effects or solve specific problems. These include:

1. Surfactants, which help stabilize dispersions
2. Viscosity agents
3. Defoamers
4. Surface modifiers
5. Stabilizers
6. Wetting agents
7. Catalysts
8. Others discussed in Ch. 7

3.3 Important Properties of Liquid Coatings

A number of important properties describe liquid coatings. Sometimes these properties are part of the specifications of the coating. Often they are not, but none-the-less are important to know, particularly when problems arise.

3.3.1   Rheology/Viscosity

Viscosity in its simplest definition is the resistance of a liquid to flow or, as the American Heritage Dictionary puts it, "The degree to which a fluid resists flow under applied force." The viscosity is usually a specification. It is frequently reported as a single measurement such as: "200–400 cps (measured by Brookfield Viscometer at 25°C, #2 spindle at 20 RPM)."

At first glance, a specification such as this might imply that the viscosity is a single measurement of a coating. If one looks closely the information in the parentheses, then the implication is that the viscosity depends exactly on how it is measured. The "25°C" in the specification implies that the viscosity is a function of temperature, which it is. The "Brookfield Viscometer" is the instrument used to measure the viscosity, and its inclusion in the specification implies that the viscosity also depends on how it is measured. The Brookfield viscosity measurement is described in more detail in Ch. 13 "Measurment of Coating Performance." Finally, the "#2 spindle at 20 RPM" defines two of the variables one can control on the Brookfield Viscometer. The spindles of a Brookfield are different designs and each has a different surface area. The more area in contact with the liquid and with the spindle, the more force will be required to turn it. The ratio of that force to the area is called the "shear stress":

$$\tau = \frac{F}{A}$$ Eq. (3.1)

where:

F = Force (dynes)

A = Area (cm^2)

τ = Shear stress (dynes/cm^2)

The "20 RPM" defines the rotational speed or velocity. The liquids being moved against this surface include not only the liquid that is in direct contact with the spindle, but also the liquid near its surface. This creates a velocity gradient, which is called the shear rate:

$$D = \frac{dv}{dx} = \frac{v}{x}$$ Eq. (3.2)

where:

D = Shear rate (sec^{-1})

v = Shear velocity (cm/sec)

x = Thickness (cm)

The viscosity is defined as

$$\eta = \frac{\tau}{D}$$ Eq. (3.3)

where:

τ = Shear stress (dynes/cm^2) from Eq. (3.1)

D = Shear rate (sec^{-1})

η = Viscosity (poise = dyne-sec/cm^2)

The main point here is that the viscosity depends upon the shear rate and stress applied by the measuring device and the temperature at which the measurement is made. This also implies that the viscosity will change depending upon how the coating is applied.

Therein lies a key to using and understanding coatings. The viscosity varies with how the coating is used. Actually, viscosity also affects how coatings are manufactured, how they are stored, how they are prepared for use, and how long their shelf life is.

The study of viscosity as a function of shear applied to the coating is called rheology. A test instrument called a Rotoviscometer can make these measurements quickly. This work will not go into deep detail on the physics and chemistry of rheology. Many texts develop the theory, measurement, and interpretation of rheology.[1]

An ideal liquid might have a viscosity that is independent of temperature, shear, and time. Some materials approach this ideal. The ideal is called Newtonian Flow. Figure 3.1 shows the viscosity versus shear rate of a Newtonian fluid. Solvents and water are nearly Newtonian.

Most coatings exhibit viscosity change with shear change. There are many practical reasons for making coatings behave in this manner. For instance, nearly everyone is familiar with house paint. It has very high viscosity as it sits undisturbed in a can. That is high viscosity at a low shear rate. However, when it is rolled or brushed, the shear rate becomes high. The viscosity drops dramatically. This allows the paint to flow out and level well on a wall or ceiling. Then after it is applied and the shear is removed the viscosity rises dramatically preventing or at least minimizing which keeps the paint from dripping or running. This type of viscosity behavior is called "shear-thinning" or pseudoplastic. Figure 3.2 shows the viscosity versus shear relationships of two coatings. Coating "B" shows a linear relationship while "A" shows a more non-linear change. Both are considered pseudoplastic.

The opposite of pseudoplastic flow is dilatant flow or "shear-thickening." Figure 3.3 shows dilatant behavior. Dilatant coatings are rare.

Thixotropic flow is a special case of shear-thinning behavior. A thixotropic coating thins with shear, but its viscosity does not return to the original value after the shear is removed. There is time dependence. Often with enough time, the viscosity will recover. Figure 3.4 shows a thixotropic coating and what is referred to as a thixotropic loop. The arrows indicate how the experiment was run. Starting from low shear, shear is gradually increased. Then, gradually the shear is removed. This type of behavior is sometimes designed into coatings with additives (discussed in Ch.7). It can be used to minimize settling in a coating formulation, increasing the time a coating can be stored.

To give a feeling for the magnitude of the shear forces, several processes and their shears rates are given in Table 3.1.

There are many ways to measure or estimate viscosity (discussed in Ch. 13). Some are easier than others. Some have more variability than others. Many plants and paint shops use a cup method which times how long it takes for a given volume of coating to drain through a hole of specific size. This work will not deal with all of these tests, but Table 3.2 allows estimation and conversion between some common devices.

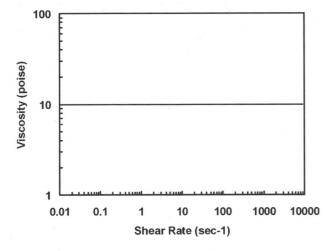

Figure 3.1 Newtonian flow.

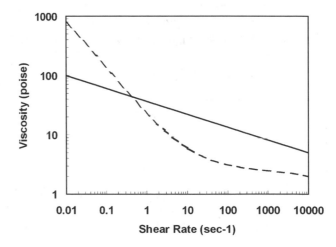

Figure 3.2 Pseudoplastic flow.

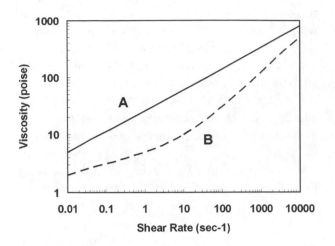

Figure 3.3 Dilatant flow.

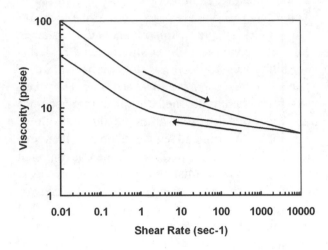

Figure 3.4 Thixotropic flow.

Table 3.1 Approximate Shear Ranges for Common Coating Processes

Process	Shear Range, sec^{-1}
Sagging	$10^{-2} - 10^{-1}$
Leveling	$10^{-2} - 10^{-1}$
Dipping	$10^{0} - 10^{1}$
Flow Coating	$10^{0} - 10^{1}$
Pumping	$10^{0} - 10^{2}$
Mixing	$10^{1} - 10^{2}$
Dispersion	$10^{2} - 10^{5}$
Spraying	$10^{3} - 10^{5}$
Roller Coating	$10^{3} - 10^{5}$
Brushing	$10^{3} - 10^{4}$

Table 3.2 Viscosity Conversion Chart[2]

Poise	cp	Parlin #7	Parlin #10	Fisher #1	Fisher #2	Ford #3	Ford #4	Gardner Holdt Bubble	Gardner Litho.	Krebs Units, KU	Seybolt Univ., SSU	Zahn #1, secs	Zahn #2, secs	Zahn #3, secs	Zahn #4, secs	Zahn #5, secs	Sears, secs
0.1	10	27	11	20			5	A4			60	30	16				
0.15	15	30	12	25			8	A3			80	34	17				
0.2	20	32	13	30	15	12	10				100	37	18				
0.25	25	37	14	35	17	15	12	A2			130	41	19				
0.3	30	43	15	39	18	19	14	A1			160	44	20				
0.4	40	50	16	50	21	25	18	A			210	52	22				19
0.5	50	57	17		24	29	22			30	260	60	24				20
0.6	60	64	18		29	33	25	B		33	320	68	27				21
0.7	70		20		33	36	28			35	370		30				23
0.8	80		22		39	41	31	C		37	430		34				24
0.9	90		23		44	45	32			38	480		37	10			26
1.0	100		25		50	50	34	D		40	530		41	12	10		27
1.2	120		30		62	58	41	E		43	580		49	14	11		31
1.4	140		32			66	45	E		46	690		58	16	12		34
1.6	160		37				50	G		48	790		66	18	13		38
1.8	180		41				54		000	50	900		74	20	14		40
2.0	200		45				58	H		52	1000		82	23	16		44
2.2	220						62	I		54	1100			25	17	10	
2.4	240						65	J		56	1200			28	18	11	
2.6	260						68			58	1280			30	20	12	
2.8	280						70	K		59	1380			32	21	13	
3.0	300						74	L		60	1475			34	22	14	
3.2	320							M			1530			36	24	15	
3.4	340							N			1630			39	25	16	
3.6	360							O		62	1730			41	26	17	

(Cont'd.)

Table 3.2 (Cont'd.)

Poise	cp	Parlin #7	Parlin #10	Fisher #1	Fisher #2	Ford #3	Ford #4	Gardner Holdt Bubble	Gardner Litho.	Krebs Units, KU	Seybolt Univ., SSU	Zahn #1, secs	Zahn #2, secs	Zahn #3, secs	Zahn #4, secs	Zahn #5, secs	Sears, secs
3.8	380										1850			43	28	18	
4.0	400									64	1950			46	29	19	
4.2	420										2050			48	30	20	
4.4	440							Q			2160			50	32	21	
4.6	460							R		66	2270			52	33	22	
4.8	480								00	67	2380			54	34	23	
5.0	500							S		68	2480			57	36	24	
5.5	550							T		69	2660			63	37	25	
6.0	600							U		71	2900			68	40	27	
7.0	700									74	3375				44	30	
8.0	800								0	77	3880				51	35	
9.0	900							V		81	4300				58	40	
10.0	1000							W		85	4600				64	45	
11.0	1100									88	5200					49	
12.0	1200									92	5620					55	
13.0	1300							X		95	6100					59	
14.00	1400								1	96	6480					64	
15.0	1500									98	7000						
16.00	1600									100	7500						
17.0	1700									101	8000						
18.0	1800							Y			8500						
19.0	1900										9000						
20.0	2000									103	9400						
21.0	2100										9850						
22.0	2200										10300						

(Cont'd.)

Table 3.2 (Cont'd.)

Poise	cp	Parlin #7	Parlin #10	Fisher #1	Fisher #2	Ford #3	Ford #4	Gardner Holdt Bubble	Gardner Litho.	Krebs Units, KU	Seybolt Univ., SSU	Zahn #1, secs	Zahn #2, secs	Zahn #3, secs	Zahn #4, secs	Zahn #5, secs	Sears, secs
23.0	2300							Z	2	105	10750						
24.00	2400									109	11200						
25.0	2500							Z-1		114	11600						
30.00	3000									121	14500						
35.0	3500							Z-2	3	129	16500						
40.0	4000									133	18300						
45.0	4500							Z-3		136	21000						
50.0	5000										23500						
55.0	5500										26000						
60.0	6000							Z-4	4		28000						
65.0	6500										30000						
70.00	7000										32500						
75.0	7500										35000						
80.00	8000										37000						
85.0	8500										39500						
90.0	9000										41000						
95.0	9500										43000						
100.0	10000							Z-5	5		46500						
110.0	11000										51000						
120.0	12000										55500						
130.0	13000										60000						
140.00	14000										65000						
150.0	15000							Z-6			67500						
160.00	16000										74000						
170.0	17000										80000						
180.0	18000										83500						
190.0	19000										88000						
200.0	20000										93000						
300.0	30000										140000						

3.3.2   Weight Solids, Volume Solids

Users of coatings need to know how much a given volume of coating they need for their particular coating job. This information is important not only in determining how much to buy, but also how much it costs if they are a processor, a seller, or a distributor of coated items.

Two measures are typically reported by coatings manufacturers. *Weight solids* is frequently a specification and is quite easy to measure. It is simply what is left of the paint on the surface after the volatiles have evaporated during the curing. The American Society for Testing and Materials (ASTM) test for this determination is D1644-01 "Standard Test Methods for Nonvolatile Content of Varnishes." This measure is not directly useful to a paint user. He needs to know the cured coating density to calculate how much dried paint he has.

Volume solids is a more useful measure than weight percent solids. It is the volume of the solid materials left after a gallon of paint's volatile components are removed. With this number on hand, one can easily calculate how much surface area can be painted with a gallon of a particular coating.

Eq. (3.4)
$$C = \frac{1604 \cdot S \cdot E}{I \cdot A}$$

where:

C = Square feet of substrate covered per gallon of paint

S = Percent volume solids

E = Transfer efficiency

I = Dry film thickness of the paint in mils

A = Part area to be coated in square feet

Volume solids is more difficult to measure. The procedure is described by ASTM D2697-03, which is the "Standard Test Method for Volume Nonvolatile Matter in Clear or Pigmented Coatings."

Typically, weight solids is measured and reported as a specified factor. Volume solids, or coverage, is reported but it is generally not measured. It is usually calculated based on the prescribed mixture of the raw materials using their densities. The underlining assumption is that there are no chemical reactions or unusual interactions. This is sometimes incorrect.

The ingredients all can affect the coating properties. The next several chapters (Chs. 4–7) look in more detail at each of the components of a fluorinated coating.

REFERENCES

1. Patton, T. C., *Paint Flow and Pigment Dispersion: A Rheological Approach to Coating and Ink Technology,* John Wiley & Sons, New York (Apr 1979)

2. Viscosity conversion chart based on charts widely available from multiple sources including, "Fine Woodworking," online extra to Vol. 169 (Mar 2004)

4.1 Introduction

The binder in fluoropolymer coatings includes all the film forming materials of the dry and cured coating. Generally binders include all polymers in the coating. Many of the physical properties of the final coating depend on the nature of the polymeric portion of the binder. The fluoropolymer is usually considered to be part of the binder. There can be other binder material such as high-temperature organic polymers or inorganic polymers.

The structures of the polymers in the binder affect the coating properties. While many fluoropolymer topcoats are nearly pure fluoropolymers, primers and one-coats are generally blends of high temperature organic polymers or inorganic polymers with fluoropolymers. One of the functions of the non-fluoropolymer binders is to provide adhesion to the substrate. More on adhesion follows in Sec. 4.2.

In those coatings that are called *thermosets*, there are *crosslinks* in some of the binder molecules. These are strong chemical attachments between molecule chains. Crosslinks are like the rungs in a ladder; they connect different polymer chains, as shown in Fig. 4.1. The crosslinks generally form after the paint is applied, by a chemical reaction frequently started by moisture, light, oxygen, or heat. The crosslinks inhibit redissolving and remelting. Most, but not all, commercial fluoropolymers used in coating have no crosslinks, though other polymer binders blended with them might.

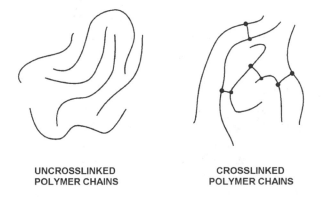

UNCROSSLINKED POLYMER CHAINS   **CROSSLINKED POLYMER CHAINS**

Figure 4.1 Uncrosslinked and crosslinked polymer chains.

4.2 Adhesion

The primary function of non-fluoropolymer binders in fluorocoatings is to provide adhesion to the substrate. The non-stick character of fluorocoatings is well known, so the problem with getting adhesion to the substrate is expected. Adhesion is related to absorption of binder molecules to that surface. Chemical bonds between the binders and the substrate will increase adhesive strength.

To form a good adhesive bond, however, the molecules of the binder must reach and wet the surface. Wetting the surface means the binder spreads out to cover the surface completely, thus displacing the air or any other substance from the interface.

The viscosity and surface tension are, therefore, important. When the binder reaches the surface it must bond to the surface with sufficient force. Surface energy and wetting are involved. High rates of wetting are often associated with low viscosity during application.[1] The binders must also wet the substrate if and when they melt.

A good description of the *exact* chemical nature of a surface is hard to find except in rare instances. Practically, all surfaces are contaminated with foreign materials. The atoms of a substrate are not homogeneously arranged and dislocations and flaws exist. All the useful metals are coated to some degree with oxide, hydroxide, carbonates, etc. Therefore, describing the chemistry of adhesion is difficult because it is not well understood. Adhesion is usually optimized experimentally by trial and error. Measuring or comparing adhesion, is discussed in Ch. 13 on coating performance.

There are polymer binders that earn the description of adhesives because they tend to adhere to many surfaces. In fluorocoatings the polymer role is similar. Besides sticking to a surface, the binder must remain attached to the surface during the coating processing and during use. To maintain adhesion during large temperature changes, it must be sufficiently elastic to withstand the dimensional changes of the substrate during coating processing without the formation of cracks. Sometimes the binder must maintain its integrity even when the coating is bent or otherwise deformed, after it has been applied and cured. It must be hard enough to withstand, to an acceptable level, the effects of abrasion and/or erosion. It also needs to maintain the properties in spite of exposure to different environments during use, such as chemical, water, or corrosive exposure.

As stated previously, many fluoropolymer one-coats and primers are blends of binders. Since the fluoropolymers do not dissolve in the solvents used in the paints, they are more like small particles distributed throughout the liquid paint. However, when the paint is baked above the melting point of the binders, the binder molecules become fluid under heat and may intimately mix. If the polymer chains intertwine to a very large extent, the mixture is sometimes referred to as an *interpenetrating polymer network*. It forms a very homogenous mixture of the two binder resins. Depending on the desired paint properties, this may or may not be desirable. If the binders start out well mixed, and one can get them to partially separate, then one can envision a coating that has a high concentration of fluoropolymer on the surface and a high concentration of the other binder at the substrate. This process is driven thermodynamically resulting in the formation of a concentration gradient area between the top and substrate surfaces, shown in Fig. 4.2. This is called *stratification*, and is the basis of many one-coat products and primers. DuPont's early products using this principle were called Teflon-S®. The surface of the finish behaves like a pure fluoropolymer, and the material contacting the substrate behaves like an adhesive.

Binders serve several functions in a paint or coating. These include:

1. Provide adhesion
2. Increase strength
3. Alter barrier properties
4. Pigment dispersion
5. Control the electrical properties
6. Improve the durability

4.3 Non-Fluoropolymer Binders

Almost any polymer used in a coating can be blended with a fluoropolymer if the formulator is clever enough. These binders might be split into two general groups, those processed above the melt point of the fluoropolymers added and those processed below the melt temperature of fluoropolymer. Usually the best properties of fluoropolymers are obtained for those processing temperatures above the melt point. The binders that are processed below the melting point are very diverse and cover all of coating, paint, and ink chemistry. This subject is too broad for discussion in this text.[2][3] This section will concentrate on those polymers used that are processed at high temperatures.

4.3.1 Polyamide/Imide (PAI)

Polyamide/imide (PAI) is one of the most common and most important binder materials. It is the basis for nearly all the cookware primers. The highly aromatic molecule, when cured, has very high thermal stability and can bind strongly to most metal substrates. It can be made from a mixture of trimellitic anhydride and methylene dianiline as shown in Fig. 4.3. PAI is amorphous and strongly colored. It is considered a thermosetting resin. The most common chemical structure is shown below in Fig. 4.3, but other amines or anhydrides could be used.

Polyamide/imide resins were originally designed for electromagnetic wire coatings. They are available from a number of manufacturers in the form of solutions or powders. The properties are summarized in Table 4.1 and Sec. 4.5.

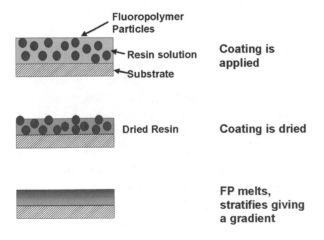

Figure 4.2 Stratification in fluoropolymer resin bonded coatings.

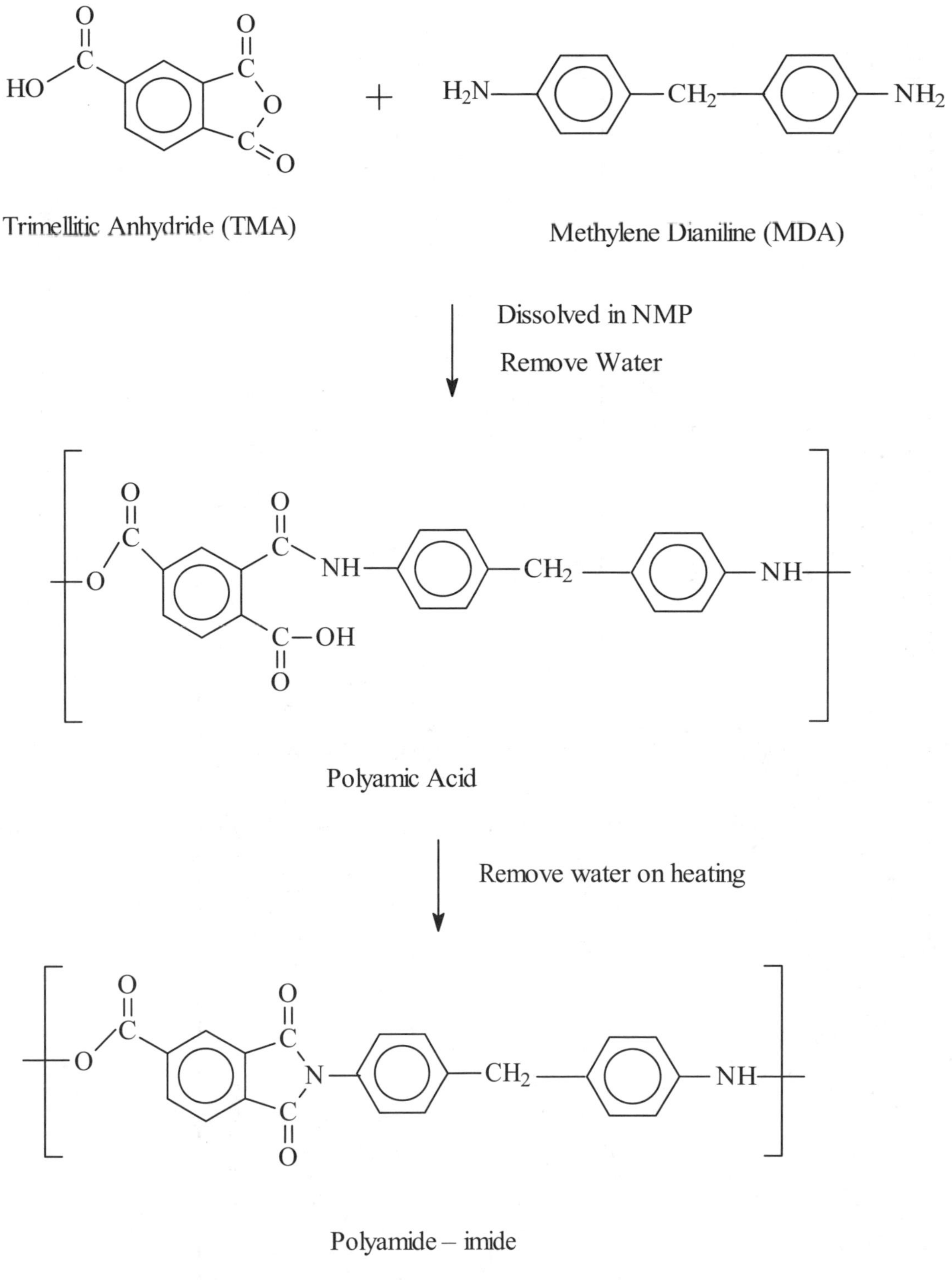

Figure 4.3 Curing chemistry of polyamide/imide polymers.

Table 4.1 Commercial Polyamide/Imide Resins

Manufacturer	Trade Name	Product Code	Form	Weight Solids	Solution Viscosity, 25% in NMP cps	MW (×10³)	Glass Transition, °C
Solvay Advanced Polymers	Torlon®	AI-10	Powder	90%	800		
		AI-30	Wet Powder	35%	35		
		4000TF	Powder	99%	7000		
		4000T-LV	Powder	99%	7000		
		4000T-MV	Powder	99%	42,000		
		4000T-HV	Powder	99%	85,000		
		AI-50	Wet Powder	35%			
Huntsman Chemical	Rhodeftal®	Rhodeftal 200	Solution	28%			
		Rhodeftal 311	Solution	24%			
		Rhodeftal 322	Solution	22%			
Toyobo	Vylomax®	HR11NN	Solution	15%		15	300
		HR12N2	Solution	30%		8	255
		HR13NX	Solution	30%		10	280
		HR14ET	Solution	25%		10	250
		HR15ET	Solution	25%		6	260
		HR16NN	Solution	15%		30	320
Bayer Polymers	Resistherm®	AL 244 L	Solution	44%	1500–3500		
		AL 336 L	Solution	36%			

4.3.2 Polyethersulfone (PES)

Polyethersulfone (PES) is an amorphous, transparent, and pale amber high-performance thermoplastic and is the most temperature resistant transparent commercially available thermoplastic resin. It has relatively high water absorption. Stable solutions can be made if solvents are correctly chosen. Chemical structure repeating units of several of the commercial polymers are shown below:

PES

PSU

PPSU

These materials are very important. They are the most common high temperature polymers used with fluoropolymers that are nearly colorless and permit light colored primers and one-coat coatings to be made. The properties are summarized in Table 4.2 and Sec. 4.5.

4.3.3 Polyphenylenesulfide (PPS)

Polyphenylenesulfide (PPS) is a semicrystalline high performance thermoplastic. It has low water absorption and is resistant to and insoluble in all organic solvents even at elevated temperatures. It has a relatively low melting point of 285°C. It has a very low melt viscosity and so flows out well. It is generally strongly colored, so only dark colors can be made

with this polymer. It adheres well to most metals. The structural repeating unit is:

Some of the commercial PPS resins are shown in Table 4.3.

4.3.4 Polyimide (PI)

Polyimide (PI) is structurally similar to Polyamide/imide. The properties are similar except that they possess even higher thermal stability. They are also much more expensive, somewhat harder to process and are not acceptable for most food contact applications. Like PAI, the highly aromatic molecule, when cured, has very high thermal stability and can bind strongly to metal substrates. It is also strongly colored. The use of polyimides in commercial fluoropolymer coatings has been limited.

The materials are known by several tradenames. Kapton® and Vespel® are Dupont well-known materials. Others include Kinel®, Upilex®, and Upimol®. These materials are normally produced in solvents like NMP or DMF by the reaction of diamines with dianhydrides. The reaction forms a polyamic acid, which is soluble. Once the polymers are fully cured to the polyimide form they are insoluble. Besides excellent high temperature properties, they also possess radiation resistance, low flammability and smoke emission, low creep, and high wear resistance. Most polyimides have moderately high water absorption and are prone to hydrolysis and attack by alkalis and concentrated acids.

The chemical structure repeating unit follows, though many monomer variations are used:

Table 4.2 Commercial PES Resins

Manufacturer	Trade Name	Product Code	Form	Comments
BASF	Ultrason® E	E 1010	Pellets	HDT 216°C
		E 2010	Pellets	HDT 218°C
		E 2020 P*	Flake	HDT 218°C
		E 3010	Pellets	HDT 218°C
		E 6020 P*	Flake	HDT 208°C
	Ultrason® S	S 2010	Pellets	HDT 181°C
		S3010	Pellets	HDT 186°C
		S 6010	Pellets	HDT 186°C
SOLVAY	Radel® (PES)	A-100		HDT 204°C
		A-200A		(PEES) Hydroquinone 25% HDT 204°C
		A-300A		HDT 204°C
		A-701		HDT 202°C
		R-5000		HDT 207°C
		R-5100 NT15		HDT 207°C
		R-5500		HDT 207°C
		R-5800		HDT 207°C
	Udel® (PSU)	P-1700 NT11		HDT 174°C
		P-1710		HDT 174°C
		P-1720		HDT 174°C
		P-1700 NT06		HDT 174°C
		P-1700 CL2661		HDT 174°C
		P-3500		HDT 174°C
		P-3703		HDT 174°C
Gharda	Gafone® PES	PES 3000		
		PES 3200		
		PES 3300		
		PES 3400		
		PES 3500		
		PES 3600		
	Gafone-S® PSU	PSU 1200	Powder	
		PSU 1300	Granules	HDT 184°C
		PSU 1400	Granules	
		PSU 1500	Granules	
	Gafone-P® PPSU	PPSU 4300		
Sumitomo Chemical	Sumikaexcel®	PES 3600P		
		PES 4100P		
		PES 4800P		HDT 203°C
		PES 5200P		
		PES 7600P		

*HDT is heat deflection temperature, a measure of where a polymer softems.

Table 4.3 Commercial PPS Resins

Manufacturer	Trade Name	Product Code	Melt Flow, g/10 min	Comments
Ticona	Fortron®	PPS 0203HS PPS 0320 PPS Powder		HDT 110°C
GE	Supec®			
Chevron Phillips Chemical Company	Ryton®	V-1 PR-11 P-6	5000 5000 380	
Solvay	PrimeF®	C-0016 C-0037	1600 3700	

4.3.5 Polyether Ether Ketone (PEEK)

Polyether ether ketone (PEEK) is a high performance thermally stable thermoplastic. It is strong, stiff, and hard, has good chemical resistance, and inherently low flammability and smoke emission. PEEK is pale amber in color. Thicker samples are usually semicrystalline and opaque. Thin films are usually amorphous and transparent, though still amber in color. PEEK also has very good resistance to wear, dynamic fatigue, and radiation. However, it is difficult to process and very expensive. Filled grades, including those designed for bearing-type applications, are also used. The chemical structure repeating unit is:

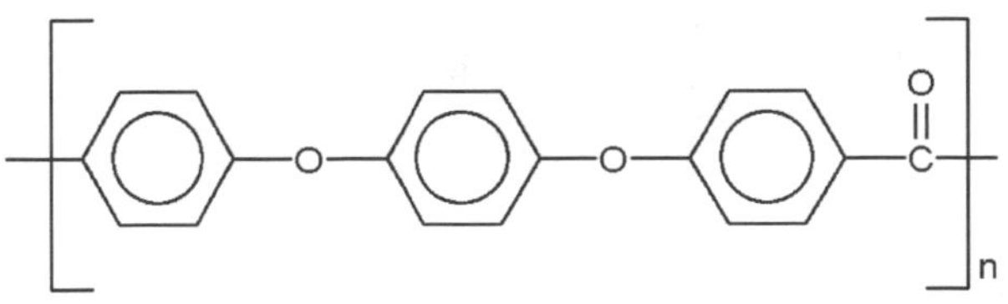

4.3.6 Polyetherimide (PEI)

Polyetherimide (PEI) is another high temperature polymer that has potential in fluoropolymer coatings applications, but it has not yet had a large impact in major commercial coatings. It is known mostly by the tradename Ultem® (Table 4.4). It is an amorphous, transparent, and amber thermoplastic with the characteristics similar to PEEK. Light colored coatings can be made from this material. It is soluble in NMP. Relative to PEEK, it is less temperature resistant, less expensive, and lower in impact strength. It is prone to stress cracking in chlorinated solvents. The chemical structure repeating unit is:

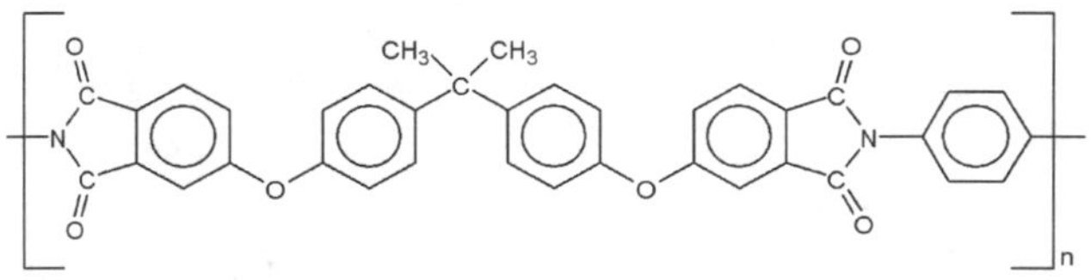

GE Advanced Materials is the primary manufacturer of this resin. They also make several variations. One is called polyetherimide sulfone, which offers higher thermal stability. The SO_2 sulfone group is more thermally stable than the CH_3-C-CH_3 group. Its structure is:

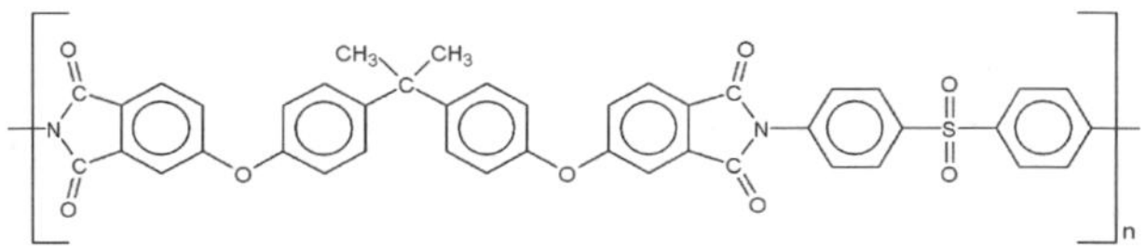

GE also makes an analog that is called a block copolymer of polysiloxane and polyetherimide called Siltem®. This material is more flexible than its Ultem® cousins.

Table 4.4 Commercial PEI Resins

Manufacturer	Trade Name	Product Code	Glass Transition Temperature, °C	Melt Flow Rate, g/10 min @ 337°C
GE	Ultem®	1000/1010	217	9
GE	Ultem®	5001/5011	227	17.8
GE	Ultem®	6050	248	
GE	Siltem®	STM1500		
GE	Ultem®	6000	234	
Mitsui	Aurum®	PL450C	250	

4.3.7   Other Less Common Binders

In this section a few additional binders that are used in some commercial products will be mentioned. Details on the chemistry of these polymers are available elsewhere.[4]

4.3.7.1 Acid

One of the first binders used for a fluoropolymer finish was a mixture of chromic and phosphoric acids with PTFE dispersion. It is commonly called "acid primer." The mixture is generally sold in two packages. One contains chromic-acid/phosphoric-acid, the other the rest of the components of the coating. The chemistry is not well understood, but it is believed that the two acids combine to form a hard inorganic glass that binds to substrate and fluoropolymer alike. The properties of acid primers can be controlled or optimized for a particular end use. This is done by changing the ratio of the acids and the PTFE dispersion.

4.3.7.2 Acrylic

Acrylics and methacrylics describe large family of chemically related polymers where polymethylmethacrylate (PMMA), (see Fig. 4.4) is the most common one. Common tradenames include Lucite® and Plexiglass®. PMMA is an amorphous, transparent, and colorless thermoplastic that is hard and stiff, but brittle. It has good abrasion, UV resistance, and

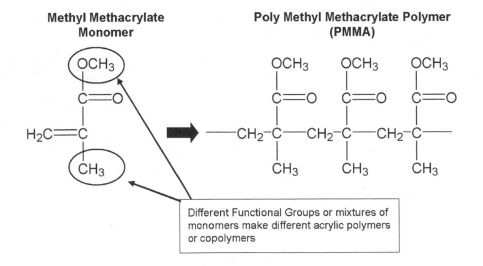

Figure 4.4 Acrylic chemistry.

excellent optical clarity, but poor low and high temperature performance and solvent resistances.

For fluorinated coatings it is used as a low-bake binder, primarily for plastic substrates.

4.3.7.3 Phenolic

Phenolic resins are thermosetting polymers formed from the reaction of phenol (C_6H_5OH) and formaldehyde or similar molecules. They became first known by the tradename Bakelite®.[5][6] They are sometimes called novolac resins. The basic structure is given:

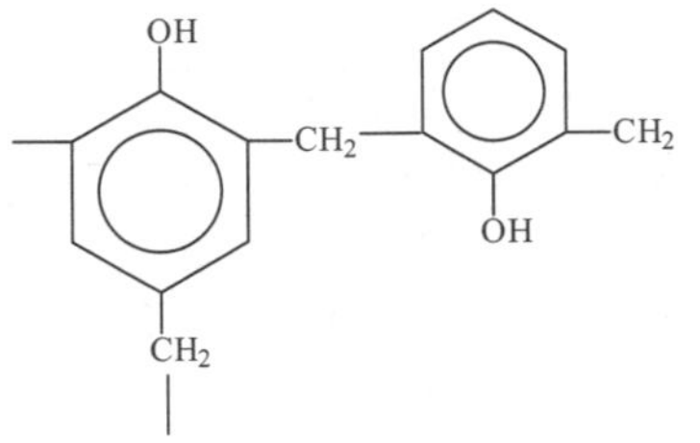

When cured, the resins are transformed from the fusible, thermoplastic state to a densely, highly cross-linked thermoset matrix. Methylene bridges join the phenol molecules in three dimensions. This strong, rigid polymer has superior resistance to a wide range of chemicals and high heat resistance.

4.3.7.4 Epoxy

Epoxy resins are well known for their adhesive, thermal resistance and physical properties. The chemistry has been widely studied. Nearly any reactive molecule will react in some way with an epoxy group.[5] Therefore, the offering of resins and crosslinking (or curing) agents is very large. A common epoxy is called diglycidyl ether of bisphenol A (DGEBA resins). A small segment of its structure is given below:

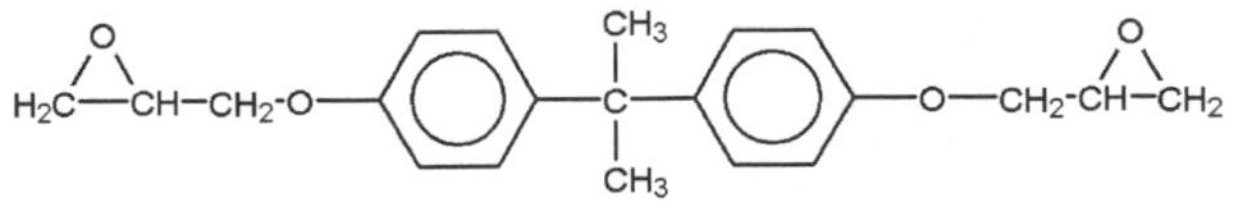

A common curing agent, hardener or crosslinker is a melamine. The simplest melamine has the following structure:

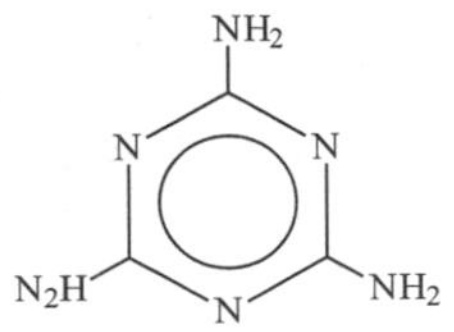

The epoxy group from the DEGBA resin can react with the amine groups to harden or crosslink the resin in the reaction shown in Fig. 4.5. There are many DEGBA resins, other epoxy resins, and many crosslinkers that will give a wide range of cured polymer properties that can cure from room temperature to a 400°F (204°C) bake. The resins are considered thermosets. The cured resins offer excellent chemical resistance, can be quite hard, and many have good thermal resistance. Some can be processed as high at 600°F (316°C) and can be used at 300°F (149°C) continuously, or see intermittent temperatures to 500°F (260°C).

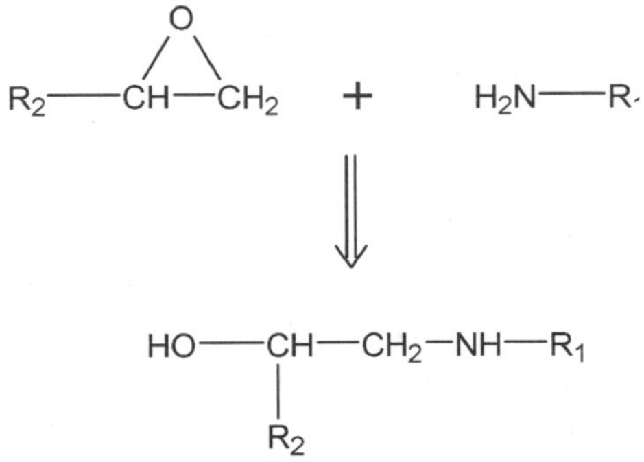

Figure 4.5 Basic epoxy amine-curing reaction.

4.3.7.5 Polyurethane

Polyurethanes are generally formed by reacting an isocyanate group (N=C=O) with an alcohol. The isocyanate group is very reactive. Coatings based on polyurethane chemistry can be made to cure at room temperature or at an elevated temperature. Room temperature curing coatings are provided in a two-package form. The two components are mixed and the end-user has a limited time to apply them, as the polymerization reaction begins immediately, though slowly. Single package polyurethanes require some heat. The heat generates the isocyanate group from a molecule. An example of the urethane reaction is described in Fig. 4.6.

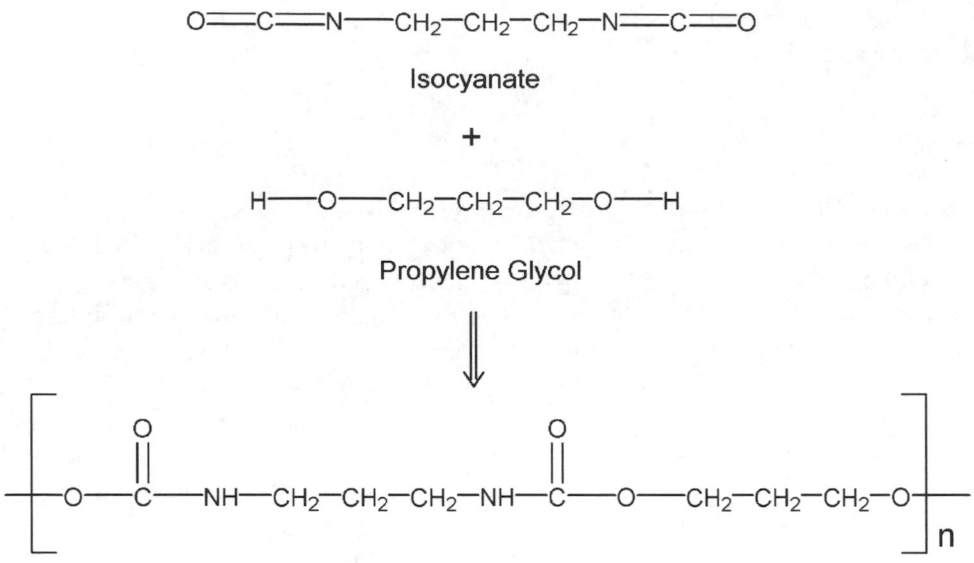

$$O = C = N - CH_2 - CH_2 - CH_2 - N = C = O$$

Isocyanate

+

$$H - O - CH_2 - CH_2 - CH_2 - O - H$$

Propylene Glycol

⇓

$$\left[O - \overset{\overset{\displaystyle O}{\|}}{C} - NH - CH_2 - CH_2 - CH_2 - NH - \overset{\overset{\displaystyle O}{\|}}{C} - O - CH_2 - CH_2 - CH_2 - O \right]_n$$

Figure 4.6 Polyurethane monomers and curing chemistry.

4.3.7.6 Alkyd

Alkyd resins are organic polyesters. They are derived from a polybasic acid, a molecule with two or more acid groups and a polyhydric alcohol, a molecule with two or more hydroxyl groups. Examples of two such molecules are given in Fig. 4.7 with the polymeric structure given in Fig. 4.8. Phthalic anhydride and glycerol are examples. An esterification reaction between the two ingredients produces an ester molecule. Alkyd resins are made to low molecular weights, generally 2,000–10,000. However, they contain unreacted hydroxyl and acid groups that can react with other molecules or with oxygen to form the solid binder.

4.3.7.7 Electroless Nickel Plating

Electroless nickel plating works without the external current source used by galvanic electroplating techniques. Electroless nickel plating is also known as chemical or autocatalytic nickel plating. The process uses chemical nickel plating baths. The

most common electroless nickel is deposited by the catalytic reduction of nickel ions with sodium hypophosphite in acid baths at pH 4.5–5.0 at a temperature of 85°C–95°C. The bath can contain PTFE. The resulting platings contain typically up to 13% phosphorus by weight and perhaps 20%–25% of PTFE by volume. PTFE powder is usually used because PTFE dispersions are unstable at 85°C. Coatings of this type have low friction, exceptional resistance to wear, and good corrosion resistance.

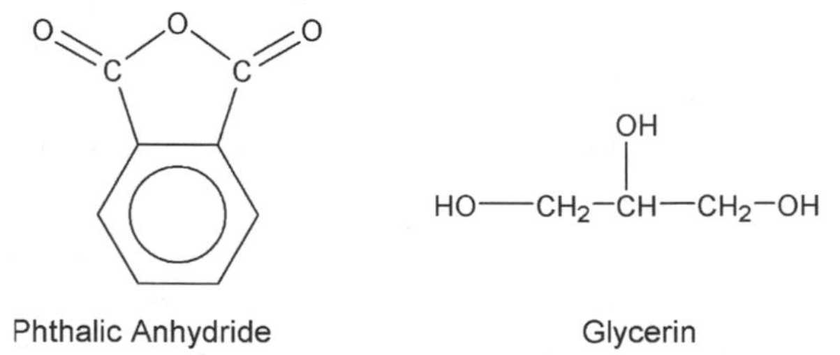

Phthalic Anhydride   Glycerin

Figure 4.7 Monomers used to make alkyds.

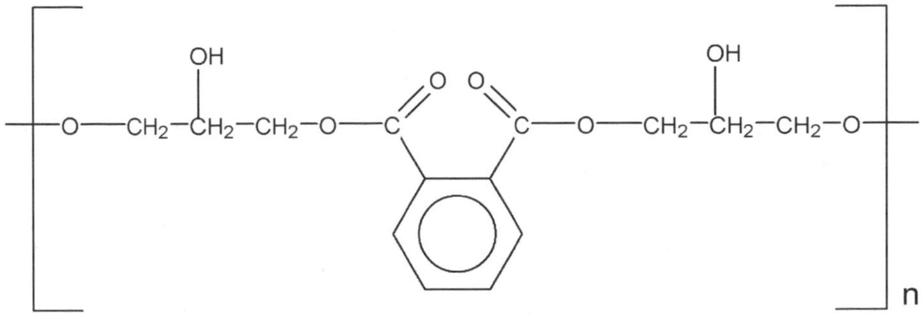

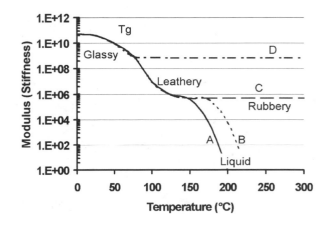

Figure 4.8 Structure of an alkyd polymer.

4.4 Effect of Temperature on Properties of Binders

Because many fluorinated coatings are used over wide ranges of temperatures, it is important to consider and understand the effect temperature variation has on the binders in the coating. This includes not only the non-fluoropolymer binders, but also the fluoropolymers. An increase in temperature increases the mobility of the polymer chain segments. Polymers also tend to expand and take up more free volume. Temperature increase, therefore, reduces the resistance to flow of the polymeric materials in the binders. The physical properties of the binders are, therefore, expected to change with changing temperature. As an example, the stiffness of each polymer incorporated in the coating changes as a function of temperature. A measure of stiffness is called modulus and its units are typically dynes per square centimeter (same as pressure). The different ways that modulus is measured is beyond the scope of this book, but the reader can refer to ASTM D747-02, "Standard Test Method for Apparent Bending Modulus of Plastics by Means of a Cantilever Beam."

Accepting that this measure can be made, Fig. 4.9 shows typical stiffness measures as a function of temperature for four hypothetical polymers:

A - Thermoplastic amorphous polymer with no crosslinking

B - Thermoplastic polymer with slight crystallinity

C - Thermoset polymer with a low level of crosslinks

D - Thermoset polymer that is highly crosslinked

Polymers A and B melt, while C and D do not since they are crosslinked thermosets.

Referring to Fig. 4.9, at low temperatures, all four polymers are relatively hard and brittle and are called glassy. In this hypothetical example they all have the same stiffness. As temperature rises, the polymers change and become less stiff. They might be called "leathery." As the temperature rises further, they lose more stiffness and become rubbery. The temperature that is half way between the glassy and rubbery states is called the *glass transition*

Figure 4.9 Polymer stiffness as a function of temperature.

temperature, noted as T_g. Polymer D, because it is highly crosslinked, is the least rubbery, or retains more stiffness or strength at higher temperature. Polymer C retains less stiffness than D. Eventually polymers A and B melt and have no stiffness at all.

There is some variation of glass transition temperature of polymers with their molecular weight. In general, higher molecular weight results in higher glass transition temperature.

The glass transition temperature is associated with various practical properties of the binder, particularly those that do not have melting points. Flexibility, such as if the coating is bent and abrasion resistance can be very different above and below the T_g.

4.5 Comparison of Properties of Non-Fluoropolymer Binders

Properties of the various high temperature non-fluoropolymer binders are summarized and compared in the following tables (Tables 4.5–4.10).

Table 4.5 Chemical Resistance

	PAI	PES	PPS	PEI	PI	PEEK
Acids – concentrated	Good	Fair	Fair		Good	Fair
Acids – dilute	Good	Good	Good	Good	Good	Good
Alcohols	Good	Good	Good		Good	Good
Alkalis	Poor	Good	Good	Fair	Poor	Good
Aromatic Hydrocarbons	Good	Fair	Good	Good	Good	Good
Greases and Oils	Good	Good	Good		Good	Good
Halogens		Good	Fair			Good
Ketones	Good	Poor	Good		Good	Good

Table 4.6 Electrical Properties

	PAI	PES	PPS	PEI	PI	PEEK
Dielectric constant @1MHz	5.4	3.7	3.8–4.2	3.1	3.4	3.2–3.3
Dielectric strength, kV.mm^{-1}	23	16	18	30	22	19
Dissipation factor @ 1 MHz	0.042	0.003	0.0013–0.004	0.0013 @ 1 kHz	0.00018	0.003
Surface resistivity, ohm/sq	5×10^{18}		10^{16}	4×10^{13}	10^{16}	
Volume resistivity, ohm-cm		10^{17}	10^{16}	7×10^{15}	10^{18}	10^{15}–10^{16}

Table 4.7 Mechanical Properties

	PAI	PES	PPS	PEI	PAI	PEEK
Compressive Strength, MPa	170-220			140		
Compressive Modulus, GPa				2.9		
Abrasive Resistance ASTM D1044, mg/1000 cycles		6		10		
Elongation at break, %	7-15	80	1.2	60	8-70	50
Hardness – Rockwell	E72-86	M88	R123	R125	E52-99	M99
Izod Impact Strength, J m-^{1}	60-140	85	75-80	50	80	85
Tensile Modulus, GPa	4.5-6.8	2.6	7.6-12.0	2.9	2.0-3.0	3.7-4.0
Tensile Strength, MPa	110-190	95	124-160	85	70-150	70-100

Table 4.8 Physical Properties

	PAI	PPS	PES	PEI	PI	PEEK
Density, g cm^{-3}	1.42-1.46	1.66	1.37	1.27	1.42	1.26-1.32
Flammability	V0	V0	V-0	V-0	V0	V0
Limiting Oxygen Index, %	44-45	46	34-41	47	53	35
Radiation resistance, Alpha	Good		Good		Good	Good
Radiation resistance, Beta	Good				Good	
Radiation resistance, Gamma	Good				Good	
Refractive Index			1.65		1.66	
Resistance to Ultraviolet	Good		Fair	Good	Good	Fair
Water absorption, 24 hours, %	0.3	<0.05	0.4-1	0.25	0.2-2.9	0.1-0.3
Water absorption, equilibrium, %	3-4		2.2	1.3		0.5

Table 4.9 Thermal Properties

	PAI	PPS	PES	PEI	PI	PEEK
Heat-Deflection Temperature, 0.45 MPa, °C		>260	>260	200		>260
Heat-Deflection Temperature, 1.8 MPa, °C	≤ -200	240	203	190	360	160
Lower Working Temperature, °C	1.0		-110		-270	
Thermal Conductivity @ 23°C, W m^{-1} K^{-1}	0.26-0.54	0.29-0.45	0.13-0.18	0.22	0.10-0.35	0.25
Thermal Expansivity, × 10^{-6} K^{-1}	25-31	22-35	55	56	30-60	47-108
Upper Working Temperature, °C	200-260	200-260	180-220	170-200	250-320	250
Glass Transition Temperature, °C	280	88	230	220	325	140

Table 4.10 Surface Properties

	PAI	PPS	PES	PI
Surface Energy, dynes/cm	~40	38	50	40

REFERENCES

1. Wu, S., *Polymer Interface and Adhesion,* Marcel Dekker, Inc., New York (1982)

2. Patton, T. C., *Paint Flow and Pigment Dispersion*: *A Rheological Approach to Coating and Ink Technology*, John Wiley & Sons, New York (Apr 1979)

3. Lambourne, R., and Strivens, T. A., *Paint and Surface Coatings: Theory and Practice*, William Andrew Publishing, Norwich, NY (1999)

4. Allcock, H. R., and Lampe, F. W., *Contemporary Polymer Chemistry*, Prentice-Hall, Inc., Englewood Cliffs, NJ (1981)

5. Lee, H., and Nevlle, K., *Handbook of Epoxy Resins*, McGraw-Hill, New York (1967)

5 Pigments, Fillers, and Extenders

5.1 Introduction

Pigments, fillers, and extenders are essentially insoluble (in solvents) fine particle sized solids that are added to a paint formulation. The pigments are usually added to a fluoropolymer coating for appearance, such as for color or sparkle. Dyes, which are soluble colorants, are used in some coatings applications, but not usually in fluoropolymer coatings. Most dyes are organic and would decompose during processing at elevated temperatures. Fillers are typically added to a coating to reduce cost, or for performance enhancement. Performance enhancements might include properties such as permeability, abrasion resistance, and conductivity. The difference between pigments and fillers is not always clearcut. One might also consider a powdered fluoropolymer as a filler, particularly if the coating is not processed above the melt point of the fluoropolymer.

Pigments and fillers are usually solid fine particles. They are added to coatings formulations for a number of reasons:

1. For appearance—this is the most obvious reason for adding pigment, but appearance factors besides color, include hiding, roughness, or gloss.

2. To alter rheological properties—to produce thixotropy or pseudoplasticity, which are described in more detail in Ch. 3.

3. For economy—to reduce the cost by adding inexpensive ingredients.

4. As carrier for active materials:

    a. Anticorrosive—to protect the substrate from chemical or environmental attack.

    b. Antibacterial—to prevent bacteria from growing on the surface.

    c. Fireproofing—to provide fire resistance.

    d. Electrical conductivity—to dissipate static electricity or provide electromagnetic shielding.

    e. Ultra violet protection—to protect against damage from the sun.

5. Reduce permeability.

6. Reinforcement—to improve physical properties such as strength and abrasion resistance of the coating.

Pigments in paint formulations can be as fine as 0.02 microns in diameter or as coarse as 100 microns. They are generally insoluble. For some uses, however, (for example, inhibition of corrosion, fungicidal action) their solubility, though small, is important. Besides the particle size, shape, and surface area, the surface chemistry of the pigment is also important.

Most pigments are mined and crushed, or precipitated as crystals, and crushed. Their surfaces are rounded by processing often in a liquid media. For example, washing and drying of a filter cake containing the pigment are common operations. Sometimes chemicals are added to pigment slurries to aid in filtration, and these remain, to some extent, stuck to pigment surfaces. The treatments to pigment surfaces affect the way they behave when added to paint formulations. Small amounts of common soluble salts also remain from the water used for washing. An aerosol process such as condensation makes a few expensive pigments. Nano-sized metal powders and metal oxides can be made this way.[1]

For pigments or additives that are made from crystalline materials, the properties can be dependent on the direction of cleavage and the crystalline form. The surfaces, therefore, differ in their affinity or absorption characteristics. The surface area of a pigment depends on the particle size and shape.

Chemistry of a pigment's surfaces helps to determine the rheological properties of a fluid coating that contain them and can also affect the mechanical properties of an applied coating. Surface chemistry is often critically important in the stability of the final coating.

5.2 Dispersion of Pigments

Pigments and fillers can be added to a liquid coating directly or "*stirred in*" as is commonly referred to in the paint industry. However, pigments as purchased are usually supplied as a filter cake or as a dry powder. The individual particles usually consist of clumps of particles, often called *agglomerates*. Before adding pigments to paints or coatings, the agglomerates of pigment are separated and sometimes made finer by breaking crystals, by a process called grinding or dispersion. Dispersion consists of three steps:

1. As the dry pigment is added to the liquid, it must be *wet* by the liquid composition. This displaces the air, water, or other contaminants on the surface of the pigment. If the solvent system does not wet the pigment particles, then a surfactant molecule or resin must be used to promote the wetting.

2. Once the particles are wetted, the next step is stabilizing, which inhibits the pigment particles from reagglomerating with each other.

3. Once the particles are wetted and initially stabilized, then the agglomerates are broken down, sometimes to primary particles, in the grinding process.

Once the particles are wetted in the millbase, then the agglomerates are separated to primary particles in the grinding process. Grinding usually refers to the breaking of hard agglomerates or large pigment particles, but it often accomplishes the dispersion process as well. The grinding process requires milling equipment that is sometimes called a *smasher*. There are several smashing type mills for dry pigments. The separation of agglomerates with minimal primary particle size reduction can be accomplished by generating a lot of shear on the pigments in a mill described as a *smearer*. For most grinding or dispersion processes used in the fluorocoatings area, the equipment is a hybrid of these. It does a small amount of grinding and a great deal of dispersing.

The most common types of mills used in the fluorocoatings industry fall into four basic types:

1. Ball or pebble milling
2. Sand and bead mills
3. Attritors
4. High speed disperser (HSD)

5.2.1 Ball or Pebble Milling

Ball mills are steel or ceramic jars (in lab size mills) or cylinders. They are usually lined with a nonmetallic liner such as ceramics to avoid metallic contamination of the millbase. Mills are mounted horizontally and are partially filled with natural pebbles, steel or ceramic balls, or cylinders. They are rotated at a rate that causes the grinding media to cascade as shown in Fig 5.1. The cascading action imparts impact, "smashing", and shear, "smearing", to the pigment particles in the liquid added to the mill. The mixture of pigment particles, additives, and liquid is commonly called a *millbase*.

There are a number of variables to control and consider when using ball/pebble mills:

1. Mill variables:
   a. Diameter of mill
   b. Rotational speed
   c. Temperature
2. Ball/pebble:
   a. Load
   b. Density
   c. Size
   d. Shape
3. Millbase charge:
   a. Relative volume
   b. Viscosity (changes with time)
   c. Density
   d. Compositional ratios

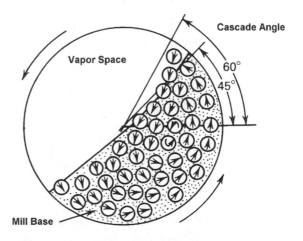

Figure 5.1 Cascading action in a pebble mill.[2]

Rotational speed is particularly important and three basic operating conditions are depicted in Fig. 5.2. When a mill is rotated too quickly, the grinding media and millbase are driven to the outside of the mill by centrifugal force. No milling action occurs at all in this condition. Slowing down the rotational speed slightly creates what is called a *cataracting* action, which is inefficient, leads to excessive media wear (which becomes part of the dispersion), and can generate foam. The optimum is the *cascading* action shown in the figure.

After the mill is run for the specified time, a sample is usually withdrawn from the mill, and its properties are checked. The properties checked are usually viscosity and fineness. If the dispersion passes the control tests, the mill must be drained. Gravity is usually used to drain a ball mill, but air pressure can be added to assist the draining for higher viscosity dispersions. Material usually remains on the mill walls and on the media surfaces. The residues can be washed out and thrown away, constituting a yield loss. Sometimes, a specific amount of solvent is added, the mill is closed, and then rotated for a short while. This wash is then added to the main part of the dispersion, which needs to be thoroughly mixed before use. Pebble milling is an old technology that was very common in the early days of fluorocoating technology (1960's), but it is rare now.

5.2.2 Shear Process Dispersion

If the agglomerates are loosely bound and the pigment/filler particles do not need to be made smaller, shear processes can complete the dispersion process. There are many dispersing mills to select from, but their underlying operating principle is the same. Nearly all can be called "internally agitated media milling." These mills operate on a single principle. A container is filled with small hard spheres, such as glass beads. The mixture to be ground and dispersed is added next. An agitator is mounted to stir the grinding media bed and mixture rapidly. The mixture of pigment and liquid medium experiences shear between the sphere surfaces that are in relative motion due to the agitator. Agglomerates that are subjected to sufficient shear are broken down into smaller fragments, thus forming the dispersion. There are three basic types of media mills:

- Horizontal media mills
- Attritor
- Sand mill

5.2.2.1 Media Mills

The principle of operation of these media mills is the same, all three use what is called the grinding media. These are small hard spherical particles. A sand mill uses sand, but other mills usually use very small glass beads, ceramic beads, or small metal shots. The media are put into a container that is generally closed. An agitator stirs the media mixture. The agitator shape varies, but it often has a spinning disk or pin design, both of which rotate at high RPM. The millbase is pumped from the bottom of a vertical media mill, or from one end of the cylinder on a horizontal media mill. A diagram of the mill chamber of a horizontal media mill is shown in Fig. 5.3. The millbase is subjected to high shear when it passes through the small spaces formed by the media. It

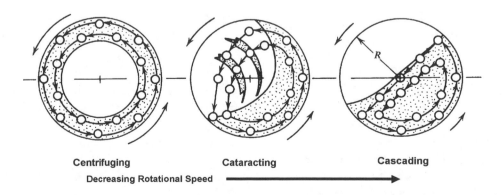

Centrifuging   Cataracting   Cascading

Decreasing Rotational Speed ⟶

Figure 5.2 Schematic diagram possible operating regimes of a ball or pebble mill.[1]

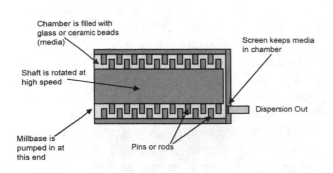

Figure 5.3 Drawing of a horizontal media mill grinding chamber.

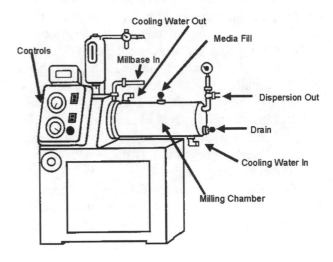

Figure 5.4 Drawing of a horizontal media mill.

can also be subjected to collisions among the media beads. The high shear will separate the agglomerates of pigment particles. The collisions with the media can break the pigment particle. These mills produce a small degree of size reduction, but a great deal of deagglomeration. The millbase can be passed through the mill multiple times to improve the dispersion quality (fineness of grind). The media mills can be run in batch, continuous, or circulation modes. Horizontal media mills, such as the one shown in Fig. 5.4, are most common in the fluoropolymer coating industry. Some common brand names of these mills are Netzsch mill (manufactured by NETZSCH Incorporated, Exton, PA) or Dynomill (Dyno-Mill, manufactured by W. A. Bachofen of Switzerland).

The operation parameters controlled or monitored during media dispersion processes include:

1. The agitator RPM is directly set on the control panel.

2. Mill temperature is monitored and can be indirectly controlled by adjusting cooling water flow, or by reducing the agitator speed. When dispersing fluoropolymer powders, temperature can be very important.

3. Production rate (mass through the mill per unit time) is controlled by setting the feed pump RPM—the faster the millbase goes through, the less residence time in the mill, which means less dispersion energy applied.

4. Media type. A special type of glass is used for food contact products. Ceramic media is common for non-food contact products.

5. Media size and shape. It affects the number and size of gaps between the media, which affects dispersion and grinding capability.

6. Amount of media load (typically 50% – 95% of the milling chamber volume). It also affects the number and size of gaps between the media, which affects dispersion and grinding capability.

5.2.2.2 *High-Speed Disperser*

One means of breaking up pigment agglomerates without using media such as glass or ceramic beads is called a high-speed disperser (HSD). A sawtooth type of mixing blade, shown in Fig. 5.5, is attached to a sturdy shaft, which is then attached to a powerful high-speed motor. This blade is put into the millbase and is operated at very high speed, thousands of RPMs. The parameter of importance is the speed at the tips of the blade, called *tip speed*. Tip speeds of 5,000 ft/min (1,524 m/min) are not unusual. If the millbase viscosity is "high," then a lot of shear is generated at the tips of the HSD. The high shear can separate pigment agglomerates. This

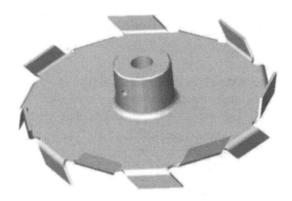

Figure 5.5 Drawing of a HSD mixing blade. (*Courtesy of Pete Csiszar.*)

Stator is stationary- Liquid is "pumped" up into assembly and out through stator holes exposing liquid to very high shear forces

Rotor fits in Stator with a narrow gap. It is attached to a shaft and rotates at very high speed

Figure 5.6 Drawing of a rotor (*bottom right*) and stator (*upper left*). (*Photo Courtesy of Melton Mixers.*)

type of disperser generally is only used for high viscosity solvent-based millbases. The higher the viscosity, the more shear is applied to the pigment agglomerate leading to better dispersion. Water-based millbases would foam up quickly due to the use of surfactants. As evidence of the amount of energy being put into the millbase, the temperature can rise quickly and often requires external cooling.

5.2.2.3 *Rotor-Stator*

An alternate type of high-speed media-less disperser is the *rotor-stator*. This equipment consists of two basic parts. Figure 5.6 shows the rotor (*bottom right*) and the stator (*upper left*) separately. The rotor has carefully machined blades that insert into the stator. The stator has holes, or slots, in its wall. The gap between the rotor's blades and the stator is very close. The rotor spins at high RPM, typically 1,200—3,600 RPM. The liquid is drawn up through the bottom of the rotor-stator assembly. The liquid is subjected to very high shear at the walls and slots in the stator. Centrifugal force expels the liquid outward from the stator. The dispersers often are inserted into tanks or other containers. These dispersers can be used in line, and are sometimes used as a pre-disperser for a media mill. As with the HSD equipment, the rotor-stator disperser works better in higher viscosity millbases.

5.3 Measuring Dispersion Quality or Fineness

Generally, the more finely a pigment is dispersed, the more efficiently it is used. Proper dispersion maximizes color strength, uniformity, gloss control, and other properties of the finished paint. The easiest method to measure fineness is also the most common. The device used to make this measurement is the Hegman Fineness Gauge, shown in Fig. 5.7. The standard measurement procedure is ASTM method D1210-96-2004 titled "Test Method for Fineness of Dispersion of Pigment-Vehicle Systems by Hegman-Type Gauge."

Figure 5.7 Hegman fineness gauge.

The Hegman gauge consists of a flat, smooth, steel block into which is machined a groove that is uniformly tapered along its length. Typically, the groove depth changes from 100 microns (4 mils) at one end to zero at the other. There are also Hegman gauges that start at 25 or 50 microns. A scale is engraved along the groove that denotes the depth of the groove at any point along its length. Scales may be in microns, milli-inches (called mils), or Hegman units. A Hegman unit equals 0.25 milli-inches, but it counts in reverse, so a Hegman reading of 8 is zero milli-inches while a Hegman unit of 0 is 2 milli-inches.

To measure the fineness, a sample is placed in the groove at the deep end and a blade is used to draw the liquid down the length of the groove. When the gauge is viewed at an angle, it is possible to note the point along the length of the groove where it becomes shallow enough for the pigment particles to protrude above the level of the liquid. The pigment particle size at this point is read from the scale. The procedure is relatively simple, but at times, it can be difficult to "read" properly, particularly if the pigment levels are low, or the particles are not strongly colored.

5.4 Dispersion Stabilization

The dispersions must be stabilized, otherwise the pigment particles will stick back together. Dispersed pigment particles are subject to continuous motions of vibration and rotation, called Brownian motion, due to random collisions with molecules of the solvent. As the particles collide or come very close, attractive forces pull the particles closer together. These intermolecular forces come from different chemical sources. There are fundamentally attractive electrostatic interactions such as ionic interactions, hydrogen bonding, and dipole-dipole interactions. Weaker attractive forces are electrodynamic interactions and are known as van der Waals, or London forces. Detailed discussion of these forces is beyond the scope of this book, but the literature on them is extensive.[3]

There are also repulsive forces that are very strong when the particles are very close, and the forces of attraction between the particle and the liquid also are repulsive (inhibit agglomeration). Whether they stick together or not depends on the difference between competing attractive and repulsive forces.

If the attractive force between pigment particles is stronger than the liquid-particle forces, then the particles will stick together after collision. This process can continue until essentially all the particles are stuck to each other. This process is called *flocculation*. The clumps of particles so formed are often called *flocs*.

Flocculation depends on forces between pigment particles and molecules of the liquid phase. Putting a coating that has a strong attraction to the liquid phase on a pigment particle is effective in preventing flocculation, if that coating also has a strong attraction to the pigment surface. Surfactants, surface-active agents, sometimes called soaps, usually have two types of chemical groups on the same molecule. In a classic case, one end is hydrophobic, or water "hating," and one is hydrophilic, or water "loving." Figure 5.8 shows the structures of the common surfactant types.

The hydrophilic end has an affinity for the pigment particle and coats the pigment particle. The hydrophobic tail on the surfactant extends out into the solvent as shown in Fig. 5.9

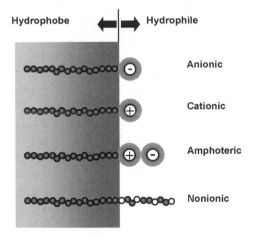

Figure 5.8 Surfactant types.

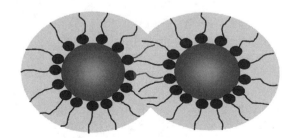

Figure 5.9 Steric stabilization.

The surface-active agent coating on the pigment particles prevents them from getting too close. Basically, it keeps the pigment particles far enough apart so that particle-to-particle contact does not occur. This is called *steric* stabilization.

Some surfactants work by using charge repulsion. Surfactants of this type, or other ionized materials, are coated on pigment particles to keep them apart by electrical repulsive forces. This is called electrostatic stabilization and is depicted in Fig. 5.10. If water-soluble ionic salts are added to the dispersion, repulsion is reduced, and flocculation may occur. The salt addition may not be intentional. Sometimes it comes with tap water that might be used for thinning or from an additive that is being used to solve some other problem. Also, changes of pH can cause a reduction in ionization of the surfactant and, hence, lead to flocculation.

Often a combined electrostatic and steric approach to dispersion stabilization is used. In non-aqueous systems the surfactant or dispersant is usually a polymer resin or other relatively large organic molecule. These dispersants come in a variety of chemical structures, but the most convenient is often a polymer with polar groups. The dispersant polymer may also be the binder polymer of the coating, although the binder of the coating is often not suitable for that function. The polymeric surfactant functions by coating the pigment and making it behave as though it was a large molecule of that polymer. However, the dispersant polymer can affect the coating performance and, in some cases, can even affect the curing chemistry of thermosets.

Besides flocculation, one other very noticeable thing can happen to a dispersion. The particles can settle to the bottom of the container. Occasionally, the particles will rise to the surface, in which case that effect is called *creaming*.

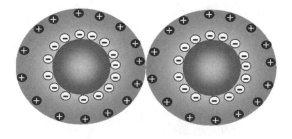

Figure 5.10 Electostatic stabilization.

5.5 Pigment or Particle Settling

The mathematical interpretation of simple settling provides a relationship that shows how formulation variables can affect the settling rates. The major force acting upon a fluoropolymer, or pigment particle in a fluid (Newtonian), is gravity. That force pulls the particle downward towards earth. However, there are opposing buoyancy and viscous components that present resistance to that movement of that particle. The viscous resistance can be calculated from Stokes' Law (Eq. 5.1).

Eq. (5.1)
$$F = 6\pi r \eta v$$

where:

F = viscous resistance (dynes)

r = radius of the particle (cm)

η = viscosity (poise)

v = terminal velocity (cm/sec)

The overall equation that incorporates gravity and buoyancy is Eq. (5.2).

Eq. (5.2)
$$F = \frac{4}{3}\pi r^3(\rho - \rho_l)g$$

where:

F = downward force (dynes)

r = radius of the particle (cm)

ρ = density of the particle (g/cm³)

ρ_l = density of the liquid (g/cm³)

g = gravitational constant (980 cm/sec²)

At equilibrium, the above two equations equal each other and the particle falls at a constant viscosity:

Eq. (5.3)
$$6\pi r \eta v = \frac{4}{3}\pi r^3(\rho - \rho_l)g$$

Solving for the settling rate, v:

Eq. (5.4)
$$v = \frac{218 r^3(\rho - \rho_l)}{\mu}$$

The exact mathematics of this equation is not as important to the user or formulator as what is learned:

1. As the particle size (r) goes up, the settling rate increases by the square of the radius of the particle. Bigger particles settle much faster.

2. The settling rate is minimized if the density of the particle and the density of the liquid are almost equal.

3. As the viscosity of the liquid increases, the settling rate is reduced.

Figure 5.11 shows the effect of viscosity on settling of a fluoropolymer non-aqueous dispersion as a function of viscosity. In this figure, the initial viscosity is lowered for each glass jar of dispersion towards the right.

Figure 5.11 Settling as a function of viscosity.

5.6   Hard and Soft Settling

Settling is often an issue with coatings users or applications. It is easily seen and generates concern about reincorporation. Some coatings are easy to redisperse, some are very difficult. Settling will almost always occur over time, to some extent, because there are density differences and gravity can not be turned off. Settling is an important issue for the formulator and it receives lots of attention.

Coatings that settle slowly frequently settle harder and are more difficult to redisperse, while those that settle rapidly settle softer and are easy to reincorporate. There are definitely exceptions to this rule of thumb, but if the instructions of the manufacturer are followed, then the problem can be manageable. Frequently, a program to redisperse coatings in storage will keep them usable longer. DuPont often suggests that their coatings be rolled every thirty days.

5.7   Functions of Pigments

5.7.1   Appearance, Color, Hiding

The primary purpose of pigmentation is to impart color and improve appearance, besides hiding the substrate. Because most applications, except cookware, for fluorofinishes have historically been for performance reasons, little attention has been paid to color. Color matching and reproducibility historically have not been important. However, in recent years color has become more important, but still has not risen to the level one sees for house or car paints. Most manufacturers prefer not to make many colors because the volume sold is small and it adds a great deal of expense to the business. This work will not delve into much detail on this subject.

The science of color can be complex. Color is the result of how visible light is:

1. Reflected

2. Absorbed

3. Refracted

The source of the visible light can affect the visual perception of color by supplying different intensities of differing wavelength between sources. A color that is viewed in sunlight will look different when viewed under a fluorescent lamp, which has stronger levels of green, blue, and violet light waves. It might also look different when viewed under an incandescent light that is stronger in the orange/red area of the visible light spectrum. The approximate spectra of these three light sources are shown in Fig. 5.12 along with Table 5.1 which breaks the spectrum into approximate colors.

Also, variation in the surface or substrate texture can lead to visual color differences. Variations in gloss can appear as different colors when viewed at different angles.

There are instruments that are used to measure color, called *colorimeters*. They measure color and quantify it in terms of units, one of which is called the Hunter "Lab" scale. The three parameters measured correspond to:

L - Lightness/darkness

a - Redness/greenness

b - Yellowness/blueness

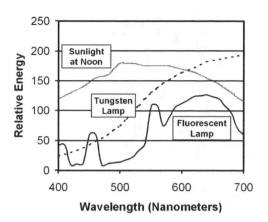

Figure 5.12 Spectra of sources visible light.

Table 5.1 Visible Light Wavelength (nm) and Perceived Color

Wavelength	Color
400-430	Violet
430-500	Blue
500-560	Green
560-620	Yellow to Orange
620-700	Orange to Red

These three parameters define any color. The details of this measurement system are available from instrument manufacturers. One explanation is made by Hunter Associates Laboratory, Inc. (www.hunterlab.com). They have an application note called "Hunter Lab Color Scale" in Vol. 8, no. 9, 1996.

Hiding and color usually improve when the pigment particles are smallest. Some pigments are better at hiding and imparting color than others. The dispersion process and formulation are quite important because they ultimately control the particle size of each pigment in the paint.

Sometimes formulations change in appearance with age or even with the spray process. This is almost always caused by pigment agglomeration or flocculation, which increases the effective particle size reducing color development and hiding. Flocculation and agglomeration were discussed in Sec. 5.4.

5.7.1.1 Gloss

The *gloss* of a surface is described as the reflection of light from the surface that is independent of color. Gloss is a property of reflected light, and as mentioned in the previous section, it can influence the visual color of a surface when viewed at various angles. Gloss can be bake-dependent for fluoropolymer coatings and can be used as an indicator of proper bake. Typically, a fluoropolymer that has not been exposed to temperatures above its melt temperature will not have a glossy surface. ASTM method D523-89(1999) Standard Test Method for specular gloss outlines the procedures for performing the test, using a glossmeter. Gloss is typically measured at a given angle, either 20°, 60°, or 85°. The gloss of a coating is affected by surface roughness and surface contamination. In coatings, pigment particles that protrude through the resin, or binder surface, cause the scattering of the light, which is visible as dullness. Gloss is sometimes a specification for coatings. Adding flattening additives such as silica can lower gloss.

5.7.1.2 Hiding

Understanding the hiding ability of a coating is important to the formulator and user. Hiding is defined as the ability to prevent seeing the substrate through the coating. Hiding power is typically measured using a modified procedure described in ASTM D344-97 (2004), "Standard Test Method for Relative Hiding Power of Paints by Visual Evaluation of Brushouts." For most paints, a black and white sticker is applied to the substrate and the paint is applied over it. If a difference between the white and black regions can be seen, then either the hiding is not good enough or the coating is too thin. Because most fluorocoatings require high bake temperatures, the sticker or its adhesive decomposes, making the hiding determination difficult. A good approach for fluorocoatings is to prepare an aluminum panel that is primed with a suitable black primer on one side. The panel can then be painted at the desired dry film thickness (DFT) and the hiding evaluated. An example of this type of study is given in Fig. 5.13.

An instrumental approach that can measure hiding is described in ASTM D2805-96a (2003), "Standard Test Method for Hiding Power of Paints by Reflectometry," but this method is rarely used.

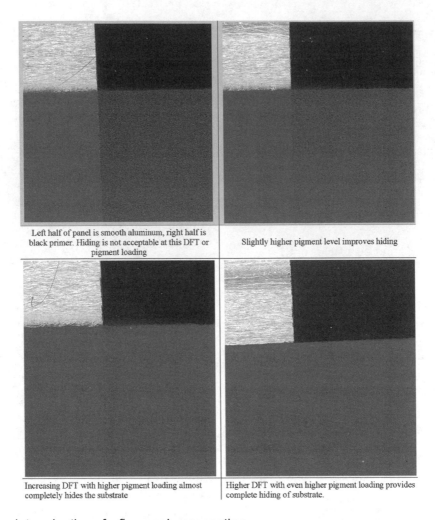

Left half of panel is smooth aluminum, right half is black primer. Hiding is not acceptable at this DFT or pigment loading

Slightly higher pigment level improves hiding

Increasing DFT with higher pigment loading almost completely hides the substrate

Higher DFT with even higher pigment loading provides complete hiding of substrate.

Figure 5.13 Hiding determination of a fluoropolymer coating.

5.7.1.3 *Types of Pigments*

The pigments used in fluoropolymer coatings, generally must be thermally stable, and are inorganic. Dyes, which are soluble organic colorants, are used in some coatings applications, but not in fluoropolymer coatings because most would be decomposed during processing. Dyes will not be discussed in this work. Historically, color has not been a property of primary importance in fluoropolymer coatings. The focus has been more on functional performance.

Carbon Blacks. The most common pigment is an organic one, carbon black. There are three major types of carbon black used in coatings:

1. Furnace process black—non-FDA applications.

2. Channel process black—FDA compliant.

3. Electroconductive blacks—Ketjen black is very common.

Because carbon black is organic, it has been known to decompose during long or high temperature bakes. It can be stabilized somewhat with phosphoric acid or phosphate/amine salts.

Inorganic Pigments. Inorganic pigments are very common, but they are not available in as many colors as the organic pigments. Historically, red iron oxide, chromium oxide (green), titanium dioxide (white), ultramarine blue, cobalt blue, and both mica and colored micas have been the dominant pigments in fluorinated coatings. One must keep in mind that there are often many variations in these pigments including particle size and pigment surface treatments. Often the surfaces are treated for exterior durability (as in exposure to outside weather) or acid or base resistance. Table 5.2 lists the most common inorganic pigments used in fluoropolymer coatings.

There are many more inorganic pigments than those listed, they are not used very often. For example, inorganic yellows include:

- Bismuth vanadate
- Nickel titanate
- Chrome titanate and other chromium-based pigments
- Cerium sulfide
- Cadmium-based pigments
- Lead molybdate
- Inorganic yellows blended with other pigments such as titanium dioxide

Of particular interest for fluoropolymer coatings are complex inorganic pigments. These were developed originally for ceramic glazes and porcelain enamels where very high temperatures are seen during processing. These pigments are metal oxide crystals in which additional metal cations are added to or replaced in the crystal structure. Many pigment

manufacturers make these now. Based on the spinel crystal structure, there are:

Copper chromites	–	Blacks
Cobalt aluminates	–	Blues
Cobalt chromites	–	Blue-greens
Cobalt chromites	–	Greens
Cobalt titanates	–	Greens
Iron chromites	–	Browns
Iron titanates	–	Browns

Based on the rutile crystal structure, there are:

Nickel titanates	–	Yellows
Chrome titanates	–	Buffs
Manganese titanates	–	Browns

The complex inorganic pigments are frequently supplied as free-flowing, one or two micron diameter powders. What makes these unique is that these can often be stirred into the coating, especially if the coating is solvent based. It is, thus, much easier to make small batches of specialized colors because the milling part of the dispersion process may be avoided.

Table 5.2 Common Inorganic Pigments Used in Fluoropolymer Coatings

Color	Pigment	CAS Number	Density, lb/gal (kg/l)	Comments
White	Titanium dioxide	13463-67-6	33.8 (4.05)	Two crystalline forms, many treatments
Red	Red iron oxide		43.2 (5.18)	
Black	Carbon black	1333-86-4	15.5 (1.86)	Many forms
Yellow	Yellow iron oxide	68187-02-0	35.0 (4.19)	
Blue	Ultramarine blue	57455-37-5	19.2 (2.30)	
Blue	Cobalt blue	1345-16-0	36.5 (4.37)	
Green	Chromium oxide	1308-38-9	43.4 (5.20)	
Metallic	Aluminum paste			
Fillers	Aluminum oxide	1344-28-1	31.6 (3.79)	Used for abrasion resistance
	Aluminum silicate	1335-30-4	21.6 (2.59)	
	Silicon carbide	409-21-2	26.8 (3.21)	Used for abrasion resistance
	Barium sulfate, barytes	7727-43-7	36.1 (4.33)	
	Graphite	7728-42-5	22.0 (2.64)	Used for dry lubrication and conductivity
	Molybdenum disulfide			Used for dry lubrication

Micas. Mica or other micaceous pigments are common to fluorocoatings. These pigments are small flakes and are commonly used in household products, makeup, and car finishes to provide a glitter of metallic appearance. This pigment is very common in fluorocoatings. Usually, they are added at trace levels to about two percent by weight of the final film. Different flake sizes provide different visual effects.

Many mica pigments are coated with a very thin layer of aluminum oxide or titanium dioxide (on the order of the wavelength of light). These materials can be highly transparent and have a high index of refraction. Light impinging on these pigments is refracted and reflected, causing shifts in the wave character of light. The refracted and reflected light can interfere with one another causing canceling out of some of the colors of white light leading, producing a color. This gives an interference color, one that changes depending upon the angle at which it is viewed and upon the thickness of the coating, also known as *flop*. The principle is shown in Fig. 5.14 along with how color shifts with increasing coating thickness.

The coating on the mica particles can be changed even further by adding some transparent color pigments to the thin coating layers making the color stronger. Gold, reds, greens, and blues can be made in this fashion.

To maximize this effect the flake pigments should all be lying parallel to the substrate. Since paint drops strike the substrate in a random process, this might seem to be a difficult objective to achieve. However, in liquid finishes one can improve planarization by film shrinkage. The process is shown in Fig. 5.15. The more shrinkage occurs, the better the planarization will be.

These materials are frequently made at a very fine particle size. An average particle size of one micron is common. These particles do not tend to agglomerate as much as other pigments. Agglomeration is the clumping up of small particles forming a larger particle as in a bunch of grapes. The agglomerates need to be broken down into separate particles when added to the coating. This is achieved by grinding or dispersing processes, which are discussed earlier in this chapter (Sec. 5.2). However, coated mica is generally stirred into a coating. Some manufacturers pre-wet the mica pigments with water to make them even easier to disperse.

The number of these types of materials available is too large to list here. The primary manufacturers are EMD Chemicals Inc., an associate of Merck KGaA, and Engelhard. EMD produces the pigments using the trade names Iriodin® and Afflair®. Engelhard produces pigments under the trade name Mearl®.

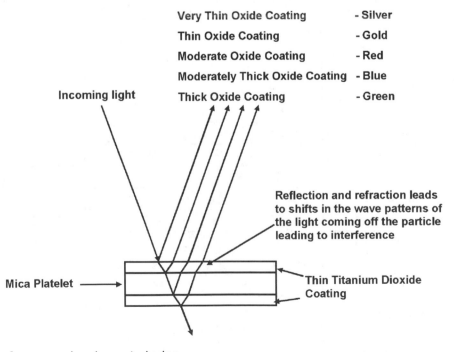

Very Thin Oxide Coating	- Silver
Thin Oxide Coating	- Gold
Moderate Oxide Coating	- Red
Moderately Thick Oxide Coating	- Blue
Thick Oxide Coating	- Green

Incoming light

Reflection and refraction leads to shifts in the wave patterns of the light coming off the particle leading to interference

Mica Platelet

Thin Titanium Dioxide Coating

Figure 5.14 Interference colors in coated mica.

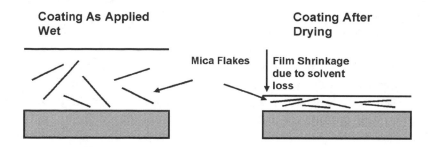

Figure 5.15 Planarization of flake pigments during film shrinkage.

5.7.2 Permeability, Barrier Properties

One function of a coating is to keep particular materials, such as water, air, carbon dioxide, acids, bases, or other materials from penetrating to the substrate, causing damage to the substrate.

In general, small molecules permeate through a film by a variety of mechanisms. The film may consist of unevenly spaced molecules with many relatively large unobstructed paths through it. These might be called pores. The permeability or porosity through pores mainly depends on the molecular size and shape of the permeant, but little on the chemical selectivity.

Another way small molecules move through a polymer film occurs when the polymers themselves absorb water (or other solvent) into its structure and swelling occurs. Solvent molecules are associated with the polymer molecules. The solvent molecules diffuse through the swollen polymer by a process of small discrete displacements, so that a particular solvent molecule need move a short distance while displacing another water molecule which moves on to the next position and so on until one leaves the opposite side of the film.

Polymers vary considerably in their ability to absorb and transmit liquids and vapors. Fluoropolymers are permeable to many small molecules. Table 5.3 gives permeation data for films of various fluoropolymers

Table 5.3 Permeation of Various Polymer Films to Small Molecules

Polymer	Carbon Dioxide	Hydrogen	Nitrogen	Oxygen	Water
	Units: $\times 10^{-13}$ cm^3. cm cm^{-2} s^{-1} Pa^{-1} at 25°C (* is at 38°C)				
FEP	10	10	1	3	13
PTFE	7	7	1	3	25
PVF	0.07	0.3	0.001	0.015	250*
ETFE	3		0.2	0.6	170*
PFA	10		1	3	13
PMMA				0.1	500
PI	0.5	1	0.03	0.1	400
PES				0.4	1200*
* This table does not include water vapor; water vapor permeation is generally much higher than liquid water permeation. Permeability generally increases exponentially with temperature. Additional and more detailed permeation data is compiled in Appendices II, III, and IV.					

and some other polymers sometimes used in conjunction with fluoropolymers.

Permeability to small molecules, particularly water and oxygen contribute to corrosion. Coatings are never completely impervious to water, so eventually water in some form will reach the substrate. The damage caused by the water depends on the adhesion of the surface coating to the substrate, or in the case of multilayer coating systems, the adhesion between the coating layers.

If corrosion does not occur rapidly, blisters can form. If there are any water soluble salts in the coating or at its boundary layers, then blisters can form due to osmotic pressure. Osmotic pressure is driven by a higher concentration of water-soluble material in the film than in the water in contact with it. The water is drawn into the coating to dilute the concentration. The salts do not permeate rapidly out of the coating, so the net effect is a build-up of water at interfaces. If the adhesion forces are not strong enough to resist the pressure build-up caused by water accumulation, blisters result in the film.

Permeation can be affected strongly by pigmentation. Mica and other platelet types of pigments are often used to make a coating more permeation resistant. These pigments are impermeable on their own, but when used at a sufficient level in a coating, they effectively increase the path for the small molecules to get from the coating surface to the substrate as shown in Fig. 5.16.

Appendix II details the permeation character of some of the pure fluoropolymers.

5.7.3 Abrasion Resistance, Reinforcement: Physical Property Improvement

Pigments and fillers are sometimes added to improve physical properties of the coating. The most common goal is improving the abrasion or scratch resistance. The coating can be made stronger by reinforcing the coating.

Nearly everyone is familiar with concrete road and building construction. Rebar is used to strengthen the poured concrete. Rebar is usually made of lattice of iron or steel bar, or a screen over which concrete is poured. A screen is generally not put in a coating (though this has been done for some very thick coatings). Random distribution of acicular pigments or fillers in the cured coating can be relied on. Acicular pigments are those that are not generally round in shape. Mica platelets, glass or metal flake, and short metal fibers (or whiskers) fall in this category and are used for film reinforcement.

Resistance to abrasion or wear can sometimes be improved by adding hard material to the generally soft fluorinated coatings. Many coatings sold on premium cookware since mid-1990s were reinforced with hard fillers such as aluminum oxide, silicon carbide, or metal flake. There is a limit to the amount added, or the non-stick properties would be lost. Particle size can have an effect of abrasion and release. Some coatings use large particles, some small, and some a mixture of large and small particles to improve their packing in the coating to enhance abrasion resistance.

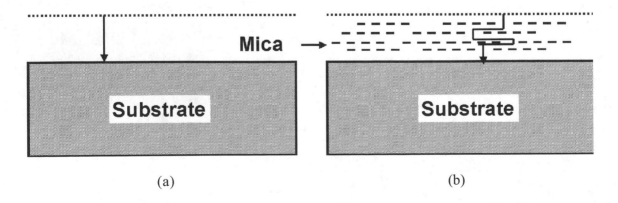

(a)   (b)

Figure 5.16 Permeation of various polymer films to small molecules. (*a*) Small molecule diffuses or permeates directly through unpigmented coating. (*b*) Small molecule diffuses or permeates along a much longer path through pigmented coating.

5.7.4 Electrically Conductive Fillers

A number of fluorocoating applications require that the coating dissipate static electricity. As explained in Ch. 11, which covers powder coatings, fluoropolymers are at the negative extreme of the triboelectric series, which means that anything that rubs against the fluoropolymer will become positively charged due to loss of electrons and the fluoropolymer will become negatively charged. The charge can cause performance problems in some applications. There are pigments that conduct electricity that will help neutralize charges that build up on the fluoropolymer surface.

The most common fillers are conductive grades of carbon. These are commonly called Ketjen blacks. They differ from conventional carbon blacks in that they are very pure and have few chemical functional groups left on the surface. They also have small particle size. Other pigments or fillers are used for this purpose, such as metal flakes, metal fibers, and certain pigments coated with conductive materials.

Formulating conductive coatings can be tricky. The fluoropolymers themselves are great electrical insulators, so the added pigments must be present in sufficient quantities to get particle to particle contact so that a conductive pathway, or circuit, leads from the coating surface to the substrate. If one plots the conductivity versus the amount of conductive pigment added to the coating as shown in Fig. 5.17, three distinct regions are found in the curve. This curve plots resistivity rather than conductivity,

because resistivity is usually what is measured by instruments. Conductivity is the inverse of resistivity. At low pigment concentrations as shown in region "A" of the figure, the coating is still in the insulation zone. The resistivity does not change on this non-conductive plateau because there is not enough particle-to-particle contact. After enough pigment is added so that particle-to-particle contact is assured, the "percolation threshold" is reached. At the percolation threshold, particle-to-particle contact extends from the coating surface to the coatings substrate. At this point, the coating begins to get conductive.

Small increases in pigment level beyond the percolation threshold causes a very rapid decrease in resistivity as shown in region "B", also known as the percolation zone. As more pigment is added, the conductivity levels off as shown in region "C", the conducting zone.

The amount of pigment needed to get to the percolation level can be very high, so high that it adversely impacts some of the other properties the coating needs such as release. To reduce the pigment amount but still get conductivity, the formulator can take advantage of the natural tendency of pigments to flocculate. This is contrary to what paint chemists usually do, which is to try to make non-flocculating dispersions. If the chemist can control the flocculation to produce strings of pigment particles rather than agglomerates then conductivity can be achieve at level much below the expected percolation threshold. This is shown in Fig. 5.18.

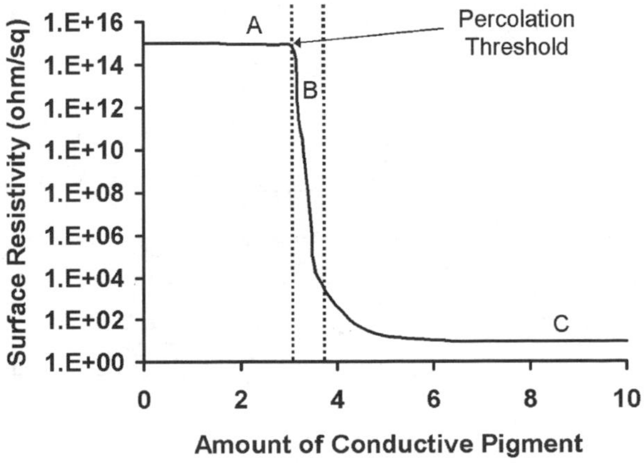

Conducting Network   Non-Conducting Network

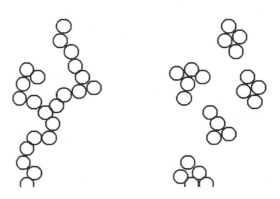

Figure 5.17 Resistivity/conductivity dependence on conductive pigment concentration.

Figure 5.18 Controlled flocculation to promote conductivity.

Dispersions of conductive forms of carbon black can be dispersed just poorly enough to allow conducting network flocculation but well enough to maintain paint stability.

An alternate way to reduce the conductive pigment loading is to use acicular pigment shapes such as rods, needles, whiskers, or platelets (such as mica). Most of these conductive pigments are not inherently conductive. They are made conductive by coating the pigments with a clear antimony doped tin oxide coating. Some of these pigments are nearly colorless and transparent. Table 5.4 lists some conductive pigments.

Table 5.4 Commercial Inorganic Conductive Pigments

Manufacturer: Product Code	Technology	Color	D50 Particle Size, μm
Milliken			
Zelec 1610-S	Sb/SnO$_2$ hollow shell	Light gray	3
Zelec 2610-S	Sb/SnO$_2$ hollow shell	Light gray	3
Zelec 1703-S	Sb/SnO$_2$ hollow shell	Light Green	3
Zelec 2703-S	Sb/SnO$_2$ hollow shell	Light Green	3
Zelec 1410-M	Sb/SnO$_2$ Mica	Light Gray	5
Zelec 1410-T	Sb/SnO$_2$ TiO$_2$	Light Gray	1
Zelec 3410-T	Sb/SnO$_2$ TiO$_2$	Light Gray	0.5
Zelec 3010-XC	Sb/SnO$_2$ particles	Blue	0.5
Ishihara			
ET-500W	Sb/SnO$_2$ TiO$_2$		0.2
ET-600W	Sb/SnO$_2$ TiO$_2$		0.2
ET-300W	Sb/SnO$_2$ TiO$_2$		0.05
FT-1000	Sb/SnO$_2$ TiO$_2$ Rods		0.13/1.68*
FT-2000	Sb/SnO$_2$ TiO$_2$ Rods		0.21/2.86*
FT-3000	Sb/SnO$_2$ TiO$_2$ Rods		0.27/5.15*
SN-100P	Sb/SnO$_2$ particles	Transparent	0.02
SN-100D	Sb/SnO$_2$ particles water dispersion	Transparent	
FS-10P	Sb/SnO$_2$ rods	Transparent	0.02/1.0*
FS10D	Sb/SnO$_2$ rods water dispersion	Transparent	0.02/1.0*
Akzo Nobel			
EC-300J	Carbon Black	Black	
EC-600JD	Carbon Black	Black	
EM Industries			
Minatec® 31CM	Sb/SnO$_2$ Mica	Light Gray	<15
Minatec® 30CM	Sb/SnO$_2$ Mica	Light Gray	<15
Minatec® 40CM	Sb/SnO$_2$ Mica + Sphere	Light Gray	<15, Mica, <10 Sphere
* Diameter/length in μm.			

5.8   Quantifying Pigment Concentrations in Formulations

It is important to know how much pigment is in a cured coating. If the amount of pigment is large, then a fall off in non-stick properties might be expected, or a rise in coefficient of friction. Abrasion resistance might go up and permeation could certainly change.

The most common measure of pigment concentration is call "P/B," pigment-to-binder ratio. This is simply the weight of the pigment divided by the weight of the binder ingredients (usually polymers) in the dry film. While this is easy to calculate, to visualize what is in a cured coating, you need to consider the densities of the binder and pigments.

The author finds it easier to think of pigment concentration in terms of volume. This is called pigment volume concentration (PVC). While this is a little more difficult to calculate, it is infinitely easier to visualize. A favorite explanation frequently goes like this:

What does it mean when it is said that a coating has a 5% PVC? To understand, imagine you have a box of 100 ping-pong balls where 5% of them are red and those represent the pigment particles. The remaining 95% are white and are the binder. One can then understand that the chance of having a red ball on the surface is small, and so the pigment at five volume percent probably will not affect the non-stick character of the coating much. However, if you had a PVC of 20%, then there would almost certainly be red balls on the top of the box. It would be easy to expect that since the pigment is not a material with good release, the non-stick character of that coating may not be as good. If the balls do not all have the same size, then the visualization is somewhat complicated, but it still works.

An important characteristic in coatings is the *critical pigment volume concentration* (CPVC). Here, a similar view can be taken, except now one can imagine the binder is a liquid. If a hundred balls (pigment) are put into a jar, the volume taken up by those pigment particles can be calculated. If a liquid is added to just fill the container enough to cover all the pigment balls, then you have the volume of the binder. The critical pigment concentration is the volume of the pigment divided by the volume of the

pigment and binder in this scenario. If a coating is formulated above the CPVC, then there is not enough binder to fill all the gaps in the pigment particles, and the coating will become much more porous. This can affect properties such as permeability and corrosion protection. The CPVC is affected by pigment shape and particle size distribution because they can affect the way the pigment particles pack. This in turn affects the open space between the particles.

Rarely in a fluoropolymer coating is the pigment content above CPVC, but there are some commercial examples in primers. Some patented formulations claim performance enhancements by mixing different particles sizes.

5.8.1   P/B: PVC

A measure of pigment content is of interest to paint formulators. It is not generally of interest to users. Pigment levels can affect many paint properties including release and non-stick, chemical resistance, corrosion performance, and hardness besides appearance.

Pigments can reduce the basic fluoropolymer properties of release, non-stick, and chemical resistance. Often through experience a formulator may develop his own guidelines for how much pigment can be added and still maintain other critical properties. For instance, in an application where release is very important, the formulator may know that if the volume of the pigment rises above 7% of the cured film, then the release is not adequate. If P/B is used, one would have to consider the density of the pigment in making the determination. A powdered metal pigment or filler is much denser than many colored pigments, so that particular application can tolerate a higher P/B with the metal filler than with the usually colored pigments.

5.9   Commercial Pigment Dispersions

There are many companies that offer various pigment dispersions. On the internet at www.kellysearch.com, a search can be made for "pigment dispersions" and dozens of companies that make stock dispersions in water or other solvent will be displayed. Many can make custom dispersions. Some of these dispersions can be used to tint fluorocoatings, but one must take care in selection. Aqueous dispersions factors that need consideration include:

- Thermal stability of pigment—needs to survive the baking conditions.

- Pigment concentration—if too much is added, it can compromise coating performance, especially non-stick and coefficient of friction.

- Viscosity—determines, in part, how easy it is to mix in.

- Surfactant type and level—this could affect the fluorocoating stability.

- pH—it should be similar to the fluorocoating; it can also affect stability.

Non-aqueous dispersions are more difficult to select because surfactants are replaced by dispersants, which are usually polymeric resins. These dispersants can affect the curing chemistry on the non-aqueous coatings particularly if the binders are thermosets. Factors to consider include:

- Thermal stability of pigment—needs to survive the baking conditions.

- Pigment concentration—if too much is added, it can compromise coating performance, especially non-stick and coefficient of friction.

- Viscosity—determines in part how easy it is to mix in.

- Dispersant type and level—this could affect the fluorocoating stability, curing, and coating performance.

- Solvent system—this can affect wet ting, surface tension, evaporation rates, solubility and all those factors described in Ch. 6 on solvents.

REFERENCES

1. Pratsinis, S. E., and T. T. Kodas, Manufacturing of materials by aerosol processes. In: *Aerosol Measurement* (K. Willeke and P. A. Baron, eds.) Van Nostrand Reinhold, New York (1993)

2. Patton, T. C., *Paint Flow and Pigment Dispersion: A Rheological Approach to Coating and Ink Technology*, John Wiley & Sons, New York (April, 1979)

3. Stone, A. J., *The Theory of Intermolecular Forces*, Clarendon Press, Oxford, UK (1997)

4. Davidson, M. W., Abramowitz, M., Olympus America Inc. and The Florida State University

5. Massey, L., *Permeability Properties of Plastics and Elastomers*, 2nd Edition: *A Guide to Packaging and Barrier Materials, PDL Handbook Series,* William Andrew Publishing, Norwich, NY (2004)

6 Solvent Systems

6.1 Introduction

The solvent systems or liquid carriers in fluorocoatings are frequently complex and carefully designed by the formulator. Generally, in regards to solvent systems, users are only concerned with how to dilute (dilute, reduce, cut, or thin are equivalent) a coating. A thorough understanding of all the intricacies of solvent formulation process is outside the scope of this book. A basic understanding will make it easier to select thinners correctly and understand its effect on coating application and quality.

Water is the most common solvent, but there are few commercial aqueous solvent systems that do not contain some other solvent, often called a *cosolvent*. Many factors are considered in determining the solvent system. Some of these are:

- The impact on rheology/viscosity of the coating or paint.
- Evaporation rates and vapor pressures.
- Boiling point.
- Solubility of polymers in the coating, both in can and as the solvent evaporates.
- Dispersion stability.
- Surface tension.
- Flash point and safety.

These factors all interact and affect coating stability, application properties, and quality of the final finish. Because of these interactions, the formulation process is complex and difficult to explain. All the factors mentioned above are discussed separately, but deeper exploration of the variable interactions is beyond the scope of this book.

Rheology and viscosity are discussed in Ch. 3. The solvent system has direct impact on the rheology of the coating and, in turn, on its application and processing.

6.2 Solids-Viscosity Relationships

Solvent systems influence many coating properties, but the most obvious one is the viscosity. All coatings' applicators or users know they sometimes need to thin a coating to lower its viscosity and make it apply in a different way. Coatings are generally manufactured at the highest reasonable viscosity to allow for application flexibility. It is nearly impossible for a user to raise the viscosity, but it is easy to reduce it. For instance, a coating might be sold at 1,400 cps viscosity, which might be ideal for a particular application technique. But the user in this example may need a lower viscosity for use with his particular application equipment. Coating manufacturers frequently supply guidance about how much thinner to add by using a chart such as the one given in Fig. 6.1. It is a plot of the viscosity as a function of the amount of a particular thinner added. A *thinner* is a particular blend of solvents. Sometimes the percent of solids is used in the x-axis as shown in Fig. 6.2. Over a narrow range of viscosity such as that in Fig. 6.1, the y-axis for viscosity can be linear, but over wider ranges of viscosity, the axis is best plotted in a logarithmic scale, and the relationship will appear more linear, as in Fig. 6.3.

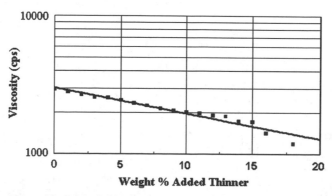

Figure 6.1 Viscosity reduction with added thinner.

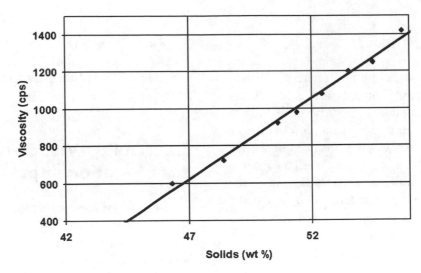

Figure 6.2 Viscosity is a function of weight solids – linear axis.

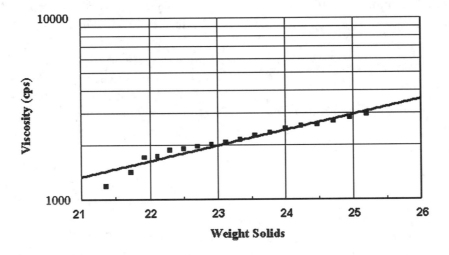

Figure 6.3 Viscosity is a function of weight solids – logarithmic axis.

6.3 Viscosity as a Function of Temperature

The measurement of viscosity is discussed in Ch. 13 on coating performance. The viscosity of a coating is temperature dependent, which is why viscosity specifications always reference a measurement temperature, usually 25°C. The temperature dependence can be quite strong. For most solvents and some polymer solutions, it can be described by Andrade's equation, shown in Eq. (6.1),

$$\text{Eq. (6.1)} \quad \eta = A \cdot 10^{B/T} \text{ or } \log \eta = \log A + \frac{B}{T}$$

where:

η = viscosity

A = constant (dependant on liquid)

B = constant (dependant on liquid)

T = temperature (absolute units, K)

A plot of log of viscosity versus $1/T$ should give a straight line.

The temperature dependence of viscosity has practical applications. Generally, a paint is warmed up to get lower viscosity. The coating can be warmed either by heating the container or by heating the application equipment. This allows the coating applicator to take advantage of viscosity reduction without thinning. When thorough temperature dependence information for a coating system is not provided, measured viscosities at two different temperatures (T_1 and T_2) can be used to estimate the viscosity at many temperatures. With the two viscosity measures (η_1 and η_2), the values for A and B in Eq. (6.1) can be calculated. B is calculated using Eq. (6.2).

$$\text{Eq. (6.2)} \quad B = \left(\frac{\log\eta_1}{\log\eta_2}\right) \bigg/ \left(\frac{1}{T_1} + \frac{1}{T_2}\right)$$

Then this calculated value for B is put back into Eq. (6.1) along with one of the temperature and viscosity measurements to calculate A as in Eq. (6.3).

$$\text{(Eq. 6.3)} \quad \log A = \log\eta_1 - \frac{B}{T_1}$$

Having determined A and B, Eq. (6.1) is use to calculate viscosity at unknown temperature. It is best to use this calculation approach for viscosity at temperatures only in the range of the two temperature measurements made. Absolute temperature (K) must be used in all the calculations.

One must keep in mind that the viscosity dependence of complex coatings does not always drop with increasing temperature.

6.4 Evaporation

One of the most important characteristics of a solvent system is its evaporation rate during application and drying. Generally, thin or low viscosity paint is easy to atomize and spray. When the coating strikes the intended substrate, it must spread out, resulting in wetting and covering of the surface. Here again low viscosity improves the spreading out of the coating and its *leveling*. Leveling is the process of flowing out to uniform thickness shortly after application. However, after the paint has flowed over and wetted the surface, low viscosity combined with gravity could cause runs, drips, or sags. A well-designed solvent system would retain low viscosity just long enough for the coating to flow over the substrate and level out to uniform thickness. The viscosity would then rise quickly to the point that the coating would not run. Coating formulators balance a solvent system with slow and fast evaporating solvents to control viscosity during the application and curing processes. Most of this refinement does not have to be done experimentally. Simulation software is available at companies that manufacture paints and coatings. This type of software allows the formulator to conduct the optimization using a computer model. An example of this type of calculation is shown in Fig. 6.4, which plots the rise in solids weight

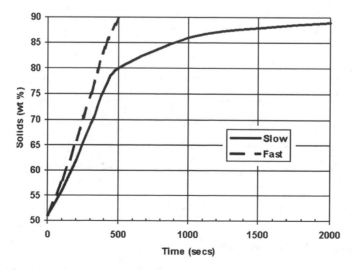

Figure 6.4 Evaporation of solvent from an applied film.

% versus time for two different solvent systems for a given coating. The computer program assumes a thin film and generates solids weight percent versus evaporation time data at a given temperature (25°C in this cxample). If a study of the viscosity versus solids weight % is conducted (as described in Sec. 6.2), and then an estimate of the viscosity of the coating versus evaporation time can be calculated.

6.5 Solvent Composition and Evaporation Time

Composition of the solvent system can also be calculated versus evaporation time, which is important because solvents evaporate at different rates. One such calculation is shown in Fig. 6.5. Such a calculation is useful when the formulator wants to replace a solvent or determine which solvents affect the coating shortly after its application, or which solvents affect the coating long after the it has been applied. The current versions of this software do not consider *azeotropes*, which are combinations of solvents that boil at different temperatures than either of the solvents alone.

6.6 Solubility

The solubility of paint materials in the coating system is particularly important in solution coatings. This behavior is quite complex, but solubility can often be calculated by the same software described in the previous two sections. A polymer has three experimentally determined solubility parameters. These parameters are called dispersive or *non-polar*, *polar* or dipole moment, and *hydrogen bonding*. These parameters are also called Hansen Solubility Parameters.[1] A polymer will dissolve in a solvent system that has similar solubility parameters. These solubility parameters can be calculated for solvent mixtures. Further, since the solvent composition can also be calculated as a function of evaporation time, the solubility parameters can be calculated as a function of evaporation time. A formulator should look for major shifts in the solubility parameters as an indication of possible problems related to polymer solubility, such as bumps in the coating caused by crystallization of one of the raw materials from solution. For example, this tool has helped the author in solving problems with a new formulation. Particles were discovered in a coating after it had dried which were not present at the time of application.

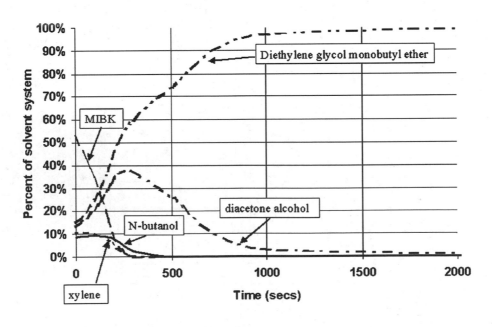

Figure 6.5 Solvent composition of an applied film.

These particles were a dissolved component of the coating system that had become insoluble during evaporation of the solvent, essentially forming crystals in the coating. An example of solubility data is shown in Fig. 6.6. This figure shows a significant shift in solubility parameters between the initial application and 200 seconds of evaporation. While the shift, in this example, may not actually cause a solubility problem, it should alert the formulator to the possibility.

6.7 Surface Tension and Wetting

Surface tension is observed as the tendency of liquids to minimize their air interfacial surface. A falling drop of water, for example, tends to assume the shape of a sphere, which has the smallest surface area per unit volume of any three-dimensional shape. It could form a perfect sphere in a weightless environment. The technical details of surface minimization are beyond the scope of this work.[2] Attractive forces among the molecules of liquid and at the interface among air and liquid molecules cause this phenomenon. The molecules within the liquid are attracted equally from all sides. Those at the surface of the drop experience unequal attractions and, thus, are drawn toward the center of the liquid mass by this net force. The surface then appears to act like an extremely thin membrane. The small volume of water that makes up a drop assumes the shape of a sphere.

The surface tension of the coating liquid impacts coating application in two ways. Lower surface tension coatings atomize into small droplets more efficiently when sprayed. More important is the effect of surface tension on the coating after it is applied to the substrate. Most metals have very high surface-energy. Liquids (of coatings) with different surface tension values affect the wetting and flow-out of the paint liquid as shown in Fig. 6.7.

When the substrate has low surface energy such as a plastic or fluorinated coating, then the surface tension of coating has to be even lower for wetting to occur. That is why it is often difficult to coat plastics or apply an additional fluorinated coating to the surface of a fully cured fluoropolymer coating. Surface tension also affects the formation of bubbles and foam, which are problematic. Lower surface tension reduces the likelihood of these problems.

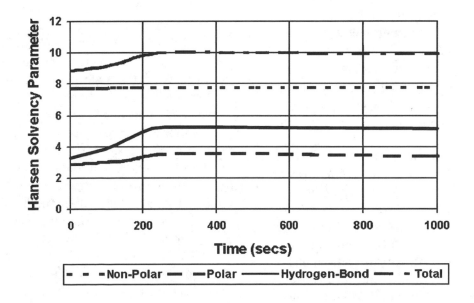

Figure 6.6 Solubility variation with time.

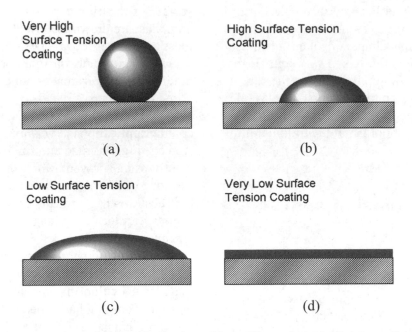

Figure 6.7 Surface tension of coating liquid affects wetting of the substrate. (*a*) A coating of very high surface tension does not wet the substrate at all. (*b*) A coating of high surface tension partially wets the substrate. (*c*) Low surface tension coatings almost completely wet the surface. (*d*) Very low surface tension coatings completely wet the surface.

6.8   N-Methyl-2-Pyrolidone (NMP)

N-methyl-2-pyrolidone, or NMP as it is commonly called, is a very important solvent for fluorinated coatings. It is the safest solvent for dissolving a number of the binder resins, such as polyamide/imide (PAI) or polyether sulfone (PES) used in fluoropolymer based one-coats and primers. There are several issues with NMP that require attention.

First, NMP is a very slow evaporating solvent. It has a boiling point of 202°C. A thin film of NMP requires over 15,000 seconds for 90% evaporation at room temperature. (See the solvent property in Table 6.2 at the end of this chapter.) If this is the only solvent used in PAI or PES based coatings, viscosity does not build up rapidly after the application and the coating will be prone to running, dripping, and sagging. The most effective fast-evaporating solvent for these resins is dimethyl formamide (DMF), but that solvent has serious safety issues.

Fortunately, fast-evaporating non-solvents can be incorporated in the formulation to provide rapid evaporation and viscosity build during application. Common examples include hydrocarbons and ketones. Even though they are non-solvents, the solubility of the total solvent system is sufficient to keep the PAI and PES in solution. The fast evaporating non-solvents are the first to leave the film, leaving NMP, which is a good solvent.

One of the advantages of NMP is its compatibility with water. It is often used in small amounts in water-based primers. NMP will absorb water from the air, especially on humid days. The evaporative cooling, due to loss of the fast-evaporating solvents, cools the coating. On warm humid days, small water droplets can condense onto the coating. It makes the coating look like it is covered with small raindrops. This situation should be avoided either by spraying in a temperature and humidity-controlled room, or by heating the parts being coated to 100°F–120°F followed by baking as soon as possible after the coating has been applied.

6.9 Conductivity

Electrostatic liquid spraying is discussed in Ch.10. It is an important application method. Conductivity of the coating material, or its capacity to carry an electrical charge, is an important factor in electrostatic spraying. Paint conductivity is primarily a function of the solvent system. A paint should be conductive enough to accept the negative charge from the spray gun, but not so conductive that the charged paint can track back to the ground through the spray gun and paint system. Water-based coatings are apt to be highly conductive, while solvent-based coatings tend to be non-conductive.

Nearly all solvent-borne coatings can be applied by electrostatic spray. The coating's electrical conductivity should be between 1.6×10^{-3} and 1.6×10^{-15} siemens/ft (5×10^{-4} and 5×10^{-6} siemens/m). Paints can be modified with polar solvent (e.g., alcohols, ketones, and NMP) to increase their conductivity. Coatings that are too conductive, such as those based on NMP can often be modified with a non-conductive solvent such as aromatic hydrocarbons (xylene).[3]

6.10 Flash Point and Autoignition

The flash point of a coating is the lowest temperature at which vapors above that liquid will burn when exposed to a source of ignition (i.e., a flame). Flash point is a direct function of the solvent system. A coating's flash point is typically that of the most volatile solvent in the solvent system. The flash point is used by regulatory agencies to specify shipping methods and containers and storage conditions and quantities. During the application of a flammable material, one must avoid many hazards that can cause fires. These details are usually described in the MSDS. However, when the product is modified or thinned, it is possible the flash point would change. A fire and explosion issue that is less known to most coating users is *autoignition* or *spontaneous ignition*. The autoignition temperature of a coating is the lowest temperature at which a material will ignite in the presence of sufficient oxygen without an external source of ignition, such as a spark or flame.

Fortunately, most common flammable and combustible solvents have much higher autoignition temperatures than flash points. These are usually in the range of 300°C (572°F) to 550°C (1,022°F). Autoignition has its most important impact on fluoropolymer coatings in the oven. Because of the high bake temperatures used, it is possible to exceed the autoignition temperature of a coating system. If the bake oven is overloaded, solvent vapors coming off could form compositions between the *Lower Explosive Limit (LEL)* and the *Upper Explosive Limit (UEF)*. If the oven operates above the autoignition temperature, then a powerful explosion could occur. Any part of the oven, such as exposed heater elements, could be above the autoignition temperature. The LEL is defined as the lowest concentration by volume of vapor in air at which the mixture will burn. The UEL is defined as the highest concentration of vapor in air at which the mixture will burn. Above the UEL there is insufficient oxygen in the mixture to support combustion. Table 6.1 shows the flammability properties of several common solvents.[4]

6.11 Summary

The solvent system is a critical part of the paint formulation as discussed in this chapter. Table 6.2 includes some of the important solvent parameters for many of the solvents used in fluorinated coatings.

Table 6.1 Flammability Properties of Several Common Solvents

Solvent	Closed Cup Flash Point, °F (°C)	Lower Explosive Limit, % v/v	Upper Explosive Limit, % v/v	Autoignition Temperature, °F (°C)
Acetone	0 (-18)	2.6	12.8	905 (485)
MEK	19 (-7)	1.8	11.5	941 (505)
Toluene	39 (4)	1.3	7	995 (535)
Isopropanol	54 (12)	2	12	797 (425)
Xylene	77 (25)	1.1	7	977 (525)
Aliphatic Hydrocarbon (150°C–200°C boiling)	~106 (~41)	0.6	6.5	500 (260)
Aromatic Hydrocarbon (160°C–180°C boiling)	~118 (~48)	1	7.5	925 (496)

Table 6.2 Some Useful Solvent Properties

CAS Number	Solvent Name	Boiling Point, °C	90% Evap Time, s	Flash Point, °C	Gal wt, lbs/gal	Hansen Non-Polar	Hansen Polar	Hansen H-Bond	MW	Surface Tension	Viscosity, cP
108-03-2	1-Nitropropane	131.6	520	35.56	8.32	8.08	5.47	8.73	89.1	29.85	0.798
104-76-7	2-Ethyl Hexanol	184.1	25730	74.44	6.93	7.8	1.6	11.45	130.23	28.2	6.52
67-64-1	Acetone	56.2	85	-17.7	6.55	7.6	5.1	7.14	58.08	23.04	0.307
64742-95-6	Aromatic Hydrocarbon	170.4	2460	43.89	7.14	8.64	0.47	9.34	127	31.87	0.842
123-86-4	Butyl Acetate	126	455	22.22	7.3	7.7	1.8	8.68	116.16	24.75	0.687
96-48-0	Butyrolactone	206	23011	98.33	9.4	9.3	8.1	11.13	86.09	38.54	1.717
123-42-2	Diacetone Alcohol	168.5	4128	57.78	7.77	7.7	4	11.58	116.16	29.63	2.797
111-42-2	Diethanolamine	269			9.07				105.14		
84-66-2	Diethyl Phthalate	295.6	1515700	93.33	9.3	8.6	4.7	13.7	222	36.99	10.588
112-34-5	Diethylene Glycol Monobutyl Ether	230.8	189965	101.11	7.93	7.8	3.4	10.4	162.23	33.95	3.551
	Ethyl Alcohol	78.4	312	3.89	6.71	7.7	4.3	9.5	46.07	21.99	1.082
107-21-1	Ethylene Glycol	197.8	123401	111.1	9.27	8.3	5.4	12.65	62.1	47.88	17.704
112-07-2	Ethylene Glycol Monobutyl Ether Acetate	190	16380	79.44	7.8	7.78	1.9	10.73	160.21	27.4	1.659
111-76-2	Ethylene Glycol-Mono Butyl Ether	171.2	7020	60	7.47	7.8	2.5	11.77	118.17	26.14	3.304
98-00-0	Furfuryl Alcohol	171		75	9.39				98.1		
56-81-5	Glycerine	290		160	10.48				92.1		
68551-17-7	Heavy Naphtha	189.9	5400	50.56	6.29	7.64	0	10.1	159	23.9	1.454

(Cont'd.)

Table 6.2 (Cont'd.)

CAS Number	Solvent Name	Boiling Point, °C	90% Evap Time, s	Flash Point, °C	Gal wt, lbs/gal	Hansen Non-Polar	Hansen Polar	Hansen H-Bond	MW	Surface Tension	Viscosity, cP
97-85-8	Isobutyl Isobutyrate	148.6	965	33.89	7.13	7.4	1.4	9.03	144.21	22.54	0.859
67-63-0	Isopropyl Alcohol	82.3	319	11.67	6.53	7.7	3	9.68	60.09	21	2.055
110-43-0	Methyl Amyl Ketone	150.1	1380	48.9	6.8	7.95	3.1	9.3	114.19	26.12	0.753
78-93-3	Methyl Ethyl Ketone	79.6	120	-7.78	6.66	7.8	4.4	7.44	72.11	23.96	0.396
108-10-1	Methyl Isobutyl Ketone	116.9	300	14.44	6.64	7.5	3	8.5	100.16	23.5	0.571
64742-88-7	Mineral Spirits-Aromatic Controlled	173.4	3500	40.56	6.42	7.86	0.04	9.56	141	23.38	1.011
71-36-3	N-Butyl Alcohol	117.7	1076	36.67	6.72	7.8	2.8	10.22	74.12	24.4	2.636
872-50-4	N-Methyl-2-Pyrrolidone	202	15120	96.67	8.53	8.8	6	10.65	99.13	30.25	1.664
57-55-6	Propylene Glycol	187.8	69000	98.89	8.62	8.2	4.6	12.89	76.1	35.47	40.366
107-98-2	Propylene Glycol Methyl Ether	120.1	690	32.5	7.65	7.6	3.6	10.31	90.12	26.5	1.7
108-65-6	Propylene Glycol Monomethyl Ether Acetate	145.8	1530	42.2	8.03	7.72	2.26	9.25	132.16	27.4	1.1
108-88-3	Toluene	110.6	249	4.44	7.18	8.8	0.7	8	92.2	27.92	0.555
102-71-6	Triethanolamine	339.9	9999999	179.4	9.32	8.3	4.4	16.9	149.19	45.24	608.54
8032-32-4	Vm&P Naphtha	127.2	530	10	6.17	7.7	0.06	8.4	116	20.36	0.632
7732-18-5	Water	100.1	1425	999.99	8.32	7.6	7.8	9.76	18.01	72.82	0.923
1330-20-7	Xylene	139.9	645	25.55	7.16	8.7	0.5	8.69	106.17	27.92	0.603

REFERENCES

1. Hansen, C. M., *Hansen Solubility Parameters, a User's Handbook,* CRC Press LLC, Boca Raton, FL (2000)

2. Hartland, S., *Surface and Interfacial Tension: Measurement, Theory, and Applications,* Marcel Dekker, New York (2004)

3. Stephens, D., and Ransburg, A. M., "Basics of Electrostatic Spray Painting," (Applicator Training Bulletin), *Protective Coatings Europe,* Vol. 5 (No. 1), pp. 60–62 & 66 (Jan 2000)

4. "Working with Modern Hydrocarbon and Oxygenated Solvents: A Guide to Flammability," American Solvent Council (Sep 2004)

7.1 Introduction

When application problems or performance shortfalls arise with a basic paint formulation, they can often be eliminated, minimized, or improved through the use of additives. These additives are usually a minor, but important part of a paint formulation. They are typically present at less than 5% by weight, but often at a fraction of 1%. Additives regularly are present at levels below that required for reporting on an MSDS. They can make or break the commercial success of a coating. Old time paint formulators sometimes call them "snake oils" or part of their "black art" of formulation.

Additives generally have specific functions, but can have unexpected positive or negative effects on other coating application or performance properties. They can be used to affect the final coating properties as cured on the substrate, the way the coating is cured, the way it is applied, and the way it is manufactured. Some remain in the cured coating, and some are volatile. Often the use of one additive corrects one problem, then requires the use of a second one to alleviate a problem the first additive caused. It is not uncommon to have several additives in a formulation.

The rest of this chapter briefly covers many types of additives that have been used in fluoropolymer coatings. The list is not exhaustive and the commercial examples are just a sample of those available. Lists of additives have been published covering perhaps one hundred general paint additives.[1] Often, the same additive is classified under several categories, and categories frequently have many names for the same function.

For fluoropolymer paint users, it may be tempting to acquire additives and modify the coatings they buy on their own. Certainly, this has been done by some with success, but modification of a coating will generally absolve the manufacturer from any responsibility for quality or performance issues.

There are many companies that offer a wide range of additives. These companies would be a good place to start a search for additives to deal with specific problems:

- Troy Chemical (http://www.troycorp.com)
- Air Products (http://www.airproducts.com)
- Degussa – Tego (http://www.tego.de/en/prod/prod2.html)
- DuPont (http://www.dupont.com)
- Lubrizol (http://www.noveoncoatings.com/CoatingsAmerica/PCIDefault.asp)
- Elementis Specialties (http://www.elementis-specialties.com)
- King Industries (http://www.kingindustries.com)
- Dow Chemical (http://www.dow.com/products_services/industry/paints.htm)
- Dow Corning (http://www.dowcorning.com/content/paintink/paintinkadditive/default.asp)
- 3M (http://www.3m.com)

7.2 Abrasion Resistance Improvers, Antislip Aids

Abrasion resistance, in general, is difficult to define, but usually it means the loss of coating by rubbing or scraping. It is sometimes called "erosion." In some coatings, fluoropolymer powder is viewed as an abrasion resistance improver because it lowers the coefficient of friction. In many cases, the additives are hard materials and could be called pigments or fillers. The materials include silicon carbide powders, glass microspheres, and aluminum oxide particles (see Table 7.1). The particle size of these additives can also affect performance.

Table 7.1 Abrasion Resistant Additives

Abrasion Resistant Additive	Knoop Hardness, kg/mm	Specific Gravity, g/cm³
Silicon Carbide – Black	2580	3.2
Silicon Carbide – Green	2600	3.2
Boron Carbide	2800	2.52
Aluminum Oxide	1175–1440	3.69
Hollow Glass Spheres		0.6–1.1
Solids Glass Spheres		2.5
Silicon Nitride	1580	3.29

7.3   Acid Catalysts

Acid catalysts are added to some resin-bonded coatings to accelerate the chemical reactions involved in curing the non-fluoropolymer resin. Epoxies that crosslink with melamine formaldehyde resins can be made to cure at lower temperatures in the presence of an acid catalyst. Common acid catalysts are p-toluene sulfonic acid (PTSA): [CAS: 104–15–4]* and dodecylbenzene sulfonic acid (DDBSA) [27176–87–0]*. DDBSA is generally preferred in food contact applications.

7.4   Acid Scavengers

Acid scavengers are uncommon in fluororesin coatings, except for those coatings that can generate hydrogen fluoride, as in some fluoroelastomer-based coatings. These additives neutralize the acid emitted during cure, and include materials like calcium carbonate.

---

* Note: CAS stands for Chemical Abstract Number. It is the largest and most current database of chemical substance information in the world. It is useful because the number describes a particular chemical structure. Because there can be many names for a particular chemical, the use of the number eliminates confusion.

7.5   Adhesion Promoters, Coupling Agents

Adhesion promoters are receiving a great deal of attention for applications on non-metal substrates such as glass and plastic. They also improve the compatibility of materials, such as glass fiber with the resins in the coating system. Adhesion promoters are frequently molecules with reactive functional groups on two ends of the molecule. One group is designed to interact strongly with the substrate or pigment and the other reactive group would react with the coating binders. The most common materials used in fluorocoatings are silanes, and titanates are used to a lesser extent.

Silanes are compounds of silicon and hydrogen of the formula Si_nH_{2n+2}. When one end of a molecule functional group is of the form $-Si-OH$, $-Si-(O-CH_3)_3$, or $-Si-(O-CH_2CH_3)_3$, that group reacts with the inorganic species such as pigments and the substrate. When the functionality at the other end is organic (vinyl-, amino-, epoxy-, methacryl-, mercapto-, etc.), that group can react with the resins in the coating, creating a coupling between the two constituents. As a result of possessing these two types of reactive groups, silane coupling agents are capable of providing chemical bonding between an organic material and an inorganic material (Fig. 7.1).

The use of tetraalkyl titanates [such as $Ti(OC_3H_7)_4$, tetraisopropyl titanate] for improved

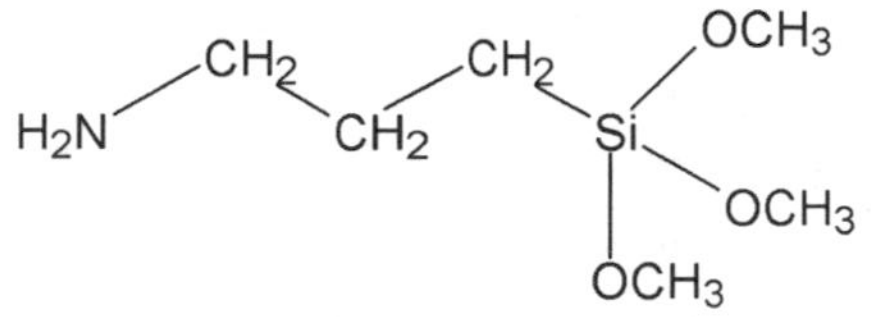

Figure 7.1 Structures of two common silane coupling agents.

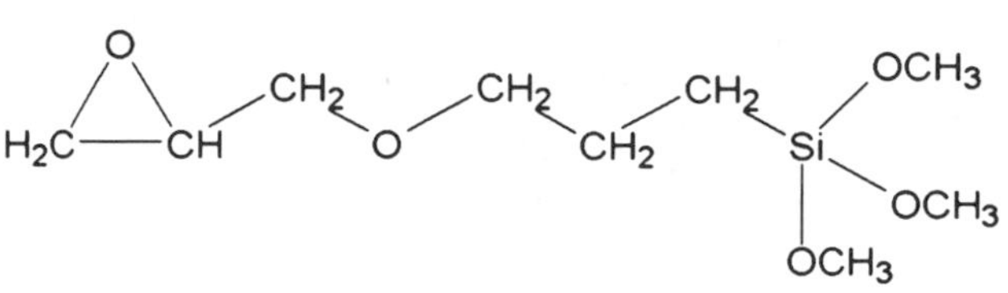

adhesion to glass is based upon the ability of the molecule to hydrolyze on the substrate surface to form a network of Ti–O–Ti bonds with the elimination of the alkyl alcohol. The Ti–O network can bond directly to hydroxyl functionality in the substrate or indirectly through hydrogen bonding.

Suppliers include:
- Dow Corning for silanes (http://www.dowcorning.com/content/silanes/silanespaint/)
- DuPont for Tyzor® titanates (http://www.dupont.com/tyzor/)

7.6 Algaecides, Biocides, Fungicides

Chemical agents that destroy or inhibit growth of algae, bacteria, or fungi on the coating have drawn increasing interest recently. Sometimes these properties are required in the final coating or in the liquid coating. Because of the high processing temperatures of fluorocarbon coatings, additives functioning in cured coatings are generally inorganic such as silver metal or silver compounds. Extremely fine nanoscale titanium dioxide has also been used, which attacks bacteria by photo-oxidation. Alkyl ammonium salts have been used in low-temperature cure systems; they attack bacteria through ionic membrane disruption. Hydantoin has also found use: it kills bacteria in a hypochlorite-like oxidation.[2]

To provide function in liquid coatings, the additives can decompose and evaporate from the film during curing. Most aqueous fluoropolymer coatings are kept at high pH levels that inhibit bacteria growth. This is done primarily using ammonium hydroxide. In fact, most fluoropolymer dispersions, from which many aqueous products are made, are modified with additional ammonium hydroxide. Bacteria growth can occur if the ammonium hydroxide concentration in the coatings drops over time. There are dozens of suppliers of these materials for in-can applications where thermal stability is not important.

For high-temperature-cured coating film application, compounds containing silver can be used. A common example is AgION® from the company of the same name (http://www.agion-tech.com/). This compound contains silver ions carried in a zeolite matrix.

7.7 Anti-Cratering Agent, Fisheye Preventer

Craters or *fisheyes* are described in more detail in the Ch. 14 on paint defects. They are frequently caused by contamination that lowers the surface tension in the paint around the contaminated area. Anti-cratering agents typically lower the surface tension of the paint, nullifying the effect of contaminant in creating surface tension gradient. Fluorinated surfactants are often used for this purpose, although other surfactants may also work. Many of the additive companies previously mentioned offer these products, but there are too many to summarize here. Fluorinated surfactants are offered by 3M (Fluorad®, http://www.3m.com) and DuPont (Zonyl®, http://www.dupont.com/zonyl/flash.htm).

7.8   Anti-Crawling Agent

Crawling is the tendency of a coating to pull back from edges of contamination on the substrate. It is similar to fisheyes, except it occurs on larger areas. If the cause can not be eliminated, it is frequently treated with the same chemicals as fisheyes.

7.9   Anti-Foaming Agent, Defoamer

Defoamers, sometimes called *bubble breakers*, are usually added for production purposes, though they are sometimes added to paints that are applied by roller or by dipping. Defoamers can be as simple as a small amount of hydrocarbon, or as complex as proprietary blends of chemicals. Some defoamers prevent the formation of the foam, others eliminate foam that has already formed. The problem is usually associated with aqueous systems, though foam or bubbles can form in non-aqueous systems, too. For aqueous coatings, defoamers usually have limited water solubility but reduce the surface tension (discussed in Ch. 7, "Solvents"). Defoamers are grouped into the following chemical types:

1. Branched chain alcohols or polyols, such as 2-ethyl hexanol.

2. Fatty acids and their esters, such as diethyl stearate and sorbitan trioleate.

3. Moderately high molecular weight amides such as distearoylethylenediamine.

4. Phosphate esters such as trioctyl phosphate or tributyl phosphate.

5. Metallic soaps such as calcium or magnesium stearate.

6. Chemicals with multiple polar groups such as di-t-amylphenoxyethanol.

7. Polysiloxanes (also good for solvent based coatings).

Generally, the types and amounts of defoamers are determined by trial and error. The following list contains some of the many companies that offer anti-foaming agents product lines:

- Air Products and Chemicals, Inc. (http://www.airproducts.com/Surfynol/default.htm)

- BYK-Chemie USA. Inc., a member of Altana Chemie Daicolor-Pope, Inc. (http://www.byk-chemie.com/language.html)

- Degussa Corporation, Tego Coating & Ink Additives (http://www.tego.de/en/prod/prod2/prod2.html)

- General Electric Co., GE Advanced Materials - silicones (http://www.gesilicones.com/gesilicones/am1/en/category/prod_category_landing.jsp?categoryId=19)

- King Industries, Inc. (http://www.kingindustries.com/coat/info/disparl/defoam/defoam.htm)

- Troy Corp. (http://www.troycorp.com/products_by_function.asp?Func=Anti-Foaming&App=Coatings)

7.10   Anti-Fouling Agent

This application is mainly aimed at preventing the attachment of marine organisms to ship bottoms. This highly specialized application of coatings is mentioned here strictly because fluorocoating technology has been studied for many years and continues to be studied in these applications.[3] Non-stick coatings generally have not worked except on fast moving boats, though the bio-fouling can be easier to remove from a non-stick coating when the boat is taken out of the water. The best additives based on organo-tin such as tributyl tin and copper are being phased out due to environmental concerns. These work by creating a surface that is toxic to marine life such as barnacles that sticks to the boat bottoms.

7.11   Rust Inhibitor, Corrosion Inhibitor, Flash Rust Inhibitor

There are two general end-uses of fluorinated coatings where rust or corrosion inhibitors are used. The first application is flash rust inhibition. Flash rusting most often occurs when aqueous products are applied to unprotected steel. Rusting starts immediately and can be very rapid. The combination of oxygen and water with the surfactants makes the attack of steel very rapid. Incorporating special additives can slow this type of oxidation. Amines added to formulations are somewhat effective, but are hazardous to handle. One of the best additives is morpholine. Flash rust inhibition is also discussed in Ch. 8 on substrate preparation.

The other application is a corrosion inhibitor. Corrosion inhibition happens after the paint is applied and cured, and is usually based on special pigments. The effective choices have been lead and zinc chromates in the past. These pigments have health and environmental problems, and are strictly controlled and avoided. Other pigment materials are available that generally do not work as well.

7.12 Anti-Sag Agent, Colloidal Additives, Thickeners, Rheology Modifiers

Anti-sag agents are often called *thickeners* or *bodying agents*. The purpose of these additives is to increase the low shear viscosity, but not affect the high shear viscosity. Adding small amounts of special small particles called colloidal particles usually increases low shear viscosity. Colloidal particles that are less than 1,000 nanometers in diameter (1 micron) are very small. Colloidal particles are usually supplied in dispersion form, called a colloidal dispersion or suspension. The colloidal particles build up a very loose structure that causes the low shear viscosity to be very high, but the structure is broken down very quickly when it is sheared. A schematic of a loose colloidal structure is shown in Fig. 7.2.

The colloidal particles need to reform the structure rapidly after shear is removed. Additionally, the additives can not detract from the final coatings properties, need to provide the anti-sag control at high temperatures, and cannot affect the high shear viscosity significantly.

A wide range of other materials is available for these purposes and can be grouped:

- *Organic thickeners*—generally degrade and volatilize during baking:
  - Cellulosics—hydroxyethyl cellulose is an example.

- Associative thickeners are polymers that are based on water-soluble polymers often based on polyethyleneglycol. An example is Bermocoll EHM 200, from Akzo-Nobel.
- Organic thixotropes—for example, Rheocin® castor-based powder for aliphatic systems from Sud-Chemie.
- Acrylics—for example, Acrysol ASE-60 thickener from Rohm and Haas.

- *Inorganic thickeners:*
  - Organoclays—commonly known as Bentonite or colloidal clay, Kaolin, or China Clay.
  - Fumed Silica—Cab-O-Sil® and Aerosil® are trade names. The surface chemistry of silica allows extensive network formation with a polar solvent or polymer molecules.

Fumed silica and organoclays can work in solvent-based coatings.

7.13 Anti-Settling Agent

Anti-settling agents are used to keep pigments or fluoropolymer particles in suspension. Anti-sag agents will do this by creating high low-shear viscosity. Pigment settling is reduced with increased viscosity as discussed in Ch. 5 on pigments. The additives used are the same as those listed in the anti-sag discussion above.

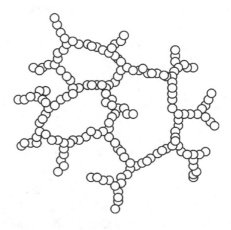

Figure 7.2 Loose colloidal structure.

7.14 Antistatic Agent, Electroconductive Additives

Antistatic agents are usually conductive pigments and are described in Ch. 5. There are also polymers available that are inherently conductive called *inherently conductive polymers* (ICPs). Polyaniline is one such polymer. These are of limited use in fluorocoatings because many of the polymers are doped with iodine and the conductivity is lost when baked above 300°F (149°C). Because these materials have limited use in fluorocoatings, they will not be discussed further, but there are several books available on the subject.[4][5]

7.15 Coalescents, Coalescing Agent, Film Forming Agent

Film forming agent is most descriptive of the function of these additives. These are very important components in many dispersion-based fluoropolymer products. These finishes, without the film-forming agent, will crack, producing unacceptable coatings. During drying and heat up of the applied coating, the agent forms a film holding the fluoropolymer dispersion particles in place along with the other components. The fluorinated coating components will, thus, form a cohesive film, which would not happen without this type of additive. Eventually, the film-forming agent decomposes and leaves the film as volatiles. Acrylic resins are examples of this type of film-forming agent. Occasionally, a decomposition catalyst is also added to aid in the volatilization. High boiling solvents such as glycerol, propylene glycol, polyethylene glycol, and butyrolactone can also function as film forming agents.

7.16 Deaerators

Deaerators are similar to defoamers, though there is a subtle difference. Deaerators prevent microfoam and pinhole formation. They are very important in high viscosity and high solids systems, especially for some application techniques. Deaerators deal more with the incorporation of air.

There are several ways air enters a coating. One of the most common is air drawn into the coating during the production process. Air is entrained when the coating ingredients are blended by mixers. If the mixer generates a vortex, then air can be drawn into the coating. If the viscosity is high, air can remain entrapped in the coating even after filtering and filling. Another source of air entrapment is during the coating application process. As droplets of coatings strike the substrate, air can be trapped under the surface, particularly if the coating viscosity is high. Most coatings need to be mixed prior to use and, if done incorrectly, air can enter the coating just like it can during production.

Sometimes air entrapment can be measured. A density measurement, (see Ch. 13 on coating performance) can yield a value lower than expected if air entrapment is significant.

When dealing with an entrained air or foam, it helps to distinguish between two types of foam, macro and micro. In both types, air is entrained in the coating material, but at different locations. Macrofoam occurs at the coating surface. Microfoam is located within the coating film. The entrapped air is prevented from rising to the coating surface by high coating viscosity.

Deaerators work by taking advantage of Stoke's Law, which states that larger bubbles rise much faster to the surface than smaller bubbles. They work by coalescing small bubbles into large bubbles that float to the surface more quickly. There are a large number of the additives available, but in general they can be classified by their chemistry:

1. Polyacrylates
2. Fluoro-modified polysiloxanes
3. Polyethers
4. Polar-modified polysiloxanes
5. Polysiloxanes

Like with most additives, experience, or trial and error often determines type and level.

7.17 Degassing Agent

Degassing agents are usually used in powder coatings. As a powder is deposited on the substrate during application, air is entrapped between the particles. That air must be released during the melting of the powder. Because melt viscosity is usually very high, the degassing process is difficult. Often a longer bake at a higher temperature will allow the air out, but when that does not work or is impractical, a degassing additive can help.

7.18 Dispersant, Dispersing Agent, or Surfactant

This subject is discussed in the dispersion stabilization section in Ch. 5 on pigments (Sec. 5.4). Common surfactants with fluorocoatings include:

1. Octyl phenol polyethoxylates (Triton®)
2. Nonyl phenol polyethoxylates (Tergitol®)
3. Sodium lauryl sulfate
4. Tridecyl alcohol polyethoxy ethanol (Serdox®)
5. Decyl alcohol polyethoxy ethanol (Dobanol®)
6. Acetylenic diol (Surfynol®)

The function of surfactants is discussed in Ch. 5. The chemical structures of the surfactants mentioned above are given in Fig. 7.3.

Many surfactants are volatile for high bake coatings and leave the final film. For others, where the surfactants remain in the cured coating, coating properties can be affected, so careful optimization of surfactant level can be important.

7.19 Flattening Agents

Flattening agents reduce the gloss or shine of a coating. The agents used in fluorinated coatings have a variety chemical compositions and particle sizes, but are mostly inorganic compounds such as:

1. Silicas: synthetic (fumed and precipitated), diatomaceous earth, silica gels
2. Clays
3. Talc
4. Carbonates

7.20 UV Absorbers and Stabilizers

These additives prevent degradation of polymers due to exposure to ultraviolet light. Absorbers and stabilizers play different roles. Absorbers absorb UV, but generally can not absorb all of it before it can attack the binder in the coating. Stabilizers, or hindered amine light stabilizers (HALS), scavenge the free radicals generated by the UV exposure that absorbers do not prevent. Free radicals can do a lot of damage to a polymer. HALS stabilizers do neutralize the radicals, but regenerate themselves. Absorbers can be pigments such as carbon black or titanium dioxide. HALS stabilizers are organic molecules, and so, have limited thermal stability and will not survive the high bakes of most fluorinated finishes.

7.21 Lubricants

One of the functions of fluoropolymer coatings is dry lubrication. Occasionally, additional lubrication is desired for high load and high temperature uses such as roller bearing surfaces. Two materials have been used in fluoropolymer coatings for this purpose.

1. Graphite is used in the form of natural or synthetic flake. A water-based dispersion called Aquadag® is also available and easy to use.
2. Molybdenum disulfide is an inorganic material used as a dry lubricant.

7.22 Moisture Scavenger

Moisture scavengers are almost always used in all non-aqueous coatings containing aluminum flake. Aluminum can react with trace water in a non-aqueous coating and produce hydrogen gas. Tightly closed containers have been known to burst when gassing occurs creating a safety issue and a potential environmental problem. The most common additive is hygroscopic silica.

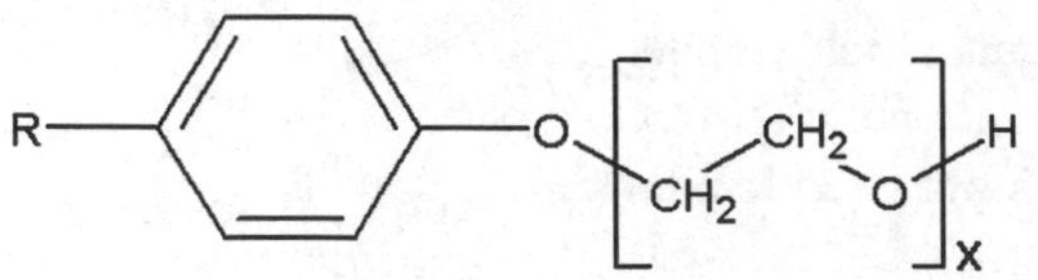

R is ——C_8H_{15} for Triton® surfactants
R is ——C_9H_{17} for Tergitol® surfactants
X varies depending on grade
and all grades are mixtures of X's

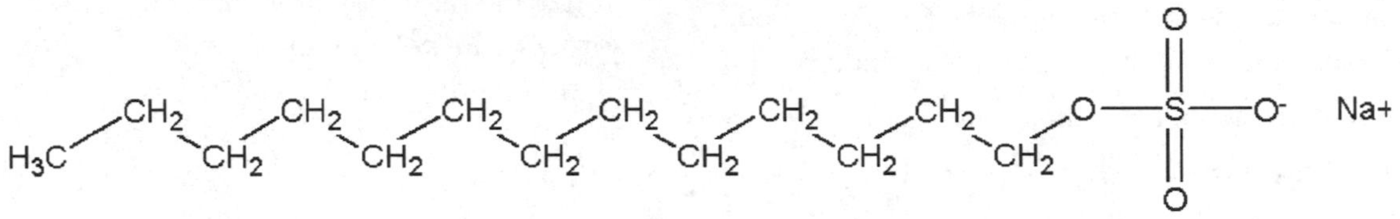

Sodium Lauryl Sulfate

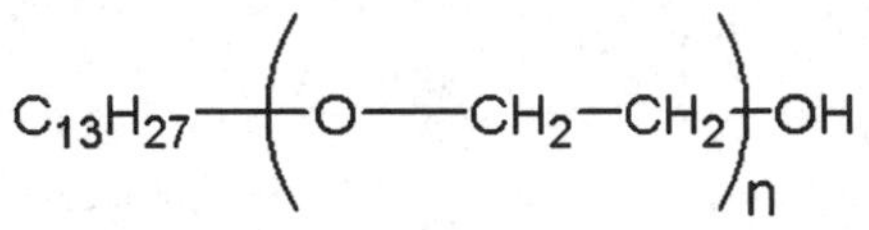

Tridecyl Alcohol Polyethoxy Ethanol (Serdox®)

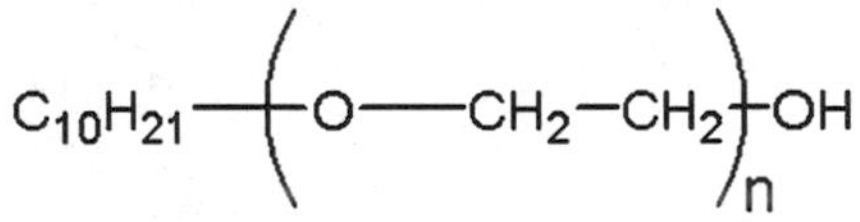

Decyl Alcohol Polyethoxy Ethanol (Dobanol®)

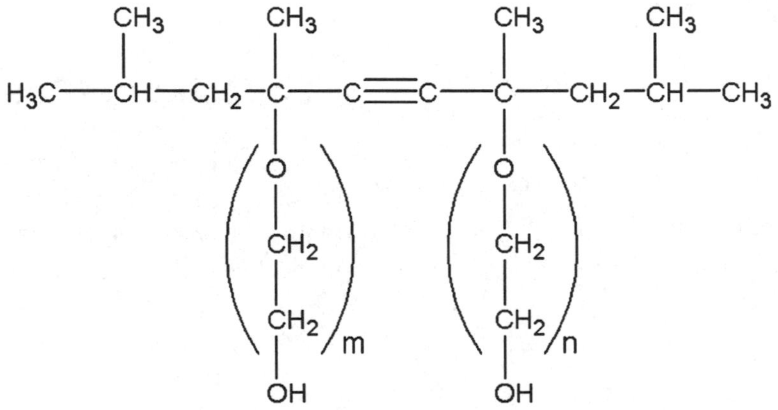

Acetylenic Diol (Surfynol®)

Figure 7.3 Structures of some common surfactants used in fluoroinated coatings.

7.23 pH Control Agent

Most aqueous dispersion-based fluorinated coatings are formulated to a basic pH above 7.0. The high pH resists bacterial attack of the organic ingredients in the coating. Amines of all types are used, as is ammonium hydroxide. Ammonium hydroxide is most effective, but it is also volatile and can diffuse out of the coating. Because ammonia is a small molecule, it can diffuse through some plastic containers. Amines are almost always present to minimize the risk of pH drift towards neutral with age.

7.24 Summary

Additives can solve problems but can also cause problems. Most formulations have several additives. During formulation development, once an additive is used, it is not possible to take it out. Others might be added to solve various problems, even those caused by another additive, but formulators tend not to remove any. The interaction of a new additive with those already in the formulation can present new problems.

REFERENCES

1. Koleski, J. V. , Springate, R., and Brezinski, D., 2003 Additives Guide, *Paint and Coatings Magazine,* 19(4) (Apr 2003)

2. Wynne, K. J., Makal, U., Ohman, D., and Wood, L., Antimicrobial Coatings Via Polymeric Surface Modifying Additives, American Chemical Society Meeting, Polymeric Materials: *Science & Engineering,* (2005)

3. Swain, G. W., Redefining Antifouling Coatings, *J. Protective Coatings and Linings,* 16:26–35 (Sep 1999)

4. Chandrasekhar, P. (Editor), *Conducting Polymers, Fundamentals and Applications: A Practical Approach,* Kluwer Academic Publishers, Hingham, MA (1999)

5. Skotheim, T., Elsenbaumer, R. L., and Reynolds, J. R. (Editors), *Handbook of Conducting Polymers,* 2nd edition, Marcel Dekker, New York (1998)

8 Substrates and Substrate Preparation

8.1 Introduction

The successful application and use of fluoropolymer non-stick and industrial coatings requires particular attention to the selection of substrates and to surface preparation. This chapter is intended to provide information and guidance in substrate selection and preparation. Ultimately, a user should consult the manufacturers' fact sheets to obtain the best results.

8.2 Substrates

Any substrate, which is dimensionally and thermally stable at the bake temperature required for the bake of a particular product, can be coated with fluoropolymer coatings. Adhesion to that substrate, however, needs to be confirmed. A variety of commercial substrates are coated with fluoropolymer coatings including those listed:

1. Various ferrous alloys
2. Cast iron
3. Steel
4. Stainless steel
5. Treated steel such as tin-plate or galvanized (zinc)
6. Non-ferrous metals and alloys
7. Aluminum
8. Cast aluminum
9. Polymeric materials
10. High temperature plastics and elastomers
11. Glass, pyroceram, and ceramics
12. Stone

At bake temperatures greater than 232°C (448°F), certain metallic substrates are unacceptable. The melting points of tin, 232°C (448°F), and lead, 328°C (622°F), are too low to permit bakes required for fluoropolymer coatings. Zinc melts at 419°C (787°F) which is below the processing temperature of many fluoropolymer coatings.

Poor adhesion to copper is the result of the copper oxide formed when copper is baked in air at high temperatures. Because of reactivity at the high baking temperatures, fluoropolymer coatings have relatively poor adhesion to magnesium and to aluminum/magnesium alloys containing more than 0.5% magnesium. Aluminum permanent mold castings and die castings are successfully coated with fluoropolymer coatings, but may show a high reject rate due to the formation of blisters caused by expansion of air bubbles in the metal during the high-temperature bake.

8.3 Substrate Preparation

Substrate preparation is aimed at several purposes:

- Cleaning
- Improving adhesion by increasing surface area
- Hardening
- Providing improved corrosion resistance
- A combination of the above

8.3.1 Cleaning

In all cases, fluoropolymer coatings should be applied over clean substrates. Normal industrial practices such as chemical washes, or solvent cleaning, and vapor degreasing can be used, but precautions must be taken to remove all residues from the cleaning process. Depending on the initial condition of the metal, it may be necessary to physically remove dirt, rust, mill scale, old paint, etc. After cleaning, the metal should be handled with clean gloves. Fingerprint contamination may show up as a stain on the finish. Residual skin oil also may cause stains, poor adhesion, or other surface defects.

Preheating metal substrates above the bake temperature required for the fluoropolymer coating is a good way to remove traces of oil and other contaminants, especially when the metal is formed by casting and is porous. With most ferrous metals, this procedure has the advantage of temporarily passivating the surface against rusting and the blue oxide formed increases the adhesion of the acid primers. In the case of aluminum and stainless steel, these advantages are not apparent and the preheating step can be omitted when the metal is clean. If stainless steel is thermally cleaned, it frequently turns a golden brown in color. Copper and brass should not be preheated in air because the resulting oxide has poor adhesion to the metal. A formic acid rinse reduces oxide formation on copper to some extent.

Grit blasting, discussed in Sec. 8.3.2.1, can also be considered a cleaning step, particularly when stripping off old coating.

8.3.2 Increasing Surface Area

The adhesion bond strength of coatings to a substrate is always increased when applied over a roughened surface. The surface area of a roughened surface is larger than that of a smooth surface. There are several common ways to increase the surface area.

8.3.2.1 Mechanical Roughening

Grit blasting is the method most commonly used to obtain good adhesion of fluoropolymer coatings. Grit blasting should precede preheating of ferrous metals to retain the protective oxide formed. With other clean substrates, the order of these two operations is not important.

Grit blasting is a relatively simple process. Hard grit is propelled by compressed air or, occasionally, by high-pressure water at the substrate needing cleaning or roughening.

Grit blast profiles are commonly measured in microinches or root mean square (RMS) by means of a profilometer (see Sec.8.4). A profilometer drags a diamond stylus across the substrate and measures the depth of the peaks and valleys.

Surface profiles in excess of 100 microinches (2.5 microns) are recommended and 200–250 microinches (5.1–6.5 microns) are frequently employed.

On hard substrates, aluminum oxide grit from #40 to #80 are commonly used at air pressures ranging from 80 to 100 psi (5.8 to 7.3 kg/cm²) at the gun. Aluminum and brass are commonly used at air pressures ranging from 80 to 100 psi (5.8 to 7.3 kg/cm²) or below. Maximum air pressures on stainless steel may exceed 100 psi (7.3 kg/cm²).

It should be noted that profiles measured by common profilometers indicate only depth of profile. They do not measure uniformity or coverage of the grit blast, nor the sharpness of the peaks. Full coverage of the grit blast is indicated by lack of gloss on the metal surface when viewed at a flat grazing angle.

There are numerous types of grit. The choice of which grit to use depends upon its intended purpose, the substrate, and the expense. Properties of common grit types are summarized in Table 8.1. Properties listed in the table include hardness (grit needs to be harder than the substrate used to roughen it). Density and bulk density are listed, as denser materials have more momentum and impart more energy to the substrate. Also, a relative cost on a volume basis is listed.

Sometimes minimal damage to the substrate is required. This could be due to the relative softness of the substrate, or because the texture or pattern machined in the substrate needs to be maintained. Plastic grit, walnut shells, or sodium bicarbonate can clean the substrate or remove the previous coating.

Occasionally, the abrasive is propelled by pressurized water. Sodium bicarbonate slurries have been used to remove fluorocarbon coatings in this fashion.

Aluminum oxide. Aluminum oxide is usually offered in a size range of 16–240 grit. It is angular in shape. It is the most popular cleaning blast media. Aluminum oxide conforms to major industrial and government standards including, MIL A21380B and ANSI B74. 12–1982.

Silicon carbide. Silicon carbide is usually offered in a size range of 16–240 grit. It is angular in shape. Silicon carbide is an extremely hard, sharp grain that is more friable than aluminum oxide. For use in blasting of extremely hard materials; it is expensive.

Sand, silicon dioxide, or silica. This is considered too smooth and uniform. It breaks down too rapidly to be useful in preparing metal substrates and is not recommended. It is cheap and occurs naturally.

Table 8.1 Grit Blast Media Properties

Media	Hardness, Moh	Density, g/cc	Bulk Density, lbs/ft^2 (g/cc)	Relative Cost, Volume Basis
Walnut Shells	1–4		40–80 (0.64–1.28)	19
Silicon Carbide	9	3.2	95 (1.52)	50
Aluminum Oxide	9	3.8	125 (2.00)	25
Glass Bead	6	2.2	100 (1.60	18
Plastic Grit	3–4	1.45–1.52	45–50 (0.72–0.80)	30
Steel Shot	6	7.87	280 (4.49)	27
Steel Grit	6	7.87	230 (3.68)	
Sand, Silica, (Silicon Dioxide)	7	2.6		11
Sodium Bicarbonate	2.5	2.16		

Glass beads. These are made from chemically inert soda lime glass. Blasting with glass beads will produce a metallurgically clean surface for parts and equipment. The beads are spheres of uniform size and hardness. Glass beads can meet OSHA standards for cleaning operations. Another advantage of using glass beads is the disposability; spent glass is environmentally friendly. This can simplify the disposal and reduce the cost. Glass beads are often used for stress relief. Mil-Spec (MIL-G-9954A) glass beads are one type that are available.

Crushed glass. Crushed glass is available in a range of sizes from coarse to very fine. Crushed glass is an excellent low cost alternative to various reclaimed blast abrasives. While it breaks down relatively easily, it is silica free with minimal iron content (2%), and produces a luster-white metal finish.

Steel grit and shot. Generally available with diameters of .007 inches to .078 inches (0.02 cm–0.2 cm), steel grit is angular in shape while shot is round. Steel grit and shot have one of the lowest breakdown rates of all blast media and can, therefore, be recycled and reused. Its density is also high which helps impart more energy to the substrate being cleaned. Steel grit is excellent for use in large blast room applications. It should not be used on stainless steel where iron impregnation is a concern. Cast stainless steel shot is available for nonferrous castings or for other items where ferrous contamination is a problem.

Walnut shells. Walnut shells are not very hard. This media is offered in sizes are 10/14, 14/20, and 20/40 grit mesh sizes (see Table 9.1 in Ch. 9). It is soft, friable, dried shells or nuts. It is sometimes called "organic" or agrishell abrasive. It is often used for removing contaminants, such as carbon deposits or old paint from delicate parts, or soft materials, such as aluminum. It is also good for blast cleaning with portable equipment.

Plastic grit. Plastic grit is another soft media typically in a size range of range 12 to 60 mesh. It is often made from recycled or waste plastic. It, like walnut shells, is typically used to clean surfaces and remove old paint without harming the substrate.

Baking soda. Baking soda blasting is unique because of its biodegradable characteristics. Clean up after use is easy because it is water soluble, and can be literally "washed" away. Baking soda is commonly used where one-pass coverage with no recovery is acceptable or desirable, and the substrate is delicate or sensitive. Typical applications for baking soda include graffiti removal, boat hulls, and large printing press rolls. It is often made into water borne slurry and propelled by high-pressure water.

8.3.2.2 Other Methods of Roughening and Cleaning

While grit blasting is the preferred metal treatment for the application of most fluoropolymer coatings, other methods of surface roughening are employed in special cases. Wheel sanding, wire brushing, and directional grinding may be used where a strong adhesive bond is not required. These operations reduce adhesive bonding in the direction of the grind. Wheel or belt sanding of aluminum previously coated with fluoropolymer must not be attempted, as violent explosions are possible.

Chemical etch. Chemical etching using acidic materials such as chromic acid, hydrochloric acid, or sulfuric acid or bases such as sodium hydroxide gives smooth peaks, without the sharp "tooth" required for best adhesive bond. Rough, as-cast surfaces are also too smooth in microprofile for strong adhesion bonding. In addition, the etching reagents require immediate rinse to stop the action and prevent salts from depositing on the surface. The rinse sometimes creates an oxidizing or rusting problem.

For reinforcement of a substrate or creating a very rough surface, a discontinuous layer of stainless steel such as 309 alloy may be applied to grit-blasted metal. Spraying molten metal onto the substrate can be achieved using a process called arc, flame, or plasma spraying. Arc spray is most common. Wires of the metal to be applied are fed into a jet of inert gas such as nitrogen. A high current flow through the wires causes them to melt. The inert gas carries the molten metal droplets to the substrate. The molten metal impacts the substrate and solidifies producing a very rough coating. However, when this procedure is used for a dissimilar metal substrates such as aluminum, there is potential bimetallic corrosion.

This type of treatment is applied more for abrasion or scratch resistance than for adhesion. When a coating is applied to such a surface the valleys are filled in with coating. Many high-end fry pans are prepared in this manner. This allows metal utensils to be used on the soft fluoropolymers. A metal spatula for example will glide across the peaks of the profile and would not scrape off the fluoropolymer coating that is deposited in the valleys. While this leaves exposed metal peaks on the coating surface, more than 98% of the surface is still fluoropolymer and, the non-stick performance is affected only marginally.

Engraving. Some cookware manufacturers engrave or machine a profile or pattern into their cookware surfaces and apply coatings over them. They are attractive and work similarly to the arc-sprayed stainless surfaces in that utensils only scrape against the high points of the machined patterns. A photo of one example is shown in Fig. 8.1.

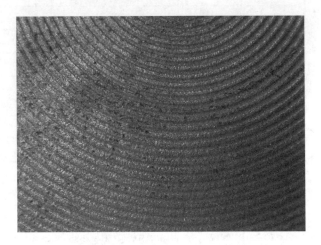

Figure 8.1 A machined bottom of a frypan.

Conversion coatings. Conversion coatings are the modified surfaces of metals resulting from specific chemical treatment. These conversion coatings, which can be applied on steel, aluminum, or most other metals, typically include zinc, manganese, and iron phosphates or chromates. The process usually involves a series of dips or sprays of the item to be coated. The baths need to be carefully maintained to function properly. This type of treatment is usually applied to large scale coating projects that run continuously. The principal function of these coatings is to promote improved adhesion of finishes, to maximize corrosion resistance, and increase blister resistance. The functionality of conversion coatings depends upon their uniformity and the integrity of the coating, both before and after application of the final finish.

Chromate conversion is a common treatment for aluminum. Sodium chromate is exposed to aluminum producing aluminum and chromium oxides on the metal surface by the reaction in Eq. (8.1). This treatment provides an increase in adhesion bond strength and corrosion protection over the untreated metal.

Eq. (8.1)

$$2Al + 2Na_2CrO_4 \rightarrow Al_2O_3 + 2Na_2O + Cr_2O_3$$

Phosphate conversion coatings for steel are common. There are several common phosphate treatments:

1. Zinc phosphate
2. 3-Stage iron phosphate
3. 5-Stage iron phosphate
4. Dried in-place coating

The zinc phosphate process typically involves:[1]

1. A cleaning bath to remove oils, other organics, and corrosion products.
2. Rinse—removes cleaning chemicals.
3. Conditioner—improves the zinc phosphate deposition uniformity.
4. Zinc phosphate bath.
5. Rinse—removes unreacted chemicals.
6. Sealing rinse.
7. Deionized water rinse.

Iron phosphates are generally easier to apply and maintain, but generally do not protect against corrosion as well as zinc phosphate. The author has found that overall manganese phosphates are best for fluorinated coatings.

Phosphates are typically applied to specific coating weights. Coating weight is generally specified in milligrams per meter (mg/m^2). Iron phosphates are typically applied at 3–7 mg/m^2, zinc phosphates at 9–28 mg/m^2. Chromates on aluminum are applied at 0.1–9 mg/m^2 and chrome phosphates at 2–28 mg/m^2.

Some of the conversion coatings are sensitive to high temperatures and will not function as expected if maximum bake temperatures are exceeded. The manufacturer should be consulted. Trade names for common chemical treatments are Bonderite®, Granodine®, Cryscoat®, Gardobond,® and Parco®. Henkel Surface Technologies (http://www.hstna.com) and Chemetall Oakite (http://www.chemetall.com) are the leading companies in providing surface treatments.

Conversion coatings are often used for automotive applications because of their corrosion resistance specifications, especially on fasteners such as nuts and bolts.

Corona, plasma, and flame treatment. Unlike most metals, non-metallic substrates such as plastic, elastomers or glass will generally need some additional means of surface treatment beyond cleaning and roughening. Corona, plasma, and flame treatment techniques are common means of imparting treatment to organic substrates. The purpose of corona, plasma, and flame treatment of a substrate is to improve the wettability and adhesion characteristics.

Corona, plasma, and flame treatments all generate different forms of plasma.[2] Plasma is a state of matter where many of the atoms or molecules are ionized, allowing charges to flow freely. It is sometimes called the fourth state of matter. This collection of charged particles containing positive ions and electrons exhibits some properties of a gas but differs from a gas in being a good conductor of electricity. The three treatments differ in the way energy is provided to produce plasma state. The energy generates atoms with a positive charge and the detached negative electrons. All are free to move about. These atoms and the resulting electrically charged gas are said to be *ionized*. When enough atoms are ionized and electricity is conducted, it is in the plasma state. Plasmas carry electrical currents and generate magnetic fields. The most common method for producing plasma is by applying an electric field to a gas in order to accelerate the free electrons.

Flame treating is easiest to understand and visualize. It is exactly what one would expect from its name. The substrate is exposed to an oxygen rich flame. It is mainly used to improve adhesion, but it can be beneficial in other ways. Because high temperatures are generated with flame treating, it can burn off dust, fibers, and residual organic matter, thus cleaning the surface for coating. The oxygen rich portion of the flame promotes oxidation of the substrate generating reactive groups. The reactive groups provide higher surface energy for better wetting and the opportunity for chemical interaction with the coating.

Corona treatment is a different form of plasma. It produces plasma by applying enough voltage across two electrodes with air space between them. The high voltage ionizes the air in the gap to produce the corona, which usually looks like a blue flame. Ozone is generated from oxygen in the air in the corona. How the corona modifies the surface is not precisely understood, but one theory states that the energy of the high-charged electrical corona breaks the molecular bonds on the surface of the substrate. The broken bonds then recombine with the free radicals in the corona environment to form additional polar groups on the surface. These polar groups have a stronger chemical affinity for coatings, which results in improved adhesion. The increased polarity of the surface also results in an increased surface energy that translates into improved wettability.

Plasma treatment is very similar to corona treatment, except gases are injected into the corona discharge to modify the chemical composition of the corona plasma and so changes the reaction with the substrate. Some substrate materials are less reactive to a traditional corona and require this special treatment.

A plasma or corona treatment is often used for coating continuous web types of materials, such as plastics and foils. Hand units have become common, however, and are now being used in some paint shops.

The coating must be applied as soon as possible after any of plasma treatments. The effects of the treatment drop off very rapidly for many substrates, often after a few seconds or minutes of exposure to air.

Many companies offer products for plasma type treatments. The following are among those that offer hand held devices:

- Enercon Industries Corp.
  (http://www.enerconind.com/treating/products/index.html)
- Surfx Technologies LLC
  (http://www.surfxtechnologies.com/index.htm)
- SOFTAL 3DT LLC
  (http://www.3dtllc.com/)

Organic substrates are sometimes treated with solvent that starts to dissolve the surface. Solvent softening may improve adhesion of a coating to a plastic surface. Spraying or dipping are two methods of applying the solvent.

8.3.3 Preventing Rust after Surface Preparation

Steel and iron rust rapidly after grit blasting. This is called flash rusting, which requires coatings to be applied immediately. Where delay is expected, or under conditions of high humidity, a solvent rinse with VM&P naphtha or toluene containing 5% kerosene may be employed. When the volatile solvent evaporates, a very thin film of kerosene remains that prevents rusting temporarily. The kerosene film may collect dust on long standing and require solvent washing before the finish is applied. It may need to be removed for some coating systems, especially on aqueous coatings.

In some instances, a water solution of an amine, such as triethyl amine can be applied to the clean metal to passivate it against flash rusting.

There are other commercial rust inhibiting formulations available that allow storage for relatively long periods of time. However, these materials would likely need to be removed before application of the coating. Chemetall Oakite provides a complete line of rust preventative products:

- Short-term indoor protection; up to eight weeks indoor rust protection:
  - Cleaner/rust preventative – water based: Oakite 443, Oakite 200–404–003, Gardoclean A 5502, Gardoclean A 5503, and Inprotect 600
  - Rust preventative – water based: NRP and CPA
- Midterm indoor protection: up to 6 months indoor protection:
  - Cleaner/rust preventative: Oakite 398 LT, Oakite 498 DFW, and Oakite 200–404–004
  - Rust preventative: Oakite Rust Proof 1 and Oakite Rust Proof 2
- Long-term indoor protection: provides greater than 1 year indoor rust protection:
  - Rust preventative: Rustproof 4002, Ryconox 20M, and Oakite HPO

Henkel Surface Technologies also provides several products.

- Turco Protectoil: medium duty, emulsifiable corrosion inhibitor.
- RI-1 Rust Inhibitor: liquid alkaline, rust inhibitor compound.
- Rust Bloc: liquid alkaline, biodegradable rust inhibiting rinse additive and cleaner.
- SF-2838M: concentrated rust inhibitor for dip and spray applications.
- PREVOX: compounded oils, water-based emulsions, and synthetic fluids for the temporary corrosion protection of coil steels and aluminum, fabricated metals, and in-process components.

8.3.4   Platings

Occasionally, the substrate is plated, electroplated, or coated with a different metal prior to coating. This is generally done at the steel manufacturing facility, not by the coater. Some of these metals are:

1. GALVALUME® steel is 55% aluminum-zinc alloy coated sheet steel. The steel is immersed in a molten aluminum-zinc alloy bath. The aluminum-zinc alloys provide corrosion protection.

2. Galvanized steel has been covered with a layer of zinc metal. During galvanizing, steel is immersed in a molten zinc bath. Zinc's natural corrosion resistance provides long-term protection, even in outdoor environments.

3. Aluminized steel is manufactured in two grades. One has a silicon-aluminum alloy coating and is best suited in an environment where a combination of heat and corrosion is involved. The second grade has a pure Al coating and has excellent resistance to atmospheric corrosion. Both are applied by hit dip process.

4. Zinc electroplating is sometimes used on fasteners for corrosion resistance.

These substrates are often processed with the other surface preparation techniques described in this chapter. When coating, one must keep in mind that the coatings, or platings, may melt at temperatures below those for which some coatings are processed. If grit blast is used on platings, care must be taken to avoid blasting through the plating, exposing the base metal.

8.3.5   Anodization

Aluminum anodizing is the electrochemical process by which aluminum is converted into aluminum oxide on the surface of a part. This coating is desirable in specific applications due to the following properties imparted by the anodization process:

- Increased corrosion resistance
- Increased durability/wear resistance
- Electrical insulation
- Excellent base or primer for secondary coatings

The process of anodizing is fairly simple. It consists of an anodizing solution typically made up of sulfuric acid. A cathode is connected to the negative terminal of a voltage source and placed in the solution. An aluminum component is connected to the positive terminal of the voltage source and also placed in the solution. When the circuit is turned on, the oxygen in the anodizing solution will be liberated from the water molecules and combined with the aluminum on the part forming an aluminum oxide coating. It is the stability of the aluminum oxide that accounts for all of the protective properties of the coating.

There are three types of anodizing:

- Chromic anodizing
- Sulfuric anodizing
- Hardcoat anodizing

Each of these has advantages and disadvantages depending on the application.

Chromic anodizing is commonly referred to as type 1 anodizing, and is formed by using an electrolytic solution of chromic acid that is about 100°F (38°C). It utilizes a chromic acid electrolyte and yields the thinnest coatings, only 0.05 to 0.1 mils (1.25 to 5 microns) thick. Chromic anodizing is often chosen when a part is complex and difficult to rinse. Chromic acid is less corrosive than sulfuric acid used in other anodizing methods. The process takes about 40 to 60 minutes. It produces a clear to gray coating, depending on the sealing and the alloy used. Chromic anodize offers a minimum of 336 hours (5%) salt spray resistance per ASTM B117 without a coating on top.

Sulfuric anodizing is commonly referred to as type II anodizing, and is formed by using an electrolytic solution of sulfuric acid at room temperature. The process will run for 30 to 60 minutes depending on the alloy used. This will produce a generally clear coating at thickness of 0.3 to1.0 mils (8 to 25 microns). It offers abrasion resistance that it is more durable than chromic anodize. Like most anodizes, corrosion resistance is excellent.

Hardcoat anodizing is commonly referred to as type III anodizing, and is formed by using an electrolytic solution of sulfuric acid at approximately 32°F (0°C). The process will run for 20 to 120 minutes depending on the alloy and the desired coating thickness. This process produces a generally gray coating. Hardcoat anodizing's great advantage is hardness and wear resistance. This anodize has a hardness on Rockwell C-scale rating of 60 to 70. The hardness makes it an excellent candidate for many applications that require low wear. It also offers good corrosion resistance.

Fluorinated coatings and primers often need formulation adjustments to optimize adhesion to anodization treatments.

8.4   Substrate Characterization

Running a substrate through a particular substrate preparation does not guarantee that the substrate preparation quality is what is expected. Ideally, there should be a quick test to verify substrate preparation quality. The primary test used in fluorinated coatings industry is a surface roughness test using a device called a profilometer. This device assigns a number to the roughness. One example of a portable unit for measuring roughness is shown in Fig. 8.2.

A profilometer works by drawing across the surface a very sharp, very small diamond stylus much like the needles found in an old phonograph. A close up of the stylus is shown in Fig. 8.2a. The stylus is attached to an arm as shown in Fig. 8.2b. The arm

moves across the substrate and the stylus rides up and down the profile. The profilometer instrument records these changes in height along the length. The data are then processed to generate a roughness number. Depending on the way the data is processed, one of several measures of roughness is output on the display. Referring to Fig. 8.3, the most common measure is called the roughness average, or R_a. Mathematically, the mean of all the measurements is calculated, which is labeled "mean line" in the figure. The absolute values of all the differences of all the surface data points from this mean are calculated along the length L. The average of these values gives the roughness average, or R_a. Mathematically the calculation is the integration described in Eq. (8.2).

(a)

(b)

Figure 8.2 Profilometer. (*a*) Close-up of the stylus. (*b*) The arm moves up and down as the stylus moves across the substrate, tracing the profile.

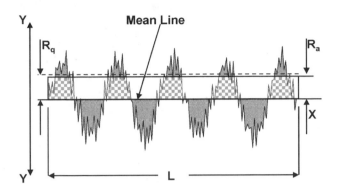

Figure 8.3 Roughness average (R_a) and root mean square roughness average (RMS or R_q).

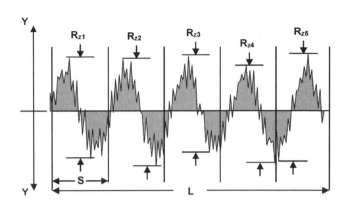

Figure 8.4 Mean roughness depth (R_z) and maximum (or single) roughness depth (R_{max}).

Eq. (8.2)
$$R_a = \frac{1}{L} \int_0^L |Y| dx$$

The root mean square roughness, (RMS) or R_q, is similar to the R_a and is noted in Fig. 8.3 and Eq. (8.3). The length the stylus moves is L, and Y is the stylus deflection (height or depth); they are indicated in Fig. 8.3.

Eq. (8.3)
$$R_q = \sqrt{\frac{1}{L} \int_0^L Y^2 dx}$$

There are two other roughness measures that a profilometer might display. These are explained with the aide of Fig. 8.4. The profilometer splits the measurement line into five equal parts. The maximum peak to valley distance in each of these parts is calculated and labeled in the figure $R_{Z\#}$. The average of these five measures is called the mean roughness depth, or R_Z, and is described in Eq. (8.4).

Eq. (8.4)
$$R_Z = \frac{R_{Z1} + R_{Z2} + R_{Z3} + R_{Z4} + R_{Z5}}{5}$$

The maximum roughness depth, or R_{max}, is just the largest of the individual $R_{Z\#}$ values.

8.5   Summary

In summary, substrate choice and preparation is critical. If not done correctly, even the best coating systems can perform poorly.

REFERENCES

1. Gardner, J., Pretreatment Shining Brighter, *Industrial Paint and Powder,* 80(9) (Sep 2004)

2. Schutze, A., et. al., The Atmospheric-Pressure Plasma Jet: A Review and Comparison to Other Plasma Sources, *IEEE Transactions on Plasma Science,* 26(6) (Dec 1998)

3. Mahr, C., Holding GmbH, Pocket Surf Flyer

9.1 Introduction

Previous chapters have covered most of the individual components that go into fluorinated coating formulations, coatings, paints, or products. Putting these components together to make a good product takes experience and experimentation, to say nothing of serendipity, which always helps.

9.2 Selecting Ingredients

Most users of fluorinated coatings rely on someone besides themselves to select the particular product for their end-use. Many applicators of the fluorinated coatings rely on the manufacturers of products for guidance. The selection of a particular coating is based on:

1. The desired function or properties that the coating will provide to the end-user.

2. The type of substrate and temperature limits (bake).

3. The application technique.

This section provides a basic explanation of why particular fluoropolymers, binders, and other components are chosen by a formulation chemist.

9.2.1 Selection of Fluoropolymer

Fluoropolymer selection depends on several property needs. The first one is *release* or non-stick. The non-stick properties of perfluorinated polymers (described in Ch. 1) are superior to the non-perfluorinated polymers. The common perfluorinated polymers include PTFE, FEP, and PFA. Generally, release ranks in the order of FEP > PFA > PTFE, but the differences are relatively small and dependent on the material that might stick to the coating. For the non-perfluorinated polymers, generally the higher the percentage of fluorine in the polymer, the better the release.

If a coating is going to be used at high temperatures, perfluorinated polymers are much more stable than the lower fluorine-content polymers. Within the perfluoropolymer group, PTFE > PFA > FEP is the order of thermal stability. Within this same perfluoropolymer group, PTFE > PFA = FEP for slip or dry lubrication. For chemical stability, the differences are minor if any. For surface smoothness, the trend follows the relative melting points of the fluoropolymers, FEP > PFA > PTFE. For toughness and abrasion resistance, PFA > PTFE > FEP.

Blends of fluoropolymers are often better performing than individual fluoropolymers. DuPont patented PTFE/PFA blends for cookware in the late 1970s.[1]

9.2.2 Selection of Binder

For primers and one-coats, the selection of the non-fluoropolymer binder is critical. The relative amount of fluoropolymer to non-fluoropolymer binder will also directly affect performance properties. The selection criteria are complex, but some simple guidelines get one started. For example, one of the hardest and best adhering binders is polyamide-imide. It will adhere to most metals. It can be used in both water-based and solvent-based formulations. Its shortfalls are strong color, slight sensitivity to light or UV exposure, and vulnerability to being attacked in strongly acidic and basic environments.

Polyether sulfone or polyetherimide are excellent choices for light-colored coatings. Adhesion properties are good on most metals, but not as good as PAI.

Polyphenylene sulfide is the most chemically resistant non-fluoropolymer binder. It adheres strongly, but it is dark colored and has a relatively low melting point and should, therefore, not be used at temperatures above 350°F (177°C). Many other binders are discussed in Ch. 4. Blends of materials can also be used.

9.3 Recipes and Formulas

It is the formulator's task to develop a recipe that the manufacturing plant can make reproducibly and within specifications. A paint recipe is a lot like a chef's recipe except that it involves much more detail. A typical recipe includes:

1. A name, production code, or product code

2. Safety information

3. Often, what equipment to use

4. What batch size to use

5. What ingredients to add, and in what order

6. How much of each ingredient to add

7. How to add

8. How fast to add

9. How to mix

10. How long to mix

11. Sometimes, the temperature is specified

12. If a dispersion is being made, the parameters for the grinding equipment

13. How to test what has been made

14. How to make adjustments

15. How to filter

16. The type of container to fill

17. How to store

Additional details on each of these steps follow.

A name, production code, or product code. The production code is what the product or intermediate is called in the manufacturing plant. It is not necessarily the same code under which the product is sold.

Safety information. Of great importance is the personal protective equipment required to meet, and preferably exceed, the standard. Some hazardous ingredients may require special clothing or special air supply. An area of the plant may be temporarily off limits to those employees not properly equipped.

What equipment to use. A large coatings manufacturing plant will have an array of mills, mixers, and filtering equipment. The mills may be of different sizes or types. The mixers can also differ in size, but could also have different mixing blades or tank configurations. Some tanks have baffles to optimize mixing, but baffled tanks are not always the best choice. Also, particular mixers or mills are frequently segregated. Preferably, food contact coatings should not be made in the same equipment as non-food contact coatings. Usually, pigmented coatings are segregated from clear coatings to minimize cross-contamination. It is especially hard to remove all traces of black pigments.

What batch size to use. Some manufacturing directions may differ from one batch size to another. For example, if one makes 100 gallons of paint in a 300-gallon mixer, the mixing instructions could differ from that of a full 300-gallon batch. A mixing blade turning at 100 RPM with 300 gallons in the tank may be specified, but if the same mixer speed is used for 100 gallons, it may splash the mixture in the tank, generating excessive foam or whipping air into the mixture. Therefore, the manufacturing directions often depend on the batch and equipment sizes.

What ingredients to add and in what order. The order of addition of ingredients can be critically important. Sometimes just swapping the order of two ingredients can be the difference between a good product and scrap. There are numerous chemical reasons for these situations to occur. Often the mixture is more shear-sensitive with one order of addition than another. Surfactants may not wet the particles they are intended to wet if they are not added at the right time.

How much of each ingredient to add. The amount of each ingredient is usually defined by weight, but occasionally it is by volume.

How to add. Sometimes materials are added by flowing down the side of the mixing tank, or through a dip tube to the mixture surface. It could even be pumped in from the bottom of the tank. This is a common practice for aqueous coatings because it minimizes foaming.

How fast to add. Each raw material is added to the mill at a specific rate. Some materials being added may be incompatible with the coating when added too quickly. The localized concentration builds up faster than mixing can take place. Adding slowly limits concentration build-up by allowing the mixer to dilute the ingredient that has the potential to cause a problem. Adding powders slowly is necessary to avoid having them clump and become more difficult to break apart and wet.

How to mix. There is usually a specific mixing blade in the equipment. The RPM is specified. Mixing too quickly can create foam, or can overshear the coating causing the formation of gel particles. If mixed too slowly, the mixture may not be uniform. The mixer RPM is changed after each addition.

How long to mix. Overmixing can create too much foam, or generate too much grit and gel; undermixing may result in nonuniformity.

Sometimes, the temperature is specified. Ingredients may be temperature-sensitive or may dissolve faster at higher temperature. The raw material or the mixing tank can be heated or cooled.

Operational parameters for the grinding equipment while a dispersion is being made. Grinding equipment such as media mills have many parameters to set. These include media type and size (for example, glass beads or ceramic beads, 2 mm or 4 mm), agitator speed, grind rate, pressures, cooling water, and mill temperature.

How to test what has been made. The plant personnel need to know when and how to sample. The quality control lab needs to know what quality control specification tests to run and what are their limits.

How to make adjustments. Not every recipe is expected to produce a product within specification "on load." Adjustments to the recipe are necessary if what has been made is slightly out of specification. These adjustments are called "hits."

How to filter. The type of filtering device and filter to use before filling the containers may be specified.

What container to fill. The container may have special properties such as a coating inside the can.

How to store. Specifics often include how-to-store or how-not-to-store instructions, such as "Do Not Freeze."

9.4 Formulating Water-Based Coatings

Water-based fluoropolymer coatings are almost always preferred for environmental reasons. Water-based fluoropolymer coatings can be made from either fluoropolymer dispersions or from fluoropolymer powders.

9.4.1 Fluoropolymer Coatings from Raw Dispersion

Coatings based on commercial aqueous dispersions are the easiest to make because all the dispersing is already done. However, some of the formulating flexibility is lost since the surfactant type and level, fluoropolymer molecular weight, particle size, etc., are mostly out of the formulator's control.

Fortunately, many commercial dispersions are available to chose from. Many of these are listed in the tables at the end of Ch. 2.

Aqueous dispersion fluoropolymer particles are very small. Their diameter is generally from 150 to 300 nanometers, or 0.15–0.3 microns. A scanning electron micrograph of a dispersion is shown in Fig. 9.1. The particles are generally very uniform in size and shape. Most are spherical, though rod-shaped dispersions are also available.

Much surfactant, about 5%–10% by weight of polymer, is used in these dispersions. This is to provide stability, but also may be left over from the concentration step of the dispersion discussed in Ch. 2. Most aqueous dispersions have very low viscosity, typically water-like, at less than 50 centipoise. There are two problems that must be addressed when using aqueous dispersions of this type in a coating formulation.

Shear stability is the first issue. Most dispersions have a limited shear stability. In fact, shear is used to destabilize the dispersion into a solid material to produce fine powder products. Therefore, the formulation needs to include ingredients that improve the shear stability of the coating products. These ingredients are often some type of solvents. The choice of these solvents is part of the art used by an experienced formulator.

When making a product based on an aqueous dispersion, the order of addition is often critically important because shear stability can change remarkably with each ingredient addition. This requires careful control of the mixing process.

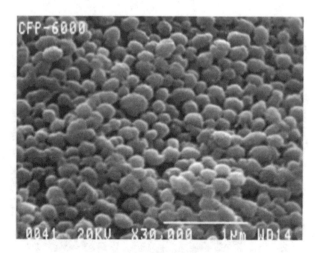

Figure 9.1 Primary PTFE particles from dispersion.

The second issue is that the small size of dispersion particles causes cracking of the coating at a fairly low dry-film thickness (DFT) as the coating is dried and baked. The DFT at which this appears is called the *critical cracking thickness* (CCT). Figure 9.2 shows a micrograph of a dispersion-based coating at the onset of cracking. There are also holes in the coating at some of the crack intersections in this micrograph.

The cracking problem can be severe. It is particularly common with PFA and FEP dispersions. Special additives can help minimize the cracking problem, or at least delay its onset to higher film thickness. To be effective, the additive needs to form a uniform film before the fluoropolymer starts to melt. Sometimes this can be done with high boiling point solvents. In the case of FEP dispersion products, glycerol, glycol, or other high boiling polyols can be added. A sufficiently high boiling solvent does not completely evaporate until after the FEP has melted at around 525°F (274°C). For PFA, glycerol can work, but because PFA has a higher melting point, about 580°F (304°C), more glycerol is needed. It does not always work well. For products based on PTFE or PTFE blended with other fluoropolymer dispersions, an acrylic resin is often added. The acrylic forms a continuous film, remains intact until the fluoropolymers melt or sinters, and then it decomposes and diffuses out of the film.

Many dispersion-based products have low viscosity, so settling needs to be considered. Also, because the viscosity is low, film builds are generally kept low to avoid running, dripping, and sagging problems.

9.4.2 Fluoropolymer Coatings by Dispersion of Powders

To make a liquid coating from powder, a dispersion must be made, which usually means some kind of grinding is required. Dispersion and grinding are discussed in Ch. 5 on pigments. The same principles apply to making fluoropolymer powder dispersions that are used to make pigment dispersions. The fluoropolymer agglomerates must be separated and stabilized with appropriate surfactant(s). In some instances, the fluoropolymer powder can be stirred into a liquid and surfactant mixture, but when using this approach, it is often difficult to break up the agglomerates without generating a great deal of foam.

There are advantages to making a coating from powder. First, the formulator has the most flexibility. The surfactant type and amount can be chosen, rather than being dictated by the dispersion manufacturer. The solids and viscosity can be controlled in a wider range.

Another advantage that powder dispersions have over standard aqueous dispersions is that higher film thickness can be applied without cracking. Large fluoropolymer particles are best for thicker films. The main problem with thick liquid coatings occurs as the coatings dry due to solvent loss during the bake. At some point there is not enough solvent left to hold the fluoropolymer particles together on the substrate. The fluoropolymer is still below its melt point, and is in a powdery state. The powder is susceptible to falling off the substrate. Vibration, air movement, or even just gravitational forces can cause this to happen. The formulator needs to include a high boiling solvent or other resin that will hold the powder together on the substrate until the polymer starts to melt. As melting starts, the fluoropolymer particles will become sticky and hold each other in place. A common solvent for this purpose is glycerol. Other glycols or polymer resins are also used. A polymer resin usually is chosen to decompose just above the melt point of the fluoropolymers. When glycols are used, it is common that they are up to ten percent by weight of the solvent system. When these eventually evaporate, a lot of white smoke is generated that will leak out of the oven through cracks or go up the exhaust stack.

One of the disadvantages of powders is that settling is more severe due to larger particle size. Higher viscosity can minimize this problem. Like in dispersion products, the order of addition is important.

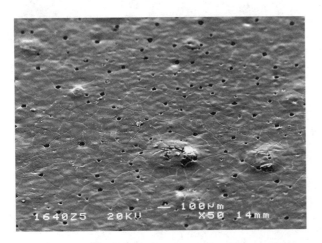

Figure 9.2 Micrograph showing the onset of cracking.

9.5   Solvent-Based Coatings

The fluoropolymers used in solvent-based coatings, with one exception that is discussed in Sec. 9.6, are always in powder form. Generally, it is non-fibrillating resin and has relatively low molecular weight. When formulating or using solvent-based fluoropolymer coatings, it is sometimes easier to think of the fluoropolymer powder as a filler or pigment. The fluoropolymer powder is dispersed into a solvent system by one of the dispersion techniques discussed in Ch. 5. When dispersing into a solvent, most surfactants could not stabilize the dispersion. Resins are usually used to stabilize the dispersion. The binder resin can be used for this purpose, but often other resins are used. These resins are called *dispersing aids.* They can affect the performance and chemistry of the coating, thus must be chosen carefully. For example, in some epoxy resin bonded coatings, a crosslinking resin such as melamine-formaldehyde or benzoguanamine-formaldehyde can be applied as a dispersing aid. These amine-based resins are known to crosslink and cure epoxy resins.

The stability of solvent-based coatings can vary widely. Generally, if after they settle, they can be redispersed, the shelf life can be very long, even many years. Manufacturers will typically specify shelf life no longer than 12–18 months to limit their liability.

9.6   Soluble Fluoropolymers

A special case of liquid fluoropolymer coatings are those rare ones that are soluble in solvent. At the time of this writing, only one high molecular weight perfluorinated polymer fits in this category: Teflon® AF.

Teflon® AF can be tailored to have narrow solubility in selected perfluorinated solvents. In spite of this solubility, the polymer remains chemically resistant to all other solvents and process chemicals. Solubilities of 3% to 15% have been observed. This solubility range permits one to solution-cast or dip ultra-thin coatings in the submicron thickness range.

Many non-perfluorinated polymers are soluble in common solvents. Polymers of this type include Lumiflon® and FEVE that are discussed in Ch. 1. These can be sprayed, dipped, or coil-coated.

Supercritical carbon dioxide has also been used as a polymerization medium for some fluoropolymers. While crystalline, high molecular weight, perfluorinated polymers are not soluble in liquid or supercritical carbon dioxide, perfluoropolyethers and many other non-perfluorinated polymers are soluble in carbon dioxide.[2] No system is commercial as of 2005, but there is promise of interesting dip application of ultrathin fluoropolymer films.

9.7   Mixing Liquid Coatings Prior to Use

As explained Sec. 5.5, coatings generally separate or settle when they are stored. Reincorporation of the settlement and mixing is critically important to the quality of the product. The manufacturers generally provide mixing instructions. Overmixing can also cause problems. Air can become incorporated, leading to defects in the coated substrate after application. Dispersion-based products frequently will generate grit or gel particles if overmixed because the dispersions have limited shear stability. The remixing instructions can be occasionally a bit bizarre. For example, one commercial coating manufacturer instructed the user to turn the container upside down and strike it with a rubber mallet until the hard settlement on the bottom broke, then roll for one hour.

Rolling is the most common way of reincorporating the settled material. A paint roller should be appropriate depending on the container size. Five-gallon pails and thirty-gallon drums are common and require a large paint roller such as that shown in Fig. 9.3. Rollers of this type can be electric-motor or air-pressure driven. Gallon containers require a smaller roller such as the one in Fig. 9.4 that can roll several cans at once. This figure shows the air-driven motor on the left-hand side of the mixer. The side panel locks in the up position to keep the containers from rolling off the rollers.

When using these rollers, the RPMs that the containers see should be measured and the rolling rate adjusted per manufacturer recommendations. Full containers are commonly rolled at 30 RPM. The RPMs should be reduced for partially full containers to reduce foaming or air incorporation.

Figure 9.3 Paint roller for large containers.

Figure 9.4 Paint roller for small containers.

Direct contact mixing with a blade is suggested for non-aqueous coatings. The procedure for doing this is quite important. A drill press or hand-held electric drill is commonly used. It is best to use an electric- or air-powered mixer, especially an electric one that has direct RPM control capability. When electric mixers are used, care must be taken about the risk of fire for coatings with low flash points.

The choice of impeller can be important, as well as where it is put into the coating containers. There are many impeller variations, several of which are shown in Fig. 9.5. Generally, the impellers are classified into two groups. One group, called *axial impellers*, push the liquid up or down along the axis of the shaft attached to the impeller. The second type is a *radial impeller* that throws the material radially outward from the impeller. The preferred impellers for fluorinated coatings are usually axial. The preferred axial impeller is the A310 Turbine or the A100 Propeller. The A310 is a patented impeller by Lightnin. The A100 Propeller is based on the common boat propeller and is made by many companies including Lightnin (http://www.lightin-mixers.com) and Chemineer (http://www.chemineer.com). Most axial impellers are available in a pump up or pump down version.

The location of the impeller in the mixing container is very important. Usually, it should be located off center by about one third of the radius of the container. It should also be at about a thirty-degree angle. This will allow the maximum RPMs to be used if the coating can withstand the shear without generating a vortex and pulling air into the coating. The impeller should be positioned at least two-thirds the distance to the bottom of the container (shown in Fig. 9.6).

After mixing, the coating material should be checked to verify whether the sediments at the bottom have been reincorporated completely. The coating should also be filtered before use as instructed by the manufacturer.

9.8   Filtering/Straining

Filtration or straining is an important part of preparing the paint for use. There really is not any difference between the terms *straining* and *filtering,* because a strainer is, in reality, a coarse filter. Filtration is necessary to remove dirt, agglomerated pigment, gels, and other contaminants that contribute to poor surface-appearance properties. This is particularly true as the coating ages or is partially used and stored for future use. The choice of filter or strainer can be very important because a good filter should remove as much of the contaminants as possible but must also not affect adhesion, color, or other formulated properties.

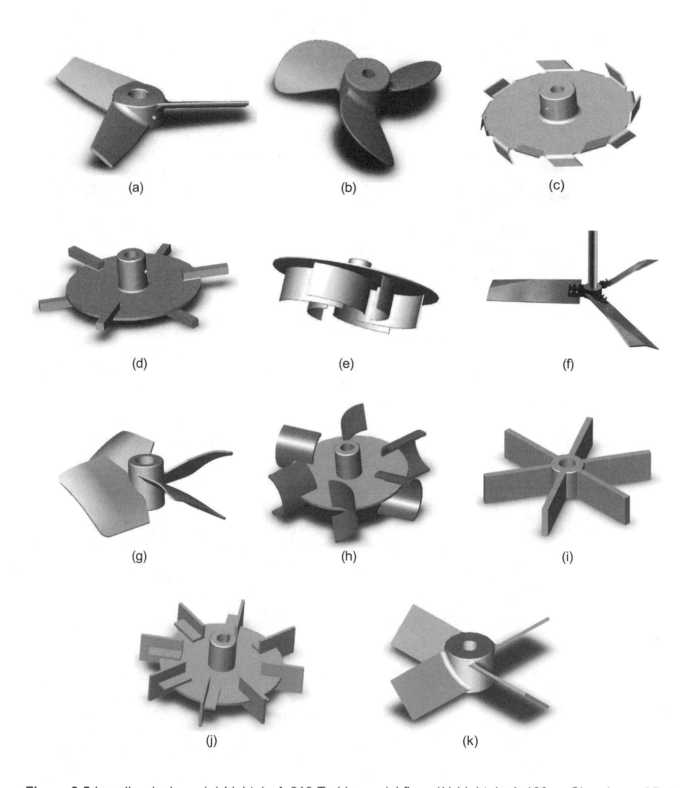

Figure 9.5 Impeller designs. (*a*) Lightnin A-310 Turbine, axial flow. (*b*) Lightnin A-100 or Chemineer AP-3 Propeller, axial flow. (*c*) Lightnin R500 Sawtooth Disperser, high-speed disperser blade, radial flow. (*d*) Lightnin R510 Bar Turbine, axial flow. (*e*) Lightnin R320 or Philadelphia Mixers CBT-6 Curved Bladed Pumper, radial flow. (*f*) Chemineer HE-3, axial flow. (*g*) Lightnin A315 Turbine, axial flow. (*h*) Lightnin R130 or Chemineer CD-6, radial flow. (*i*) Six paddle blade, radial flow. (*j*) Rushton Turbine, radial flow. (*k*) Lightnin A200 or Chemineer P-4 Pitch Blade Turbine, axial flow. *(Figures courtesy of Pete Csiszar.[3])*

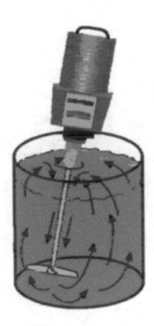

Figure 9.6 Mixer location in the container.

Normally, the manufacturers of coatings provide information about proper filtration or straining. This entails specification of a mesh or micron size for a screen. There is often confusion between mesh and microns, the most common measures applied to filters or sieves. Mesh numbers and microns are two very different measures, but they are related. A micron is a unit of measurement and it is the size of the holes in the filter media. One micron equals 1/1,000th of a millimeter. Basically, a filter defined by a micron range size should remove everything larger than the opening size, and allow passage of everything smaller. Therefore, the smaller the micron size of the filter, the smaller the particles that will be removed.

Filters, sieves, and strainers that have been manufactured from woven wires are assigned a mesh number to indicate the number of wires. Normally, weaves consist of vertical and horizontal wires woven in a simple over and under pattern. The mesh number is the number of wires in one linear inch. A 100-mesh filter therefore has 100 vertical wires per inch and 100 horizontal wires per inch. The larger the mesh number, the more wires are used per inch and the smaller the openings between the wires. As a mesh number increases, the equivalent micron range size decreases. Various wire diameters can be used to make the screen. Of course, the diameter

of the wire will affect the hole size. A 100-mesh size screen with small diameter wire will have a larger hole and micron rating than a 100-mesh with a larger diameter wire. There is a relationship between standard filter screen mesh size and micron size. The standard is slightly different in USA and Europe and it is summarized in Table 9.1.

An important detail about the filtering process is that caution should be taken to minimize splashing and foam generation. Often splashing and foam will generate grit and gel that occurs after the filtration process, thus affecting the quality of the final coating.

9.9　Shelf Life

As paints age, they develop quality problems. Settling, gel or grit formation, and bacterial contamination can occur with aging. Most manufacturers report an expected shelf life for each paint type. This is typically reported under ideal storage conditions, such as temperature limits along with periodic remixing. It is important to follow those recommendations. It is also important to note that rated shelf life usually applies to an unopened container. Once the container is opened, it can lose solvent and become contaminated. Also, a partially filled container will not mix the same way as a full container would.

9.10　Commercial Producers and Their Product Lines

This section lists some of the major manufacturers of fluorinated coatings. It does not list all of their products by names and codes. Since those listed change constantly and manufacturers phase out old products and introduce new ones, the list would not remain useful for long. Instead the manufacturers frequently group their products into lines, differentiating them by chemistry, application method, or end-use. Where manufacturers have not supplied information for this book, information has been gathered from public literature such as sales brochures and Material Safety Data Sheets (MSDS), internet, and the author's experience. Although the information is believed to be accurate, it may contain inadvertent errors, or the products may have changed with time.

Table 9.1 Relationship Between Mesh and Micron Filter Sizes

Europe		U.S.A.	
Mesh No.	**Micron Rating**	**Mesh No.**	**Micron Rating**
20	920	20	864
30	570	30	508
40	410	40	381
60	260	60	229
80	178	80	178
100	142	100	152
120	122	120	117
150	109	150	104
180	91	180	84
300	45	300	46

In this section, *manufacturers* means those companies that make and sell coatings. There are several companies that make and use their own coatings, but do not offer them for sale outside their shop. There are even more companies that buy coatings from major manufacturers but apply their own trade names and trademarks to them in order to shield their process from competitors.

9.10.1   Acheson Colloids

Acheson Colloids (www.achesoncolloids.com) (Table 9.2) is a small coating manufacturer that specializes in automotive applications. It is a National Starch and Chemical Company, whose parent company is ICI, PLC. Their main interest is corrosion protection, friction reduction, abrasion and erosion resistance. The coatings are generally applied by spray or dip spin. Their primary trade name for their coatings is Emralon®.

9.10.2   Whitford Liquid Products

Whitford Corporation is based in Fraser, Pennsylvania (www.whitfordww.com). Whitford trademarks (Table 9.3) include:

- Whitford®.
- Xylan®: These coatings come in one-, two-, and three-coat versions for a wide variety of applications.

- Xylac® identifies high-temperature decorative materials most often utilized as exterior coatings for cookware, which are generally not fluorinated coatings.
- Dykor® describes Whitford's fluoropolymer solutions and dispersions, primarily PVDF. Whitford acquired Pennwalt's Kynar® 200 Series of PVDF dispersions in 1988. (Kynar® is a registered trademark of ARKEMA.) The product line includes powder coatings.
- Xylar® identifies Whitford's inorganic coating materials.
- Ultralon® was an ICI Americas product line but Whitford purchased it in 1990.
- Excalibur® is Whitford's coating system for stainless steel applications. It is reinforced externally with a stainless-steel alloy that is arc-sprayed onto the surface of the pan.
- Eclipse® is a reinforced three-coat system aimed at superior resistance to abrasion.
- QuanTanium® is Whitford's line that is reinforced with titanium.
- Quantum2® is another one of Whitford's reinforced coating lines.

9.10.3 Weilburger Coatings (GREBE GROUP)

Weilburger Coatings has manufactured non-stick coatings in Germany since about 1990. It is a GREBE GROUP Company. They focus on markets in Europe, India, Korea, and South Africa. Their market focus is cookware, bakeware, and household electrical appliance manufacturers.

The latest range of GREBLON non-stick coatings has been categorized under the brand names GREBLON ALPHA, GREBLON BETA, GREBLON GAMMA, and GREBLON. No product information was available for Weilburger coatings.

- GREBLON ALPHA® is a two-coat finish offering release properties even with baking mixes containing high levels of sugar.
- GREBLON BETA® are one- or two-coat systems offering non-stick properties for premium bakeware.
- GREBLON GAMMA® is a low cost, one-coat non-stick coating for high volume bakeware lines.
- GREBLON® is a low-cost, easy-to-clean, non-stick system for promotional bakeware.

9.10.4 Akzo Nobel

Akzo Nobel's trade names include:
- Skandia® Marrlite Plus
- Skandia® Marrlite
- Skandia®
- Skandia® Stratos
- Skandia® Tech

No product information was available for Akzo Nobel's (www.an-nonstick.com) products.

9.10.5 DuPont

DuPont (www.Dupont.com or www.teflon.com) is based in Wilmington, Delaware. Their products are shown in Table 9.4.

9.10.6 Mitsui-DuPont Fluorocarbon Liquid Products

Mitsui-DuPont Fluorocarbon Co. (MDF) is a joint venture company between Mitsui and Dupont. It produces liquid and powder coatings in Japan. MDF coating products are shown in Table 9.5.

Table 9.2 Achesion Colloids Products

Series	Technology	Cure Temperature	Service Temperature	Typical End-Uses or Applications
Emralon® 305	PTFE, resin bonded – phenolic, non-aqueous	60 min @150°C (302°F)	•Cont: (300°F) •Int: (325°F)	•Spray cr dip-spin •Threaded and unthreaded fasteners, brackets, meshers
Emralon® 329	PTFE, resin bonded – thermoplastic, non-aqueous	Air dry, 2 hr	•Cont: 82°C (180°F) •Int: 116°C (240°F)	•Spray, dip-spin •Moving parts of wood, metal, fabric, plastic, or rubber •Solenoid plungers, piano action parts, window guides, tools
Emralon® 330	PTFE, resin bonded – phenolic, non-aqueous	60 min @150°C (302°F)	•Cont: 130°C (275°F) •Int: 150°C (300°F)	•Spray or dip •Fasteners, carburetor parts, rubber parts, conveyor belting, lock mechanisms, bearings, gears
Emralon® 333	Blended FP, resin bonded – polyamide-imide, non-aqueous	10 min @150°C (302°F)	•Cont: 232 (450°F) •Int: 260°C (500°F)	•Spray •Saw blades, garden tools, snow shovels, washer, springs, carburetor shafts, spray gun parts, business machine parts
Emralon® 334	PTFE, resin bonded – polyamide-imide, non-aqueous	10 min @150°C (302°F)	•Cont: 232°C (450°F) •Int: 260°C (500°F)	•Spray, coil coating •Carburetor shafts, links & levers, business machine parts
Emralon® 8301–01	PTFE, resin bonded – polyurethane, aqueous	30 min @100°C (212°F)		•Spray or dip-spin •Drapery hardware •Aluminum surface panels •Paper handling equipment •O-rings •Nylon parts

(Cont'd.)

Table 9.2 (*Cont'd.*)

Series	Technology	Cure Temperature	Service Temperature	Typical End-Uses or Applications
Emralon® 8370	PTFE, resin bonded – polyurethane, aqueous, two package	5 min @149°C (300°F)	• Cont: 135°C (275°F) • Int: 191°C (375°F)	• Spray • Vinyl extrusions, rubber extrusions (EPDM), rubber molded parts, door, trunk and sunroof gaskets, weather stripping (squeak suppression)
Emralon® TM-001A	PTFE, resin bonded – acrylic, aqueous	30 min @150°C (302°F)	• Cont: 150°C (300°F) • Int: 175°C (350°F)	• Spray • Rubber o-rings, flexible diaphragms, non-rigid impellers, valve seals, gaskets, elastomers
Emralon® TM-008	PTFE, resin bonded – alkyd, aqueous	10 min @ 200°C (392°F)	• Cont: (300°F–400°F)	• Spray or dip • Fasteners, seat belt components, seating mechanisms, lock mechanisms, slides, rails
GP 1904	PTFE, resin bonded – phenolic, nonaqueous, with graphite	60 min @150°C (302°F)	• Cont: 149°C (300°F) • Int: 163°C (325°F)	• Spray or dip • Fasteners, gears, valves, lock mechanisms, carburetor parts, bearings
TW-014B	PTFE, resin bonded – polyurethane, aqueous, with silicone	Air dry		• Spray • Anti-itch and squeak for styrofoam blocks and plastics used in automotive industry
TW-014C	PTFE, resin bonded – polyurethane, aqueous, with silicone and UV tracer	Air dry		• Spray • Anti-itch and squeak for styrofoam blocks and plastics used in automotive industry

(*Cont'd.*)

Table 9.2 (*Cont'd.*)

Series	Technology	Cure Temperature	Service Temperature	Typical End-Uses or Applications
TW-020	PTFE, resin bonded – polyurethane, aqueous, two package	Air dry	• Cont: 135°C (275°F) • Int: 191°C (375°F)	• Thermoplastics, TPV/TPE glass runs and molded parts, rubber extrusions (EPDM), rubber molded parts, door, trunk and sunroof gaskets, weather stripping
TW-023	PTFE, resin bonded – polyurethane, aqueous, two package	Air dry	• Cont: 135°C (275°F) • Int: 191°C (375°F)	• Spray • Rubber extrusions (EPDM), rubber molded parts, door, trunk and sunroof gaskets, weather stripping
TW-005A	PTFE, resin bonded – polyurethane, aqueous	5 min @ 150°C (302°F)	• Cont: 135°C (275°F) • Int: 191°C (375°F)	• Spray • Vinyl extrusions, rubber extrusions (EPDM), rubber molded parts, door, trunk and sunroof gaskets, weather stripping (squeak suppression)

Table 9.3 Whitford Coating Products

Series	Technology	Food Contact	Minimum/Maximum Cure Temperature	Continuous and Intermittent Service Temperature Range	Film Thickness Range, Application Methods, Typical End-Uses or Applications
Xylar 2xx	Two-package, "chromic acid primer," PTFE	No	• Min: 30 min @ 343°C (650°F) • Max: 15 min @ 400°C (750°F)	• -40°C (-40°F) to 260°C (500°F)	
Xylan 10xx	Resin bonded, PAI, PTFE, non-aqueous and aqueous (1052 contains MoS$_2$, 1058 uses FEP)	No	• Min: 20 min @ 220°C (430°F) • Max: 5 min @ 345°C (650°F)	• Cont: -195°C (-320°F) to 260°C (500°F) • Int: -195°C (-320°F) to 285°C (545°F)	• 20 ± 5 µm/coat • Rotary actuators, carburetors, bearings, seat belt clips, garden tools, actuators, bearings, sealing rings, valve springs, pistons
Xylan 12xx	Resin bonded, thermoset (PAI), PTFE, aqueous	No	• Min: 30 min @ 180°C (-355°F) • Max: 5 min @ 260°C (500°F)	• Cont: -50°C (-58°F) to 200°C (392°F)	• 20 ± 5 µm/coat • Saw blades, industrial files, threaded fasteners, lock mechanisms, bushings, bearings. shear blades, valves, valve bodies, fasteners
Xylan 1220	Resin bonded, thermoset (PAI), FEP, aqueous	No	• Min: 10 min @ 275°C (525°F) • Max: 5 min @ 400°C (750°F)	• Cont: -195°C (-320°F) to 205°C (400°F)	• Industrial molds
Xylan 1230	Resin bonded, PTFE	No	• Min: 30 min @ 100°C (212°F) • Max: 10 min @ 120°C (250°F)	• Cont: -30°C (-22°F) to 100°C (212°F)	• 20 ± 5 µm/coat • Rubber o-rings, gaskets.
Xylan 13xx	Resin bonded, PPS, PTFE, aqueous	No	• Min: 15 min @ 375°C (705°F) • Max: 5 min @ 400°C (750°F)	• Cont: -20°C (-4°F) to 245°C (475°F)	• 22.5 ± 2.5 µm/coat • Chemical processing equipment, pump components, impellers, valve bodies

(Cont'd.)

Table 9.3 (*Cont'd.*)

Series	Technology	Food Contact	Minimum/Maximum Cure Temperature, °C (°F)	Continuous and Intermittent Service Temperature Range	Film Thickness Range, Application Methods, Typical End-Uses or Applications
Xylan 14xx	Resin bonded, thermoset, PTFE, non-aqueous and aqueous (1425 includes MoS_2)	No	• Min: 15 min @ 205°C (400°F) • Max: 5 min @ 275°C (525°F)	• Cont: -20°C (-4°F) to 180°C (355°F) • Int: -20°C (-4°F) to 230°C (445°F)	• 17.5 ± 2.5 µm/coat • Threaded fasteners for building, chemical process, oil and off shore industries, hinge pins, piston casing, compressors; • Rotary actuators, carburetors, bearings, seat belt clips, actuators
Xylan 15xx	Resin bonded, thermoplastic, PTFE, non-aqueous	No	• Min: 30 min @ 220°C (430°F) • Max: 5 min @ 275°C (525°F)	• Cont: -40°C (-40°F) to 220°C (428°F) • Int: -40°C (-40°F) to 250°C (480°F)	• 20 ± 5 µm/coat • Cooling fans, light fittings, personal care products, radomes, I-O drives
Xylan 16xx	Resin bonded, multi-package, thermoset, PTFE, non-aqueous and aqueous	No	• Min: 30 min @ 120°C (250°F) • Max: 5 min @ 180°C (355°F) • Varies by code	• Cont: -40°C (-40°F) to 150°C (302°F) • Int: -40°C (-40°F) to 200°C (390°F) • Varies by code	• 20 ± 5 µm/coat • Automotive EDPM extrusions, sponge seals, body seals, window seals, piston skirts
Xylan 17xx	Aqueous	No			• Copy rollers, heat sealing bars, CPI vessels
Xylan 18xx	Resin bonded, thermoplastic, FEP or PTFE, non-aqueous and aqueous	Yes	• Min: 15 min @ 375°C (705°F) • Max: 15 min @ 400°C (750°F)	• Cont: -40°C (-40°F) to 205°C (400°F)	• 12.5 ± 2.5 µm/coat • Industrial molds, garden tools, metal sheet that is drawn or post-formed

(*Cont'd.*)

Table 9.3 (*Cont'd.*)

Series	Technology	Food Contact	Minimum/Maximum Cure Temperature	Continuous and Intermittent Service Temperature Range	Film Thickness Range, Application Methods, Typical End-Uses or Applications
Dykor 2xx	Three-coat system, PVDF	No	• Min: 275°C (525°F) • Max: 290°C (555°F) until complete melt flow is achieved	• Cont: -60°C (-76°F) to 200°C (390°F) but depends on chemical environment.	• 100 ± 25 µm/coat • Chemical processing equipment, heat exchangers, flu pipes, valves, pipe fittings, pumps, tanks, reactor vessels, sucker rods, oil well tubing, couplings, fan drive clutch discs
Xylan 4070	PES	Yes	• Min: 5 s @ 290°C (555°F) • Max: 10 s @ 300°C (570°F)	• Cont: -50°C (-58°F) to 200°C (392°F)	• 5–10 µm in total in 1–2 coats • Coil coating • Domestic bakeware and appliance exteriors
Xylan 4080/8800	Two-coat system, resin bonded, PES, PTFE	Yes	• Min: 5 s @ 350°C (660°F) • Max: 10 s @ 370°C (700°F)	• Cont: -40°C (-40°F) to 230°C (445°F) • Int: -40°C (-40°F) to 260°C (500°F)	• 5–10 µm in total in 1–2 coats • Coil coating • Domestic bakeware interiors, coffee heater plates, roasting pans
Xylan 51xx	Resin bonded, thermoset, (epoxy/phenolic), PTFE	No	• Min: 20 min @ 220°C (430°F) • Max: 5 min @ 315°C (600°F)	• Cont: -195°C (-320°F) to 260°C (500°F) • Int: -195°C (-320°F) to 285°C (545°F)	• 5–7 µm/coat • Dip-spin • Threaded rasteners, small components, screws
Xylan 52xx	Resin bonded, thermoset, PTFE, non-aqueous and aqueous	No	• Min: 30 min @ 180°C (-355°F) • Max: 5 min @ 260°C (500°F) • Varies by code	• Cont: -50°C (-58°F) to 200°C (390°F) • Varies by code	• 6–8 µm/coat • Dip-spin, threaded fasteners, roofing, appliances

(*Cont'd.*)

Table 9.3 (*Cont'd.*)

Series	Technology	Food Contact	Minimum/Maximum Cure Temperature, °C (°F)	Continuous and Intermittent Service Temperature Range	Film Thickness Range, Application Methods, Typical End-Uses or Applications
Eclipse System 7050/7252/7353	Three-coat system, internally reinforced	Yes	•Min: 5 min @ 425°C (800°F) •Max: 5 min @ 440°C (825°F)		•Cookware
Quantum 7115/7120/7320	Two- or three-coat system, internally reinforced, PTFE	Yes	•Min: 5 s @ 350°C (660°F) •Max: 10 s @ 370°C (700°F)	•Cont: -40°C (-40°F) to 230°C (445°F) •Int: -40°C (-40°F) to 260°C (500°F)	•Coil coating •Bakeware interiors for aluminum or tin-free steel
Quantum2	Two- or three-coat system, internally reinforced with ceramic materials	Yes	•Min: 5 min @ 425°C (800°F) •Max: 3 min @ 435°C (815°F)	•Cont: -195°C (-320°F) to 260°C (500°F)	•Cookware
Xylan 7910/7930		Yes			•Curtain coating •Interior top-of-stove cookware
Xylan 81xx	Non-aqueous, PAI and PTFE (food contact version of Xylan 10xx)	Yes	•Min: 5 min @ 315°C (600°F) •Max: 5 min @ 345°C (650°F)	•Cont: -195°C (-320°F) to 260°C (500°F) •Int: -195°C (-320°F) to 285°C (545°F)	•20 ± 5 μm/coat •Food chutes, sweet molds, circular knife blades
Xylan 8254/8257	Two-coat, PAI based primer, PTFE topcoat	Yes	•400°C (750°F)		•Food processing machinery, restaurant equipment, warming trays, and bakeware
Xylan 8255/8256/8257	Two- or three coat, PAI-based primer, PTFE topcoat				•Top-of-stove cookware, bakeware, pyroceram, electric kitchen appliances (griddles, waffle irons, sandwich makers)
Xylan 83xx	Resin bonded, PPS, PTFE		•Min: 15 min @ 375°C (705°F) •Max: 5 min @ 400°C (750°F)	•Cont: -20°C (-4°F) to 245°C (475°F)	•22.5 ± 2.5 μm/coat •Cookware, domestic appliances

(*Cont'd.*)

Table 9.3 (*Cont'd.*)

Series	Technology	Food Contact	Minimum/Maximum Cure Temperature	Continuous and Intermittent Service Temperature Range	Film Thickness Range, Application Methods, Typical End-Uses or Applications
Xylan 8470	Resin bonded PTFE	Yes			• Food molds and bundt pans.
Xylan 8500	Resin bonded, thermoplastic (PES?), PTFE, non-squeous (food grade analog Xylan 15xx)	Yes	• Min: 15 min @ 260°C (500°F) • Max: 5 min @ 275°C (525°F)	• Cont: -20°C (-4°F) to 190°C (375°F) • Int: -20°C (-4°F) to 220°C (430°F)	• 20 ± 5 μm/coat • Domestic bakeware, kitchen utensils, appliance components
Xylan 8541	Resin bonded, thermoset, PTFE	Yes	• Min: 15 min @ 260°C (500°F) • Max: 5 min @ 275°C (525°F)	• Cont: -20°C (-4°F) to 175°C (345°F) • Int: -20°C (-4°F) to 200°C (390°F)	
Xylan 86xx	Resin bonded, thermoplastic, PTFE (some use silicone in place of PTFE)	Yes	• Min: 15 min @ 260°C (500°F) • Max: 5 min @ 290°C (550°F)	• Cont: -40°C (-40°F) to 250°C (480°F) • Int: -40°C (-40°F) to 275°C (525°F)	• 17.5 ± 2.5 μm/coat • Domestic cookware
Xylan 88xx	Resin bonded (PES), thermoplastic, PTFE or FEP	Yes	Coil • Min: 20 s @ 350°C (660°F) • Max: 20 s @ 370°C (700°F) Spray • Min: 15 min @ 375°C (705°F) • Max: 5 min @ 400°C (750°F)	• Cont: -40°C (-40°F) to 230°C (450°F) • Int: -40°C (-40°F) to 260°C (500°F)	Coil • 8–10 μm/coat coil • Domestic bakeware interiors Spray • 22.5 ± 2.5 μm/coat • Heat sealing bars, post- or pre-formed cookware, domestic appliances, sandwich toasters, grills, waffle irons

Table 9.4 DuPont Coating Products

Series	Technology	Food Contact	Cure Temperature	Service Temperature	Typical End-Uses or Applications
851-line	PTFE topcoats, aqueous		385°C–430°C (725°F–805°F)	260°C (500°F)	• Heat exchangers • Automatic soldering equipment • Molds • Carburetor shafts, linkages • Cruise control parts • Filters • Cryogenic applications • Aerospace applications
850-line	PTFE acid primers, aqueous, two-package with VM-7799 acid accelerator		230°C–280°C (446°F–536°F) force dry	260°C (500°F)	
856-line	FEP topcoats, aqueous	Yes	370°C–400°C (700°F–750°F)	200°C (392°F)	• Chemical equipment (impellers, mixing tanks, valves, pumps) • Biomedical equipment • Heat sealing bars • Shoe molds • Textile dryers
958–line	FEP or PTFE one-coats and primers, non-aqueous, resin bonded				• Automotive gasoline filler tubes • Sprinkler ball valves • Fan blades, housings, garden tools • Automotive fasteners • CPI fasteners • Boat propellers • Saw blades • Packaging equipment, conveyors • Fuel injectors, saw blades • Best dry lubricant
954-line	FEP or PTFE one-coats, non-aqueous, resin bonded				
954-1xx	Moderate bake versions		260°C (500°F)	150°C (302°F)	
954-2xx	Low bake versions		175°C (350°F)	150°C (302°F)	
954-5xxxx	Aqueous versions				

(*Cont'd.*)

Table 9.4 (*Cont'd.*)

Series	Technology	Food Contact	Cure Temperature	Service Temperature	Typical End-Uses or Applications
420-line	Resin bonded, self priming one coats	Yes	400°C (752°F)	260°C (500°F)	• Coffee plate warmers, assorted food-processing utensils • Iron sole plates, portable electrics, sandwich makers
959-line	FEP one-coats and primers, non-aqueous, resin bonded	Yes	400°C (752°F)	260°C (500°F)	• Primer for FEP and PFA topcoat • Coffee plate warmers, assorted food-processing utensils • Iron sole plates, portable electrics, sandwich makers
857-line					
855-line	PTFE, FEP, PFA and blends, two- and three-coat systems for office machine fuser rollers				• Copier and laser-beam printer fusers • Some static dissipating
459-line	PTFE or blends, resin bonded primers				• Cookware
456-line	PTFE or blends, midcoats and topcoats				• Cookware
699-line					

(*Cont'd.*)

Table 9.4 (*Cont'd.*)

Technology	Product Code	Cure Temperature	Service Temperature	Typical Applications
PTFE	TOPCOATS (Aqueous)			
	851-214 Green 851-221 High-build gray 851-224 High-build green 851-255 High-build black 852-201 Clear 852-202 High-build clear	385°C–430°C (725°F–806°F)	260°C (500°F)	• Heat exchangers • Automatic soldering equipment • Molds • Carburetor shafts, linkages • Cruise control parts • Filters • Cryogenic applications • Aerospace applications
	ACID PRIMERS*			
	850-300 Clear 850-314 Green 850-321 Gray	230°C–280°C (446°F–536°F) force dry	260°C (500°F)	

*Used with VM-7799 acid accelerator

(*Cont'd.*)

Technology	Product Code	Food Contact	Cure Temperature	Service Temperature	Typical Applications
PTFE	PRIMERS				
	850-Acid primers				
	958-203 Black 958-207 Green				
	959-203 Black 959-205 Brown	Yes Yes			

(*Cont'd.*)

Table 9.4 (Cont'd.)

Technology	Product Code	Cure Temperature	Service Temperature	Typical Applications
	Self-Priming ONE-COAT blends of fluoropolymer with other resins			
	(Solvent)			
Teflon®-S	954-100 Unpigmented	(390°F–555°F)	150°C (302°F)	• Automotive gasoline filler tubes
	954-101 Green			• Sprinkler ball valves
	954-103 Black	260°C (500°F)	150°C (302°F)	• Fan blades, housings, garden tools
	954-201 Low-bake green			• Automotive fasteners
	954-203 Low-bake black			• CPI fasteners
	954-407 Low-bake flat black	175°C (350°F)	220°C (428°F)	• Boat propellers
	958-203 Black			• Saw blades
	958-207 Green	315°C–345°C (600°F–650°F)	260°C (500°F)	• Packaging equipment, conveyors
	958-303 Dry lubricant black			• Fuel injectors, saw blades
	958-306 Blue	260°C–345°C (500°F–650)	150°C (302°F)	• Best dry lubricant
	(Aqueous)			
	954-50003 Black	175°C (350°F)		
	954-50007 Green			

(Cont'd.)

Technology	Product Code	Food Contact	Cure Temperature	Service Temperature	Typical Applications
	Self-Priming ONE-COAT blends of fluoropolymer with other resins				
	(Solvent)				
Teflon® One Coat	420-104 Gray	Yes	400°C (752°F)	260°C (500°F)	• Coffee plate warmers, assorted food-processing utensils
	420-106 Metallic gray	Yes			
	420-109 Metallic black	Yes			
	959-203 Black	Yes	345°C (653°F)	215°C (419°F)	• Iron sole plates, portable electrics, sandwich makers
	959-205 Brown	Yes			
	(Aqueous)				
	857-503 Black	Yes	400°C (752°F)	260°C (500°F)	

(Cont'd.)

Table 9.4 (*Cont'd.*)

Technology	Product Code	Food Contact	Cure Temperature	Service Temperature	Typical Applications
	855-021 Blue (Primer)		425°C–430°C (797°F–806°F)	260°C (500°F)	Copiers Printers
	855-401 Silver (Midcoat)				
	855-402 Black (Midcoat)		425°C–435°C (797°F–815°F)	260°C (500°F)	
	855-500 Clear (Topcoat)				
	(Electroconductive)				Commercial food
	855-023 Black (Primer)		425°C–435°C (797°F–815°F)	260°C (500°F)	
	855-101 Black (Midcoat)				
	855-103 Black (Topcoat)				
PTFE/PFA Patented Blends	(Industrial Supra®)				
	459-780 Blue (Primer)	Yes	425°C–440°C (797°F–824°F)	260°C (500°F)	
	456-186 Pewter (Midcoat)	Yes			
	456-187 Black (Midcoat)	Yes			
	456-480 Clear (Topcoat)	Yes			
	(Ceramic Reinforced)				
	857-101 Black (Primer)	Yes			
	857-202 Black (Midcoat)	Yes			
	857-301 Clear (Topcoat)	Yes			

(*Cont'd.*)

132

Table 9.4 (*Cont'd.*)

Technology	Product Code	Food Contact	Cure Temperature	Service Temperature	Typical Applications
	TOPCOATS (Aqueous)				Chemical equipment (impellers, mixing tanks, valves, pumps)
					Biomedical equipment
					Silicone wafer mfg. equip.
PFA	857-210 Clear	Yes	370°C–400°C (700°F–752°F)	260°C (500°F)	Molds
					Laundry dryers
					Copier, printer rolls
					Paint spray cups
					Light bulbs
	PRIMERS				
	420-703 Black	Yes			
	850-Acid primers	Yes			

Table 9.5 MDF Liquid Coating Products

Product Line	Technology	Film Thickness Range, Application Methods, Typical End-Uses or Applications
EN-500xx	PFA, aqueous	
EN-510xx	PFA, aqueous, green	Wear resistant
EN-540CL	PFA, aqueous, clear, smooth surface	OA stripper finger
EN-700CL	PFA, aqueous, clear, high build	Up to 70 microns, mold release, food processing industry
EN-700GN	PFA, aqueous, green, high build	Up to 70 microns, mold release, CPI
EN-700GY	PFA, aqueous, gray, high build	Up to 70 microns, mold release, CPI
EN-700BK	PFA, aqueous, black, high build	Up to 70 microns, electroconductive
EN-710CL	PFA aqueous, clear, high build	Up to 70 microns, high build and wear resistant, OA fuser roll
SL-800BK	PFA, aqueous, powder slurry	Super high-build, Electroconductive, 500 microns/coat, 2,000 microns by multiple coat for CPI
SL-800LT	PFA, aqueous, powder slurry, light tan	Super high-build, 500 microns/coat, 2,000 microns by multiple coat for CPI
SL-900CL	PFA/PTFE, aqueous, clear, powder slurry with adhesion promotion	Midcoat for CPI
PR-902xx	PFA primers	Silver (AL), red brown (BN), yellow colors (YL), for OA fuser roll and stripper finger
PR-910xx	PFA heat resistant primers	Silver (AL), electroconductive black (BK), yellow colors (YL), for OA fuser roll and stripper finger
PR-914AL	PFA heat resistant primer, black metallic	Rice cookers

REFERENCES

1. Concannon; T. P., and Vary, E. M., US Patent 4,252,859, assigned to DuPont (Feb 24, 1981)
2. Young, J. L., and DeSimone, J. M., Frontiers in Green Chemistry Utilizing Carbon Dioxide for Polymer Synthesis and Applications, *Pure Applied Chemistry,* 72(7):1357–1363 (2000)
3. Drawings of impellers courtesy of Pete Csiszar, Mixing Consultant, e-mail: pcsis@telus.net
4. http://www.lightnin-mixers.com

10.1 Introduction

There are dozens of ways to apply liquid coatings. To cover these application methods in detail would require a separate volume. This book briefly describes major liquid application techniques for fluoropolymer coatings, with emphasis on where they can be used and what properties of the liquid coatings are needed.

10.2 Liquid Spray Coating Application Technologies and Techniques

Paint application techniques can be divided into two basic types, *spray application* methods and *bulk application* methods. *Spraying* is painting one part at a time, while bulk implies coating many parts at one time or coating continuously. There are many variations within each of these two groups and the rest of this chapter discusses the techniques and formulation considerations for each method. Liquid spraying, such as from a spray can, is something nearly everyone has some familiarity with, so liquid spray application is discussed first.

10.2.1 Conventional Spray Coating

The most common application of fluoropolymer coatings is conventional spray. It is used across many industries and technologies. Liquid paint is atomized by high-pressure air, typically 20–60 lb/in² (1.4–4.2 kg/cm²), escaping through a narrow orifice. But as simple as the technique seems to be, the theory of spraying is not well developed, and the entire process is quite complex. A description of generally accepted mechanisms of spraying is provided.

A stream of the liquid paint is directed into a fast-moving stream of compressed air. The velocity of the air approaches the speed of sound. The air stream elongates the stream of liquid into thin threads or sheets. The threads break spontaneously into droplets, driven by surface tension. The process of droplet formation is affected by viscosity and by elastic forces if the liquid contains dissolved polymers. The droplet size increases with decreasing air pressure (air velocity) and with increased flow rate of liquid paint. Usually the weight of air pumped through a spray gun is about equal to that of the liquid, but the volume of air is much larger.

The paint atomization occurs within a centimeter or so from the spray gun. The shear rates are great at this point and the liquid droplets are quite fine. Solvent composition of the droplets changes rapidly due to evaporation from the high surface area of the finely atomized droplets. The concentrations of low boiling solvents can be dramatically reduced compared to the bulk paint composition. The coating droplets are usually cooled by evaporative cooling. The momentum of air and liquid leaving the gun is transferred to the relatively motionless ambient air, creating a turbulent mixture. The mixture continues to travel in the direction the gun is aimed but its forward velocity falls off rapidly and carries more and more air as the distance from the gun increases. The turbulence grows more intense and the velocity of the atomized cloud decreases as the distance from the gun increases.

Spray guns often have auxiliary nozzles (called the "fan") which shape the atomized cloud. The auxiliary nozzles have minimal effect on the dimensions of the atomized liquid droplets.

Figure 10.1 shows a drawing of a typical hand-held spray gun. Automatic guns that are used in high volume applications have a similar design except that the handle is eliminated and the trigger is operated automatically.

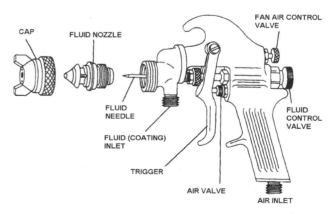

Figure 10.1 Schematic of a conventional hand-held spray gun.

The ability of a conventional spray gun to produce fine droplets depends on the rheological characteristics of a coating material at high shear rates. Viscosity rises with increasing concentration of dissolved polymers. As the concentration of dissolved polymer in the paint formulation is increased, the droplets become larger. As the particle size increases, the loss of volatile solvents from the particle by diffusion from the inside and evaporation from its surface, decreases. The large droplets, therefore, arrive at the substrate with a high content of volatile solvent.

The viscosity of the coating as it arrives at the substrate can be low. Low viscosity coatings tend to run and sag readily. If the viscosity is too high, then flow and leveling may not occur. Leveling is the process of smoothing out the surface of the wet paint. Much can happen to the wet coating after it has arrived at the substrate. The solvent composition can change dramatically. In some cases, materials will crystallize out of solution, or dispersions will become unstable, creating quality problems.

This application method offers many advantages:

- It is very common and very flexible

- It is inexpensive

- It offers "easy" application of thin films

Its primary disadvantage is that overspray (paint that does not deposit on the substrate) is severe, resulting in low transfer efficiency. Typically, less than 40% of the liquid paint deposits on the substrate, leaving 60% as waste.

When developing formulations for this application method, one usually:

- Aims for wide viscosity latitude

- Pays careful attention to solvent evaporation rates

- Keeps surface tension low, allowing the substrate to be wet by the coating

- Rheology control is formulated into the coating, shear thinning coatings are preferred

10.2.2 High-Volume, Low-Pressure Spray Application

To improve application or transfer efficiency, a system called *high-volume, low-pressure* (HVLP)

atomization has been developed. Instead of using a small amount of high-pressure air to atomize the paint, large amounts of low-pressure air are used. A "sonic venturi" converts high-pressure compressed air to low pressure. Typically, the air pressure for the atomizing air is 5–10 psi (0.35–0.7 kg/cm²).

Advantages offered by HVLP application include:

- Less overspray

- Less atomization

- Higher transfer efficiency, but wetter films

- Meets California Air Quality Standards

- Easier application of moderately thick films

Disadvantages include:

- Sometimes it is difficult to apply thin films

- Less shear during atomization can lead to appearance differences

When formulating coatings for HVLP application, solvent evaporation rates and coating rheology are important and may need optimization.

10.2.3 Electrostatic Spray Application

When a fine-wire or fine-point high voltage electrode is placed near a stream of liquid (in a spray gun as described earlier), the liquid is broken down into fine droplets. The droplets are electrically charged with the same polarity as the fine wire or fine point. However, for practical electrostatic spraying of liquid coatings, the liquid coating is usually fed out to the edge of a rotating disk-like or bell-like surface, as shown in Fig. 10.2. Fast rotation of the disk or bell will produce a fine spray by shearing the

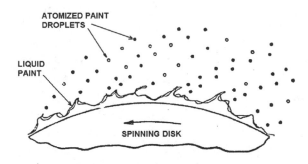

Figure 10.2 Diagram of atomization at the edge of a spinning disk.

liquid by air, similar to conventional spray guns. A high-voltage electrode is placed near the disk to charge the paint particles. However, in the presence of an electric field, fast rotation is not required and slow rotation is sufficient. The purpose of the rotation is the distribution of the coating at an even thickness that leads to uniformly atomized droplet sizes.

The droplets of coating are all charged negatively by passing through ionized air produced in a high-voltage discharge from a negative sharp point or wire electrode. The amount of charge picked up by the droplets depends on the particle diameter, the dielectric constant, and the conductivity.

When the substrate to be coated is grounded, it becomes the end point of a strong electric field. Spray particles are directed toward the substrate by an electric field terminating on the substrate. Because the spray particles are electrically charged, they are strongly attracted to the grounded substrate. When properly used, the loss of paint particles caused by missing the substrate can be minimized. The charge on each spray particle must be high and the electric field must be strong enough to overcome the competing effects of air currents, which can be strong in an industrial spray booth.

The electrical conductivity of the liquid coating affects the application process. As each charged spray particle arrives at the substrate, the conductivity of the liquid particle allows its charge to leak off to the substrate. If the coating is a poor conductor, its outer surface will retain its charge and act to reduce the electric field intensity that directs additional particles to it. High conductivity and high dielectric constant yield both an increased charge per particle and good application on the substrate. However, high conductivity can lead to poor electrostatic spraying (as opposed to deposition) and to leaks of high-voltage electricity across insulators and through pipes conducting paint to the spray equipment, such as occurs in water-based systems. A conductivity balance is required, so a coating material resistivity of 10 ohm-cm is generally targeted.

If it is desired to recoat, by an electrostatic mechanism, an article that has been coated, the conductivity of the dry coating must be raised to an acceptable minimum level.

Advantages of electrostatic liquid application include:

- High transfer efficiency: 50%–70%

- Rotational atomization efficiency approaches 90%
- More uniform thickness

Disadvantages of electrostatic liquid application are:

- Deep cavities are difficult to coat due to the Faraday Cage effect
- Safety: electrical shock and fire hazard
- Metallic paints often apply poorly
- Masking due to paint wrap-around

Formulation considerations for electrostatic application are:

- Water-based systems can be sprayed with special equipment called *voltage blocking equipment*
- Solvent conductivity affects application
- Flash point

Other liquid spray techniques have been occasionally used with fluorocoatings. For example, there is an ink jet application that requires a shear-stable coating with no particles larger than two microns. Also of interest is supercritical carbon dioxide spraying. However, since these are rare, they are not discussed in this work.

10.3 Liquid Bulk or Direct Coating Application Techniques

All bulk applications of liquid paint involve direct contact of the liquid to the substrate without atomization. These processes are generally called *meniscus-coating techniques,* where *meniscus* refers to the solid-liquid interface. The most basic of the bulk coating techniques is dip application. Quite simply, the part to be coated is dipped into the liquid paint, withdrawn, and allowed to drain. The Tallamadge Withdrawal Theory predicts the film thickness produced by this process.[1] The equation that describes this is called the *Landau-Levich equation:*

Eq. (10.1) $$h = R \cdot \frac{(\eta \cdot U_w)^{2/3}}{\gamma^{1/6} \cdot (\rho \cdot g)^{1/2}}$$

Thickness (h) depends on:

- Withdrawal velocity (U_w)
- Surface tension (γ)
- Viscosity (η)
- Density (ρ)
- Gravitational constant (g)
- Constant (R) depending on units of measurement

Assumptions:

- Newtonian fluids (viscosity does not vary with shear rate)
- One-dimensional flow
- Considers inertial, gravitational, viscous, and capillary forces
- No evaporation
- The equation has been found valid when $(\mu \cdot U_w / \sigma) < 8$

This equation is useful for understanding the affect of application variables such as viscosity and withdrawal rate. Raising viscosity and withdrawing the dipped items more quickly result in thicker coatings.

Practically, it is important to have:

- Very clean substrates, not only for surface defects leading to substrate wetting problems, but also to keep contamination from the paint reservoir (dip tank).
- Contamination-free coating.
- No bubbles on the coating surface.
- Withdrawal at a very uniform rate.

The end effects (i.e., drips at the bottom) need to be manually removed to minimize its defect. If the coating has dense particles such as metal powders, then gravity will affect these particles more than the rest of the coating and some separation can occur.

Dip processing has been used to coat tool blades, screws and bolts, wire and tubing (inside and/or outside). Various procedures can be used:

- The coated item can be withdrawn directly from a fixed coating bath.
- The item being coated can be fixed and the coating bath can be lowered.
- The coating bath can be drained at a constant rate.

10.3.1 Dip Coating

Dip coating is simple, but getting a quality coating can be difficult. As discussed in the previous paragraphs, the physical properties of coating, such as viscosity, density, and solids, are influential, as is the rate of withdrawal.

There are several practical ways dip coatings are applied. They partially depend on the size of the parts and the number being coated. Large numbers of small parts such as garden shear blades or fasteners with threads to be coated are typically hung from an overhead chain. The chain is loaded with the parts automatically or by hand. The parts can be cleaned by dipping in a solvent bath or by passing through a hot oven. Occasionally, parts are dipped in other treatment baths such as phosphating baths. The parts are dipped into a constant level paint bath as shown in Fig. 10.3 and removed slowly. The conveying line must move very smoothly to obtain the most uniform coverage. A drip of coating material often remains at the tips of the item being coated. This is removed by letting the bottom edge graze a wire that removes most of the drop. It is necessary to:

- Monitor and remove surface bubbles that sometimes form.
- Take care to avoid contamination.
- Monitor viscosity changes due to evaporation.
- Agitate the dip tank if the coating tends to settle.
- Monitor the coating quality if materials are shear sensitive and coating bath is agitated. (Many aqueous systems are shear sensitive.)

Continuous flexible substrates such as wire are often coated by dipping. The equipment is somewhat simpler as shown in Fig. 10.4. Constant level is not as important in this case. Sometimes the wire is passed through a die that removes excess paint and can improve coating uniformity. The same concerns described above need to be monitored for coatings applied to wire.

The main advantage to dip coating is its high application efficiency. The main disadvantage is the drip marks.

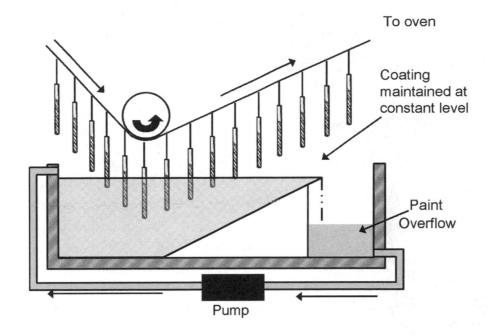

Figure 10.3 Schematic of a dip-coating line.

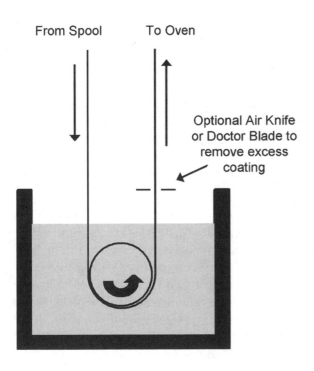

Figure 10.4 Schematic of a continuous dip-coating line for wire.

10.3.2 Dip-Spin Coating

Dip-spin coating is a process for coating large numbers of small parts that are impractical to spray. The most common parts coated by this technique are screws and springs. Typical equipment is shown in Fig. 10.5.

The process is comprised of the following steps:

- The parts are loaded in a basket that is attached to an overhead motor.

- The basket is lowered into a less-than-half-filled container of coating material (or the container is raised up to the basket).

- The parts are soaked for a specific time.

- The basket is raised above paint level, but still in the paint container.

- The basket is spun in one direction. Centrifugal force removes excess paint which is thrown to the container wall and reused.

- The basket is spun in the opposite direction. The parts shift and reorient, allowing excess paint that may have been trapped to be thrown to the container wall and re-used.

- Parts are dumped onto trays or belts.

- Parts are cured, then recoated as necessary.

Figure 10.5 A typical dip-spin machine. *(Photo courtesy of P. Ronci Machine Co.)*

For some parts such as washers, normal dip-spin will leave the parts stuck together because of the large flat surfaces. Paint can also be trapped in recesses such as in small cups. One can construct a fixturing device that can be loaded with parts and keep them separated (e.g., for fastener and washers) or oriented in a specific direction (such as small cups or cans) to minimize trapping paint in the recesses.

The most common parts coated with fluoropolymer coatings with this technique are automotive bolts and roofing nails. Roofing nails are very long screws, usually six to eighteen inches long, that are used on multilayer industrial roofs. The coating lets the screw penetrate the roof more easily and provides corrosion protection. The screws can frequently be seen on the underside of a roof in an industrial plant.

Dip-spin offers several advantages:

- Very high throughput—one can coat thousands of parts at a time.

- Application efficiency > 95%—the excess paint is reused.

- The process can be automated.

Disadvantages include:

- Poor film build control and uniformity.

- Contact points—can not avoid part-to-part contact, but defects are minimized by using multiple coats.

- Need to "make-up" fast solvent loss—fast evaporating solvents are lost relatively quickly, leading to viscosity rise in the paint reservoir. Viscosity adjustments need to be made on a regular basis.

Formulation considerations:

- The coating needs chip resistance from handling; parts are usually dumped out of paint baskets.

- Solvent evaporation control—quick evaporation helps increase film build per coat.

- Rheology control—strong shear thinning behavior improves application control.

10.3.3 Spin-Flow Coating

Spin-flow coating is most commonly used in the manufacture of silicon wafers and computer chips. The one major application in fluoropolymer coatings is the coating of aluminum disks that are then formed into deep drawn pans used in rice cookers and bread makers. Basically, paint is deposited onto a flat, planar and horizontal sheet. It is then spun at a specific RPM and for a specific time. The excess paint is thrown off the substrate and a layer of liquid paint remains that is of very uniform thickness.

There are four separate steps to the spin-coating process.

1. Deposition of coating onto the substrate, which can be done in any imaginable way, but is commonly done with a transfer tube or nozzle that pours the coating onto the center of the flat substrate piece. An excess of coating material is applied. The coating is passed through a submicron filter to eliminate the larger particles that could generate coating defects.

2. The substrate is ramped up to its preferred rotation speed. This step throws off much of the coating from the substrate surface by centrifugal force. Eventually, the coating reaches a uniform

thickness controlled by the rheology and the spinning speed. The excess paint can be reused if contamination is eliminated and cleanliness is maintained.

3. In the third step, the substrate is spinning at a constant rate. Rheological forces dominate gradual fluid thinning. The remaining film thickness is quite uniform, although initial evaporation of volatile solvents (leading to an increase in viscosity) can limit any further film thickness reduction. Edge effects are often seen because the coating forms droplets at the edges of the substrate and are thrown off. Depending on the surface tension of the liquid, viscosity, and rotation rate, there may be a small band of increased coating thickness around the outer edge of the substrate.

4. The fourth step is when the substrate is spinning at a constant rate and solvent evaporation dominates the coating-thinning behavior. The final thickness can be estimated under ideal conditions.[2]

After curing or drying, the coating process can be used to apply other layers. The substrate is finally baked and then can be formed into the end product. Because most products are post-formed, the coating needs to be flexible enough to be stretched and bent into its final shape and still maintain the required performance properties. Clean contamination-free substrates and coatings are important to maintain optimum coating quality.

10.3.4 Curtain Coating

Basically, curtain coating applies paint by allowing the substrate to pass through a "waterfall" of paint. The substrate being coated is either continuous or flat. The technique is like spin flow, but without the spin. Gravity and coating rheology control the coating thickness. Like spin flow, after curing or drying, the coating process can be used to apply other layers. The substrate is finally baked and then can be formed into the end product. Because most products are post-formed, the coating needs to be flexible enough to be stretched and bent into its final shape and still maintain the required performance properties.

This method offers high application efficiency and production rates. The paint needs stability to shear and resistance to foaming. A common problem is the stability of the contact line of the paint on the substrate. An unstable contact line can result in air entrapment under the coating, which leads to non-uniform coating thickness and other defects.

10.3.5 Coil Coating

A coil-coating process is used to coat continuous flat substrate. A continuous substrate is called a *web*. It usually consists of cold rolled steel, aluminized steel, or aluminum. The coated sheet can be cut up and post-formed by bending, folding, or pressing into the needed shapes. This is very common for fluoropolymer finishes for coating items such as bread pans, cake pans, and cookie sheets.

Coil coating is a continuous and highly automated process. The coating line requires a large capital investment. A schematic of a coating line is shown in Fig. 10.6. A large quantity of metal can be coated in a short period of time. For fluoropolymer coatings, the line may run from 25 to 100 feet per minute (7.6–30.5 m/min). Most companies contract out metal coating to merchant coil-coaters that own coating lines.

Description of coil coating:

1. A coil of metal is unwound and then cleaned or pretreated, usually by a dip process. Substrate treatment might be a light etching or the application of a phosphate treatment. The metal is dried after cleaning or treatment. The continuous metal strip is then coated with a primer. The application is by roller as depicted in Fig. 10.7.

2. The pickup roll transfers the coating liquid from the pan to the applicator roll. The coating is continuously pumped into the pan while the overflow recycles back to the supply reservoir, where it is remixed and filtered. The direction of the rotation of the applicator roll plays a part in determining the quality of the applied coating. Reverse roller coating, where the applicator roll turns in the opposite direction of the strip, is most common for

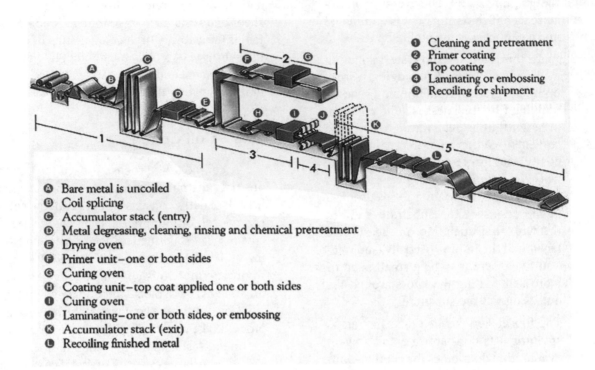

① Cleaning and pretreatment
② Primer coating
③ Top coating
④ Laminating or embossing
⑤ Recoiling for shipment

Ⓐ Bare metal is uncoiled
Ⓑ Coil splicing
Ⓒ Accumulator stack (entry)
Ⓓ Metal degreasing, cleaning, rinsing and chemical pretreatment
Ⓔ Drying oven
Ⓕ Primer unit–one or both sides
Ⓖ Curing oven
Ⓗ Coating unit–top coat applied one or both sides
Ⓘ Curing oven
Ⓙ Laminating–one or both sides, or embossing
Ⓚ Accumulator stack (exit)
Ⓛ Recoiling finished metal

Figure 10.6 Schematic of a coil-coating machine. *(Drawing courtesy of National Coil Coaters Association, www.coilcoating.org.)*

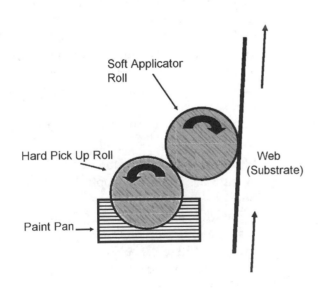

Figure 10.7 The coil-coating process.

fluoropolymer finishes and is shown in Fig. 10.7.

3. The paint sheet then enters an oven in which the coating is baked at high temperatures, often 750°F (399°C) for 20 to 30 seconds.

4. The strip exits the oven and is cooled with air and water.

5. A majority of fluoropolymer applications require two coats (primer and topcoat) so the metal must pass through a second coater, oven, and quench station.

6. The fully painted sheet can be inspected before it is rewound.

The main advantages of this application technique are its very high throughput and greater than 95% application efficiency. It also offers very precise control of film build. Most coatings are solvent based so VOCs can be high, but volatile gases are incinerated. Its main disadvantages are the very high capital cost and that coatings must be cured in very short times (ten seconds to two minutes) and must be post-formable. One of the problems with appearance that is difficult to control is a defect called "chicken tracks." This is a hint of stripes in the coating that are not meant to be there. It is particularly a problem with metallic-appearing coatings.

The largest commercial examples of fluoropolymer products coated in this fashion include home ovenware or bakeware such as pie tins, bread pans, and cookie sheets. One unusual application is automotive brake dampers. A brake damper is a component of a disk brake that minimizes the squeal. It is a multilayer laminate that has an elastomeric material on one side and metal on the other. When these are cut from coated coiled material, they are stacked up. The parts stick to each other due to the elastomer face of the laminate. Robots are used to assemble the brakes and problems develop when the parts stick to each other. By coil-coating the exposed metal side with a DuPont low-curing-temperature product that provides release, the parts do not stick together and the problems with robotic assembly are minimized.

There are other ways to apply a coating to a continuous substrate. Various roller techniques include rotogravure, which is a lot like coil coating, except that the printing or application roll is engraved with a pattern. The pattern holds the paint until it is transferred to the substrate. Knife coating has also been done with fluorinated coatings. Here the web passes under a special blade of metal that is set for a precise gap. A pool of liquid coating is continuously applied in front of the blade. The blade applies a precise amount of liquid paint to the metal. An air-knife can also be used to blow a thin stream of air at the web and hold back the excess paint.

10.3.6 Roller Coating

Roller coating uses the principle of coil coating but is designed to coat individual flat parts such as round disks that are then formed into frying pans. Coating is applied to the substrate that moves between two rollers. The amount of the coating applied is controlled by the gap or separation between the applying roller and the doctor roller as shown in Fig. 10.8. The scheme puts a precise amount of liquid coating on the applicator roll, which then applies it to the substrate.

The whole system is shown in Fig. 10.9. Multiple coating heads may be used with drying steps in between the heads to apply topcoats on the primer or just to build up paint thickness.

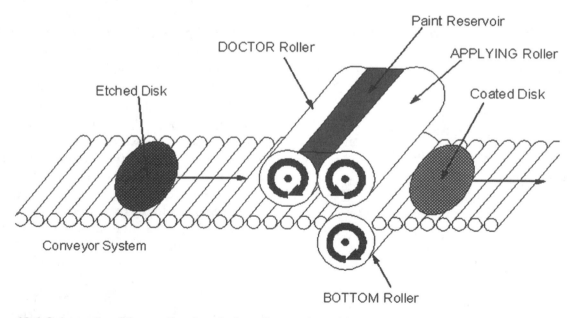

Figure 10.8 Schematic of the coating head of a roller-coat machine.

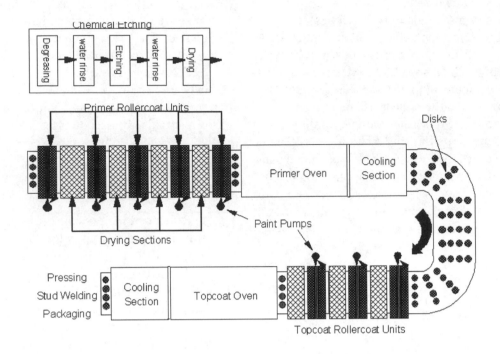

Figure 10.9 Schematic of a complete roller-coating line.

The advantages are very high throughput with greater than 95% application efficiency and precise control of film build. Disadvantages include:

- Moderate capital cost
- Short bake times are required, < 2 minutes
- Usually thin wet film build per station, multiple stations
- "Chicken tracks"

Formulation requirements:

- Coatings must be post-formable
- Shear stability
- Slow solvent systems
- Short cure cycles

10.3.7 Pad Printing

Occasionally a part needs to be coated only on a specific area. For a hypothetical example, consider a disk, shown in Fig. 10.10, that needs a fluorocoating only in the center and there are thou-

sands of these to be coated. One approach would be to mask the area where coating is not desired and then spray the exposed area. The mask would be removed and the part baked. This approach wastes materials and requires a great deal of labor. A process called *pad printing* is a very efficient and rapid way of applying a coating in this scenario.

The first step in pad printing is to make a "cliché." A cliché is basically a printing plate with the image of the desired coating area etched or engraved into it (see Fig. 10.9). Next, the etched image is filled with coating and the excess is removed by a doctor blade as shown in Step 1 of the figure. The solvents in the coating begin to evaporate from the surface and it becomes tacky.

In Step 2, a printing pad, which is a soft pliable elastomer such as silicone rubber, is pressed onto the cliché. As the pad is lifted away, the coating sticks to it (Step 3). A new surface of the coating is now exposed to air and the solvent begins to evaporate, making that surface tacky. The pad then moves to the substrate that needs the coating and is pressed onto it as shown in Step 4. The coating releases from the pad and is transferred to the substrate where it

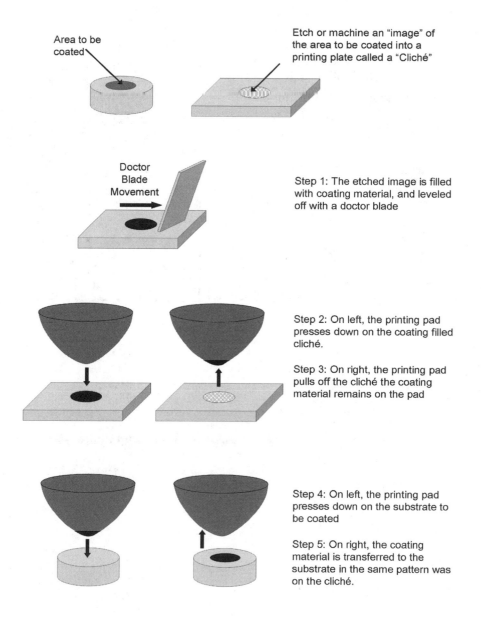

Area to be coated

Etch or machine an "image" of the area to be coated into a printing plate called a "Cliché"

Doctor Blade Movement

Step 1: The etched image is filled with coating material, and leveled off with a doctor blade

Step 2: On left, the printing pad presses down on the coating filled cliché.

Step 3: On right, the printing pad pulls off the cliché the coating material remains on the pad

Step 4: On left, the printing pad presses down on the substrate to be coated

Step 5: On right, the coating material is transferred to the substrate in the same pattern was on the cliché.

Figure 10.10 The pad printing process.

the pad and is transferred to the substrate where is it subsequently cured. The whole process can take place in a second or two and, with automation, many pieces can be coated per minute.

Coatings designed for this process are usually high in viscosity with some rapid-evaporating solvents. They are generally resin-bonded systems and are one-coats, though it is feasible to apply multiple coats by this process.

The advantages of applying coatings this way go beyond the high efficiency of applying coating only where it is wanted. It is possible to apply carefully controlled dry film thickness. The substrate can also be curved or irregularly shaped. A limiting disadvantage is that areas larger than one square inch are difficult to coat well by pad printing. Variations of this technique such as rotary gravure pad printing are not discussed in this work.

10.4 Summary

Table 10.1 summarizes the application efficiency of several of the major liquid painting methods.

There are rarely used, specialized, non-spray application techniques that have been operated with fluoropolymer finishes. Examples of these techniques include air-knife coating and silk-screen coating. Often the coating needs modification or special formulation to be applied by these processes. With coating applications, if one has an application idea, a formulation can usually be developed for that idea.

Table 10.1 Comparison of Application Efficiency of Various Liquid Coating Techniques

Application Technology	Approximate Application Efficiency
Air atomization	30%
HVLP	48%
Electrostatic, hand	64%
Electrostatic, automatic	80%
Dip and flow	85%
Coil and roller	>95%
Pad printing	>95%

REFERENCES

1. Tallmadge, J. A., and Gutfinger, C., Entrainment of Liquid Films, in: *Industrial and Engineering Chemistry,* 59(11):18–34 (1967)

2. Meyerhofer, *J. Appl. Phys.,* 49 (1978)

11.1 What is Powder Coating?

Powder coating is a way of applying a dry paint to a surface. Most people have used liquid paints. They may have applied them with a brush, a spray can, or even with their fingers. Powder coatings are dry because there are no liquid solvents. The dry powder is applied to the item (substrate) to be painted. Then the powder is turned to liquid by melting. That powder in its molten state subsequently flows out to cover the substrate; it coalesces and sometimes it crosslinks. The end result is a painted object.

The Powder Coating Institute publishes one of the best references,[1] although there are numerous others.[2]

11.2 Spray Powder Coating Process

There are several ways to apply fluoropolymer powders. Just like with liquid finishes, they can be classified as spray techniques or bulk application techniques. The spray techniques are called *electrostatic spray application* and *hot flocking*. The bulk application techniques are usually based on a fluidized bed.

Electrostatic spray can be used to apply thin or thick films. A thin film might be as low as 25 micrometers. Thick films might be applied as high as 2.5 mm, though typically thick films are 1.0–1.5 mm. The equipment used is diagrammed in Fig. 11.1. A large number of manufacturers make this type of equipment.[3][4] The process starts with fluidization of the powder. This is commonly done in a hopper or a vibratory feeder.

A hopper, shown in Fig. 11.2, is usually a cylinder with a porous bottom or fluidizing plate. Compressed clean dry air passes through the plate and into the powder. The air flows uniformly through the fluidizing plate and mixes with the powder, increasing its volume. The fluidization serves two purposes. One purpose is to get the powder to flow easily, allowing it to move from the hopper to the spray gun. The second function is to break up any loosely agglomerated powder particles. If one were to put his hand into this fluidized powder, there would be very little resistance to movement. It has the feel of a dry fluid. In the fluidized bed, it appears that the powder is boiling. In some cases, especially with finer powders, the bed is stirred or vibrated to aid in fluidizing uniformly.

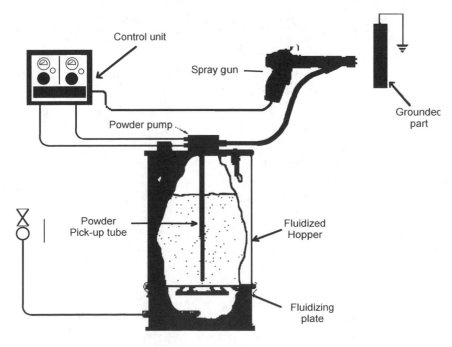

Figure 11.1 Powder coating application equipment.[1]

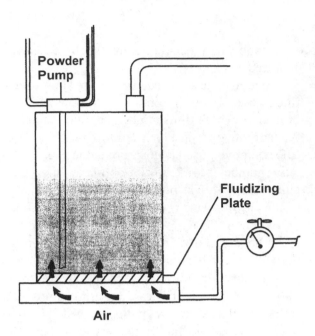

Figure 11.2 Fluidized hopper.[1]

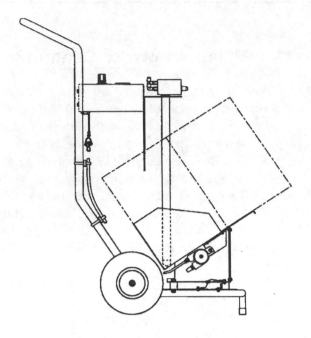

Figure 11.3 Vibratory box feeder.[3]

An alternate approach to fluidization is with a device called a *vibratory box feeder*. Figure 11.3 shows an example of one. These devices fluidize the powder directly in the box that powders are commonly provided in. It works by fluidizing the powder right around the pick-up tube with compressed air while the box is vibrated. Good fluidization is required to get consistent coating application.

The next step in the process is to move the powder in its fluidized state to the spray gun. This is done using a device called a *powder pump*. It is sometimes called an *injector* or a *venturi pump*. The powder pump usually sits on top of a dip tube that is inserted into the fluidized powder. Compressed air is injected into the top of the dip or pick-up tube as shown in Fig. 11.4. The top of the pick-up tube is called the *pump chamber*. By injecting through a narrow opening or nozzle, aerodynamic turbulence is created. The high-velocity air injected here creates a vacuum, known as the venturi effect, that draws powder from the fluidized bed up through the pick-up tube. The powder gets mixed with more air in the turbulent pump chamber and transported to the hose that leads to the spray gun. Some equipment provides for additional air to be added after the venturi throat to allow supplementary control over the flow of the powder.

The powder is transported through the supply tubing to the spray gun. If the powder passed through the gun without further interaction, it would be sprayed into the air as a cloud. There would be no physical reason for the powder to deposit on the substrate to be coated. To make this attraction happen, a charge must be applied to the powder. This is done by two methods called *corona charging* and *tribocharging*.

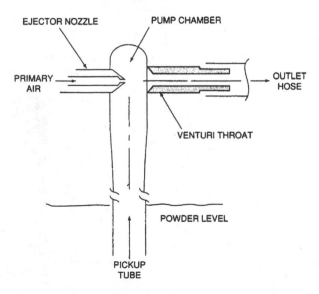

Figure 11.4 Powder pump.[1]

11.2.1 Corona Charging

A high-voltage power supply is attached to the spray gun in a corona-charging system. These power supplies provide adjustable voltage (typically 0–100,000 volts) or controlled current. When the high voltage is applied to a charging electrode in the gun, a strong electric field is created between the charging electrode and the grounded attractor electrode, shown in Fig. 11.5. This strong electric field ionizes the air creating what is called a *corona*. Normally 30 kV will ionize clean dry air, but lower voltages can be used, especially when particles are present as in powder coating. Ions are created and electrons are emitted. The electrons interact with oxygen in the air to form negative ions. (Nitrogen molecules in air can produce positive ions.) The oxygen negative ions collide with the powder paint particles and transfer the electrons to those particles giving them a negative charge.

Once the powder particles are charged, they are blown out of the gun towards the substrate to be coated. If the substrate is grounded, the powder particles will be attracted along lines of electric force to the substrate as shown in Fig. 11.6.

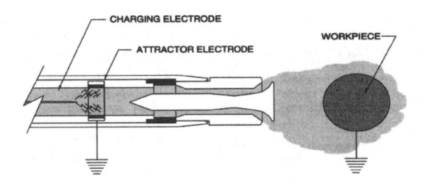

Figure 11.5 Corona charging in a spray gun.[1]

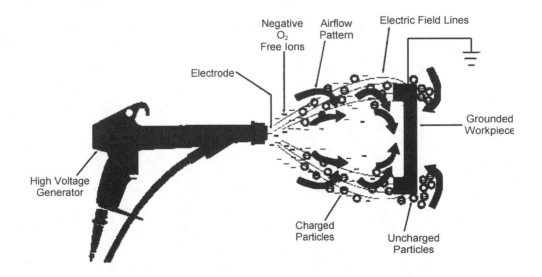

Figure 11.6 Powder from a corona-charging spray gun is attracted to ground.[1]

11.2.2 Tribocharging

Tribocharging is the process of electricity generation when two different materials rub against each other. Some materials easily give up or accept electrons from other materials under friction. A triboelectric series lists materials that give up electrons in order from easiest to hardest. For many materials, the dielectric constant determines the position of that material on the triboelectric series, as shown in Table 11.1. Additional materials such as rabbit fur have been added based on experimentation. The further apart the two materials are that rub against each other, the more charge is transferred.

To take advantage of this process for powder coating, a spray gun can be constructed without a high-voltage power supply. For most coating materials, the ideal material of construction for the gun is PTFE since it is positioned at one extreme of the triboelectric series. However, for fluoropolymers, a gun constructed of nylon is best. Nylon is the engineering plastic that is furthest away from PTFE on the triboelectric series, thus providing the best charging of a fluoropolymer powder. A schematic of a tribocharging powder gun is shown in Fig. 11.7. The interior of the gun is constructed of nylon such that there is maximum contact between the powder coating and the gun. Grounding of the gun is important because the powder flows continuously over the same nylon surfaces, removing electrons. Those electrons must be replaced from the ground for the gun to continuously tribocharge the powder and to avoid dangerously large voltage build up.

Tribocharging is affected more by the weather than corona charging. Relative humidity can affect the amount of charge transferred to the powder. Likewise, if the powder coating contains a lot of moisture, problems may arise.

Table 11.1 Triboelectric Series[1]

Material	Dielectric Constant at 1 MHz
+ (Electron Donor)	
Rabbit Fur	
Glass	
Human Hair	
Poly Ether Sulfone (PES)	3.7
Nylon 6,6	3.4
Wool	
Cotton	
Steel	
Silk	
Polyimide (PI)	3.4
Polybenzimidazole (PBI)	3.2
Polyether ether ketone (PEEK)	3.2
Polyethylene terephthalate (PET)	3.0
Polycarbonate	2.9
PVC	2.9
Polystyrene	2.7
Acrylic	2.6
Polypropylene	2.4
Polyethylene – High Density	2.3
Silicon	
PTFE	2.0
- (Electron Acceptor)	

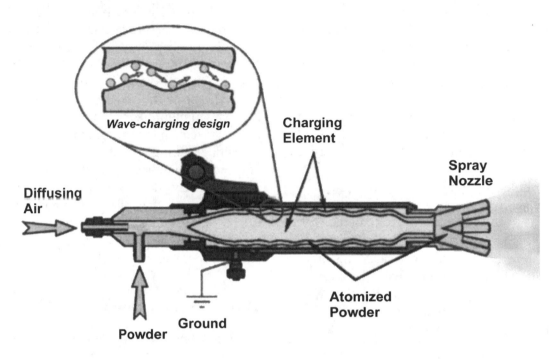

Figure 11.7 Schematic of a triboelectric charging gun.[1]

11.2.3 Powder Coating Advantages and Limitations

Powder coating has advantages over liquid coatings, but it also has limits. Most of the advantages are environmental and economic. Powder coatings generally contain no volatile organic compounds (VOC), so there are no atmospheric emissions. This reduces the cost of permits and environmental compliance. Spray-booth air can be filtered and returned to the room, reducing cooling and heating costs. *Overspray*, the powder paint that missed the part being coated, is usually considered nonhazardous waste or, at worst, solid hazardous waste, which reduces disposal costs. Overspray can sometimes be collected and reused, reducing material costs. Ovens do not need to remove solvent, so less air turnover is needed which reduces energy costs.

There are several common problems reported by powder users. Equipment and coating manufacturers can help solve them, but it is usually quicker to solve on location. A few comments on these problems follow.

Poor charging or poor attraction to the substrate often occurs because the substrate is not adequately grounded. First, check the hangers, which are usu-

ally coated and insulated from previous use. For corona guns, adjust the ionizing voltage. Usually, the further the gun is away from the part, the higher the voltage is required to get attraction. However, it is usually best to start low and work towards higher voltages.

As powder is applied to the substrate, a limit is reached where additional powder starts to repel itself because of build up of like charge on the surface. This occurs even though the part is grounded because neutralization of that charge takes time and the powders themselves are not usually conductive. Also, some charge retention is desired to help hold the powder onto the part being coated. Care must be taken when moving a freshly powder-coated part because the powder can fall off. The film thickness limit for powder coatings per coat applied electrostatically is dependent on the powder coating material, part geometry, and voltage on the powder. It is typically 50–100 micrometers of dry film thickness.

A characteristic of electrostatic coating with both charging processes is that the charged powder will wrap around the back of the substrate, especially as the powder builds up on the front of the part being coated. This is usually referred to as *electrostatic wrap*, and is shown in Fig. 11.6. Electrostatic wrap can be an advantage or disadvantage.

Powder coatings have trouble penetrating deep depressions. The electric field lines do not penetrate into these tight areas and the charged powder can not penetrate and deposit there. This is observed in almost all ninety-degree corners and is commonly referred to as a *Faraday Cage Effect*, shown in Fig. 11.8. While this effect can not be eliminated entirely, it can be minimized using triboelectric equipment.

Other problems seem to occur with the equipment. Surging powder at the gun is a common complaint. Surging can be caused by contaminated or moist fluidizing air or powder-pump air. Pinched or excessively long hoses sometimes cause this problem. The hoses should be conductive, otherwise the powder will tribocharge as it flows through them between the fluidized bed and spray gun. Poor fluidization in the hopper is often a cause. Occasionally an applicator will shake the bed while spraying. This should not have to be done. A vibrator can be attached to the fluidized bed, or adjustments to the airflow to the bed can help. A dry flow additive can also be added to the powder. This material is discussed in Sec. 11.5.1.

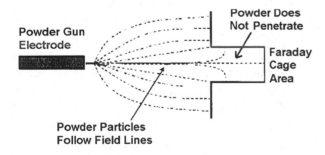

Figure 11.8 Faraday Cage Effect.[1]

11.3 Thick Film Coatings

For some applications, high film build is desired. Doing this with powder coating requires many coats and many bakes. This gets more difficult the thicker the coating gets because fluoropolymers insulate the grounded part from ground. The problem is further exaggerated because of the electrical insulation nature of pure fluoropolymers. Problems have been

observed with electrostatic attraction on multiple coats. The powder literally looks like it is repelled instead of attracted to the substrate. This is caused when the electrostatic charge used to apply the first powder coat is not dissipated off to ground during baking. That leaves the surface charged with the same charge as the powder, so it is repelled. The best way to minimize this problem is by using the minimum ionizing voltage setting on the corona gun, or waiting overnight for the residual charge to relax or bleed to ground before applying another coat.

When heavy dry powder layers are applied, electrostatic force is the only force that keeps the powder on the part. Heavy electrostatic coats are subject to falling off the substrate from jarring, vibration, and even, occasionally, airflow.

11.3.1 Hot Flocking

Some applications require very thick fluoropolymer coatings. There is a more efficient way to apply thick powder films and that process is called "hot flocking."

Powder is one of the best ways to coat a large impeller or mixing blade with a thick fluoropolymer coating. Usually, the process starts with a cold electrostatic powder-coating step that applies a thin film. The part (impeller, in this example) is baked above the melt point of the powder for a long enough time to get a smooth continuous coating. The part is quickly removed from the oven and powder is applied while the part is hot and above the melt point of the powder. Usually, the electrostatic voltage is turned off for hot-flocking coats. As the powder contacts the hot part, it melts. More powder can be applied until the melting stops. This is easily visible because the molten coating is usually quite glossy. When the melting stops, the part is put back in the oven to melt and flow out this coat. It can be pulled out of the oven and hot flocked again and again until the desired film build is achieved. For PFA, a common material for CPI applications, one can apply up to 0.2–0.4 mm per coat, mostly dependent on the mass of the part being coated.

Figure 11.9 shows the thermal history of a 6.25-mm thick steel panel being coated with PFA. The panel has a thermocouple attached to allow the temperature to be monitored even while it is in the oven. The first "tooth" shows the temperature after the panel has been primed and powder coated with about

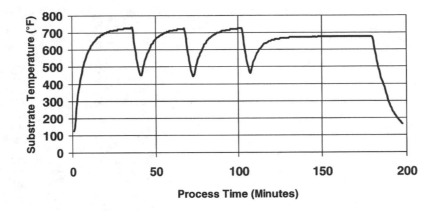

Figure 11.9 Hot flocking of PFA. *(Courtesy of DuPont Company.)*

50 micrometers of PFA. The melt point of PFA is about 581°F–590°F (305°C–310°C) and even a 6.25-mm steel panel cools quickly, One must, therefore, be ready to powder coat immediately when the panel is pulled out of the oven. About 0.2–0.4 mm of additional powder was applied on top of the first coat. The part was put back into the oven and rebaked. It was then flocked again and again. The final bake usually takes an extended time to allow the coating to completely flow out. Frequently, it is also at a lower temperature, as shown in this example, to minimize thermal degradation.

While the same powders can be used for hot flocking, usually larger particle sizes are used. This is partly due to the fact that most people turn off the electrostatic high-voltage supply. Larger particles have more momentum to carry them to the part.

11.3.2 Special Problems with High-Build Coatings

11.3.2.1 Decomposition

High-build coatings for both powder and liquid are susceptible to bubbling caused by thermal degradation. The degradation that occurs during high-temperature processing can be minute and, in thin film coatings, the decomposition gases can diffuse through the coating and escape. For thick film coatings, though, more material decomposes, and it has to diffuse through a much thicker film. This fre-

quently does not occur fast enough and the result is a coating that contains bubbles. Figure 11.10 shows a panel exhibiting thermal degradation that resulted in bubbling.

When this occurs there are limited choices to try to deal with the bubbles. One can try to lower the baking temperature, perhaps leaving it at temperature longer. One can also slow the temperature ramp-up or bake in stages. A good starting point is to set a long bake just below the melting point of the fluoropolymer. This is the point at which the coating is most porous. Then, the temperature is raised to just above the melting point and held before going to the final baking temperature. Sometimes adding a thermal stabilizer to the coating can reduce the decomposition.

Figure 11.10 Bubbling caused by thermal degradation in a thick fluoropolymer coating.

11.3.2.2 Sagging

Sagging or dripping of molten coating is another high-build problem. As the temperature rises above the melt point, gravity pulls on the melt. The higher above the melt point, the lower the viscosity of the melt becomes. An example is shown in Fig.11.11.

To minimize this problem, one must lower the baking temperature and shorten the bake time. A higher molecular weight fluoropolymer can also be used because melt viscosity rises with increasing molecular weight. Some users actually rotate the parts being coated in the oven to average out the gravitational forces.

11.3.2.3 Shrinkage

Another problem that occurs with thick fluoropolymers is shrinkage (Fig. 11.12). This is particularly true with PFA. Shrinkage is more pronounced on sharp edges. There is little one can do to stop this. Some applicators claim special bake schedules will minimize the problem. Lowering the baking temperature can help, and avoiding sharp edges minimizes pull back.

11.4 Bulk Application: Fluidized Bed Coating

Fluidized bed coating can be an efficient way to powder coat an item, once one figures out exactly how to get it to work. The fluidized bed is constructed much like the fluidized bed used in the powder application equipment as shown in Fig. 11.13. The difference is that there is no powder pump or dip tube. The part is dipped directly into the fluidized bed. Usually the part is heated to a temperature above the melting point of the polymer. For fluoropolymers, this is a very high temperature and can create safety or handling problems. As the part is "bathed" by the fluidized powder, it starts to melt and sticks to the part. This will continue until the part is below the melt temperature of the powder coating. The coated part needs to be reheated to permit the coating to remelt, flow out, and form a smooth surface. Additional coats can be applied if necessary.

Figure 11.11 Sagging in a thick ETFE coating.

Figure 11.12 Shrinkage in a thick-fill fluoropolymer coating.

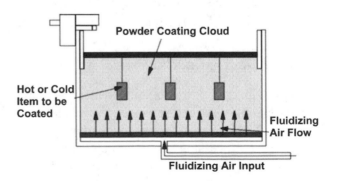

Figure 11.13 Fluizided bed coater.

One of the problems with this process is that if the part is complex with thin and thick sections, the thicker sections will frequently build up much more coating because those sections cool more slowly. It is not unusual to have film build differences of a factor of four. A complex part can also trap pockets of powder, leading to excessive film build, which can form bubbles during the rebaking.

There are many advantages to fluidized beds. It is one of the best ways to coat small parts. It offers high efficiency and low emissions. There is no Faraday Cage effect because electrostatic forces are not involved.

Fluidized beds are easy to construct allowing most applicators to build their own. Sources of porous membranes through which air passes, but powder does not, include Atlas Minerals and Chemicals, Inc., in Mertztown, Pennsylvania. Atlas makes polyethylene membranes. Eaton Products International in Birmingham, Michigan, offers glass bead plates.

Large powders (60–100 micrometers) generally fluidize better than smaller ones. Typically, 35-micrometer electrostatic powder-coating grades work well. One might have to stir the powder or attach a vibrator to the coater to get a stable fluid bed. Air powered vibrators are available from Martin Engineering Company of Neponset, Illinois.

A different approach for thin-film coating was refined by researchers at DuPont; it uses a cold part with no electrostatics. A part is primed with a liquid primer. It is immersed in the fluidized bed while still wet and the fluoropolymer powder sticks to the wet primer, holding it in place until it is baked.

This approach has been proven to be quite feasible, experimentally. A panel was primed by a hand-held air-spray gun. The panel was dipped into a bed that was fluidized using 20-micrometer fluoropolymer particles. After the part was baked, it had an excellent appearance. The fluoropolymer DFT was about five micrometers, and it was very uniform; even the sharp edges looked well covered. Measurements showed that 35-micrometer powder produced a DFT of 7.5 micrometers, and 60-micrometer powder produced a DFT of 13 micrometers. The results are graphically depicted in Fig. 11.14.

The reasons for the increased dry-film thickness obtained as larger PFA powder particles are applied is readily apparent in the micrographs in Fig. 11.15. What is rarely recognized, is that an applied powder before melting is very open. Under a microscope, it appears to contain 75% air. When those particles melt and flow together, the PFA layer is formed. Because the excess particles are blown off, this is not readily visible on these panels. The PFA

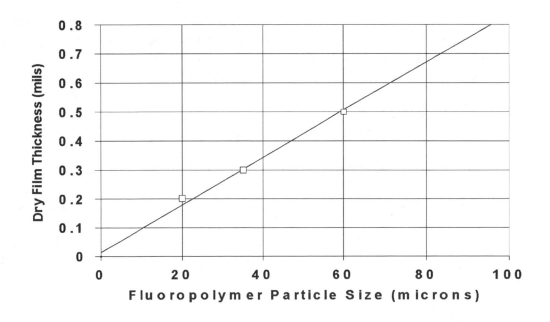

Figure 11.14 Cold fluizided bed application.

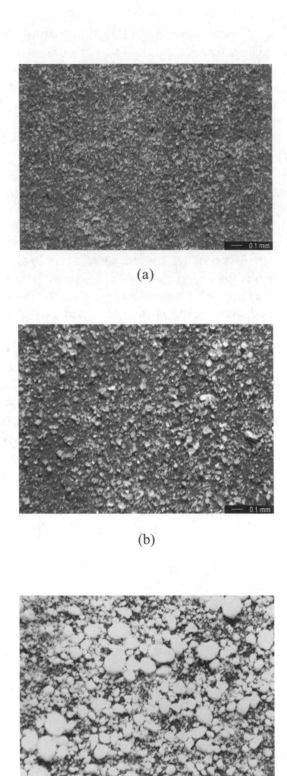

(a)

(b)

(c)

Figure 11.15 Micrographs of different size PFA particles applied on wet primer by fluidized bed: (a) 20 μm, (b) 35 μm, (c) 60 μm.

particles do not pack closely on the primer—there is a lot of space—and that is why the particles melt down to about one-fifth of their average particle size.

This technique of applying powder without electrical charge offers a big advantage. The edges and corners are covered more uniformly than is generally obtained even when coating powder electrostatically. The Faraday Cage effect is also eliminated so corners are coated.

Another way to coat from a fluidized bed involves applying an electric charge to the particles in the bed. In this case, the parts to be coated can be cold or heated, but they must be grounded. The steps are:

- Two high-voltage DC electrodes charge the fluidizing air.

- As the air fluidizes the powder, charge is transferred to the powder.

- The charged powder is then attracted to the grounded parts.

- The parts are baked to flow out the powder.

11.5 Commercial Powder Coating Products

There are many grades of fluoropolymers for powder coating. Many of the companies that make fluoropolymers produce powder-coating grades. Theoretically, any melt processible fluoropolymer can be made into a powder coating. Some coating companies buy granular fluoropolymers and grind them to a particle size suitable for powder coating.

11.5.1 Preparation of Powder Coating

Powders for electrostatic coating application usually average about 35–90 micrometers in diameter. Finer particle-size coatings are available, but fluidization is more difficult and less consistent, leading to a more difficult-to-control coating process. The finer grades are generally designed for thinner and smoother coatings not usually required in CPI applications. Larger particle sizes, especially those over 100 micrometers, have less surface area per unit weight and do not hold electrostatic charges well enough for coating.

Particle morphology or shape affects the ease of fluidization, the ability to hold charge, and how the particles pack onto the surface being coated. Shape varies greatly and is affected by the manufacturing process. The micrographs in Fig. 11.16 show the same pure molecular weight PFA prepared three different ways. Differences in the ways different PFA powders apply may be attributed in part to particle morphology.

Fluoropolymer powder coatings are not always pure fluoropolymers. There could be additives such as fillers, pigments, and stabilizers blended with the fluoropolymers. Most of these products are just dry blends. The different powders are put into a dry powder blender and blended to uniformity. The dry blending can cause a problem, particularly if the powders differ in density by a large amount. An example of density-blending problems can occur if a metal powder or flake is blended with fluoropolymer powder. Furthermore, since the dry blends are mixtures of discrete particles of different materials, each particle has a different ability to fluidize and a different ability to take up electrostatic charge. In fact, some particles may not accept charge at all. This manifests itself as a separation of materials during electrostatic application. Dry blended pigments often do not deposit on the part being coated in the same concentration as they are in the bulk powder because there is no driving force except momentum to attract them to substrate. Some companies make powder coatings that encapsulate the pigments within the fluoropolymer particles through special processing.[6]–[8] The materials are often called *encapsulated*, implying that the filler or pigment is completely surrounded by fluoropolymer. These materials, therefore, have distinct advantages over the dry blended powder-coating analogs.

One additive common to many powder coatings is a dry flow agent. It is always dry-blended and is commonly called *fumed silica*. Common trade names include Cab-o-Sil® and Aerosil®. Fumed silicas are submicrometer particles that are nearly perfectly round. When mixed with larger particles at less than one percent by weight, they coat the outside of those particles, though not one hundred percent of the area. They act like little ball bearings in between the particles, enabling them to flow against each other much more easily. The effect is dramatic. A diagram of this is shown in Fig. 11.17.

Fumed silica imparts an additional benefit. Depending on the silica chosen, it can also absorb water, thus keeping the powder coating dry. Fumed silica can also improve the ability of a powder to tribocharge.

There are numerous commercial powder coating materials. Many are made by companies for internal use only and are not sold except as applied. They are advertised and somewhat known in the industry, but physical data on the powders are not generally available. Where possible, the best-known materials are included in Tables 11.2–11.6 based on their base fluoropolymers.

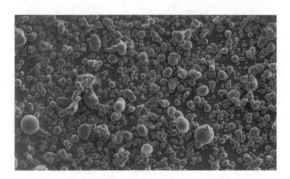

Figure 11.16 PFA powder coatings made by different processes. *(Courtesy DuPont Company.)*

13.6.1.3 Instron Peel Test

There are a number of other tests to measure adhesion bond strength. Most require gluing a metal probe to the surface of the coating. These are described in ASTM standard D4541-02, "Standard Test Method for Pull-Off Strength of Coatings Using Portable Adhesion Testers." Since fluorinated coatings often have excellent non-stick properties, the bond between the coating and the glue fails before the bond between the coating and the substrate, making these tests worthless.

One additional adhesion test worth mentioning is ASTM D1876-01, "Standard Test Method for Peel Resistance of Adhesives (T-Peel Test)." The test requires very thick coatings and an expensive Instron machine, but the data can be very useful. Basically, the Instron pulls the coating and substrate apart and measures the force required to do that.

The preparation of the test sample is important. The test panel or foil must be prepared to allow easy separation of the coating from the substrate for at least an inch (25 mm) at one end. Covering that end with a silicone coating or a piece of Kapton® film, then applying the coating over the entire panel, can accomplish this. For this test, the coating must be thick, generally 40 mils or more (1 mm). Thinner coatings may break before the adhesion of the coating to substrate fails. Figure 13.15 shows a sample with the layers separating while being pulled by the jaws on the Instron. The width of the test strip is measured; usually it is cut to one inch (25 mm). The force in pounds per inch (kg/cm) to pull apart the two strips is a measure of the adhesion of the coating to the substrate.

13.7 Environmental Exposure Testing

13.7.1 Salt Spray

The most common environmental type test is what is commonly called the *salt spray* or *salt fog* test. This test is considered to be most useful for measuring the relative corrosion resistance of related coating materials rather than providing absolute performance measures. Given that, there are still many specifications based on minimum performance levels at a given exposure time in this test. Nearly every automotive end use carries a salt spray specification of a minimum number of hours without red rust.

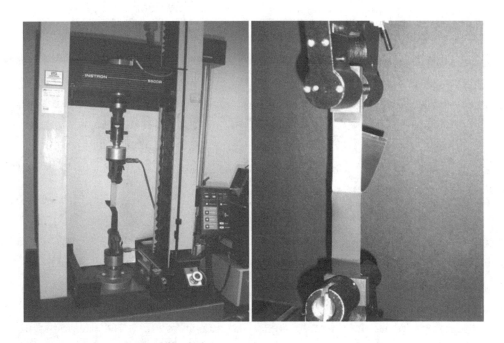

Figure 13.15 Instron peel-strength adhesion test.

The test provides a controlled corrosive environment that represents accelerated marine type atmospheric conditions. The equipment set-up, conditions, and operational procedures are described in detail by the ASTM standard B117-03, "Standard Practice for Operating Salt." It does not describe the type of test specimen and exposure periods nor does it describe how to interpret the results. There are other ASTM practices to help with evaluation of exposed panels.

The salt spray test is typically run by preparing test panels under controlled conditions. That means controlled surface preparation, film thickness, and bake. The panels are exposed at a somewhat elevated temperature to a mist of 5% salt water. The panels are examined periodically, typically after 24, 48, 96, 168, 336, 500, and 1000 hours, and evaluated for defects such as degree of rusting, blistering, and chalking. Sometimes the panels are scribed with an "X," cutting through the coating to the substrate. This is to evaluate creep or filiform corrosion.

The basic procedure for running a salt spray study:

1. Prepare replicate panels for each coating and each application condition being studied.

2. Label each panel. The writer prefers to use 4" × 12" panels with a scribed "X" on the bottom half of the panel.

3. The panels are put into the salt spray cabinet in random order as in Fig. 13.16.

4. At the prescribed time intervals, all the panels are removed from the cabinet and rinsed under running water and patted dry. Initially they can be inspected relatively quickly unless they are poor performers. The evaluation is done per ASTM standard methods.

5. Put the panels back in the cabinet and continue the test. Sometimes the experimenter may select a panel to keep after a specific number of hours. By having many replicate panels, removal of a panel at given intervals allows the test to continue.

6. Repeat Steps 4–6 until the test is complete.

The ASTM methods for evaluating panels being tested for corrosion are listed:

- D714-02, "Standard Test Method for Evaluating Degree of Blistering of Paints." Ratings vary depending on size and number of blisters. For multilayer coatings, blistering may form between any of the layers, so that must also be noted.

- D4214-98, "Standard Test Methods for Evaluating the Degree of Chalking of Exterior Paint Films."

- D610-01, "Standard Test Method for Evaluating Degree of Rusting on Painted

Figure 13.16 Salt spray testing cabinet. Panels are loaded as pictured *(left)* with the edges protected by tape from corrosion. The cabinet on the right shows the computer control panel. The cover is open to access the test panels.

Steel Surfaces." Frequently this is based on the percentage of the surface area that has red rust.

- D2803-03, "Standard Guide for Testing Filiform Corrosion Resistance of Organic Coatings on Metal."
- D1654-92 (2000), "Standard Test Method for Evaluation of Painted or Coated Specimens Subjected to Corrosive Environments."

13.7.2 Kesternich DIN 50018

Another corrosion test of interest for construction and automotive applications is the Kesternich Test. It is best know under the European test method, DIN Standard 50018, or ISO 3231. This test method is a severe measure of corrosion resistance. The components or panels to be tested are prepared in a Kesternich Test Cabinet. Two liters of distilled water are placed in the bottom of the cabinet and it is sealed. After sealing, sulfur dioxide is injected into the cabinet and the internal temperature is set to 104°F (40°C) for the cycle. Every 24-hour cycle begins with eight hours of exposure to this acidic bath created in the cabinet. Next, the test specimens are rinsed with distilled water, and dried at room temperature for sixteen hours. The test specimens are examined for surface corrosion (red rust) at the end of each cycle. Cycles are typically run until a specific amount of red rust is noted. The ASTM practice for this is G87-02, "Standard Practice for Conducting Moist SO$_2$ Tests."

13.7.3 Atlas Cell

Atlas Cell is a test that allows the estimation of the resistance of a coating in contact with a chemical at a given temperature. The picture in Fig. 13.17 describes the test. The coating is applied on the inside of the panels, which close the glass pipe. They are held in place by clamps with a chemically resistant gasket such as Goretex®, EPDM, or other heat- and chemical-resistant elastomer, insuring a leak-free seal. The test liquid is heated. The heater can be controlled at a particular temperature or the liquid can be run at reflux with a condenser returning the liquid to the cell. The coating is exposed to both a liquid and a vapor phase. It also sees a temperature gradient across the coating and the panel. The effect of the gradient is a very important part of this test. Even many thick coatings will blister within a week when exposed to these conditions.

Most often the liquid is just boiling water. With precautions taken for leaks and spills, other liquids may be used. Sulfuric, nitric, hydrochloric, and hydrofluoric acids have been used. Strong bases such as sodium hydroxide have also been studied with Atlas Cells. Visual inspection is done daily at the start of the test, but can be done less often as the test progresses. Most of the time, after 720 hours, the test is stopped. Often the panels blister as shown in Fig. 13.18. If the panel is not severely blistered, then adhesion can be tested.

The Atlas Cell test is a key test for chemical equipment applications such as tanks, mixers, and reactors. The ASTM procedure is C868-02, "Standard Test Method for Chemical Resistance of Protective Linings."

Figure 13.17 Atlas Cell during testing.

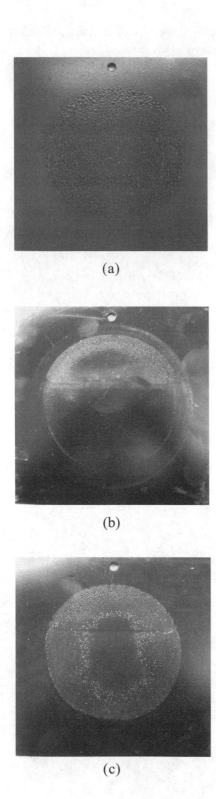

(a)

(b)

(c)

Figure 13.18 Sample Atlas Cell panel failures. (a) Blisters form uniformly over entire face of the panel. (b) Blisters form primarily in the vapor phase exposed area. (c) Blisters form on the outer edges first and move towards the middle.

13.8 Coefficient of Friction (CoF)

One of the unique properties of fluoropolymer coatings is what is often called *dry lubrication*. Many customers or users of coatings ask for the coefficient of friction for the coating. This is usually of little use because the nature of friction is not well understood.

When any two surfaces contact each other, there is a frictional force that resists motion. Picture a one pound cube (1 kg) of polished steel sitting on a flat smooth slab of PTFE. That steel cube exerts a force on the PTFE and the PTFE exerts an equal and opposite force on the steel cube. If a very small force is applied to the block, it will not move. That is because frictional force resists that motion, and it is larger than the force being applied to move the block. Because both materials are at rest, the frictional force is considered static frictional force. Physicists have developed an equation describing this static frictional force. It is shown in Eq. (13.9).

Eq. (13.9) $f_S \leq \mu_S N$

In Eq. (13.9), the static frictional force, f_S, *is* less than or equal to the static coefficient of friction, μ_S, and the normal force, N. The static coefficient of friction depends on the material pairs, so presumably aluminum and steel against PTFE have different coefficients of friction. As this equation indicates, the static frictional force is nearly independent of contact area. That is, a cube, cylinder, pyramid, or slab of any shape but of the same weight exerts the same amount of force on the PTFE and has the same static frictional force. The normal force in this example is provided by gravity, while in an actual experiment, tensioning devices such as springs could provide it.

Similarly, there is an equation for *kinetic* frictional force that looks similar to static force, with different coefficients, leading to Eq. (13.10).

Eq. (13.10) $f_K = \mu_K N$

In Eq. (13.10), the kinetic frictional force, f_K, is equal to the kinetic coefficient of friction, μ_K, and the normal force, N. The kinetic coefficient of friction likewise depends on the material pairs. As this equation indicates, the kinetic frictional force is also nearly independent of the contact area.

Usually μ_S is greater μ_K and both values are less than one. However, for PTFE, the static coefficient of friction is 0.1 or less, which is smaller than the dynamic coefficient (0.24). The concepts of static and kinetic friction were actually discovered by Leonardo da Vinci, two hundred years before the theories were fully developed by Isaac Newton.[2]

The coefficients are measured with specialized instruments. The ASTM practice is D1894-01, "Standard Test Method for Static and Kinetic Coefficients of Friction of Plastic Film and Sheeting," and can be applied to coatings.

Both coefficients of friction depend on many other variables besides the contact materials. Surface finish, temperature, humidity, and contamination can affect the coefficients. When velocities get very high, the friction coefficients can change.

13.9 Abrasion/Erosion

The field of *tribology* includes the analysis and study of friction, wear, abrasion, erosion, and lubrication. Customers or users frequently ask, "How abrasion-resistant is this coating?" This is a very difficult question to answer because it needs a great deal of qualification. The loss of coating during use is called *abrasion* or *erosion*. The conditions the coating is exposed to during use directly affect the loss of coating. Conditions would include such variables as velocity, pressure, and temperature, as well as what the contact surface is made of and whether the locus of contact is a point, a line, or an area. Also of concern is the abraded coating debris and whether it is removed or it remains in the contact area.

There are a large number of standard commercial tests. Selecting the one that correlates to a particular end-use is difficult. Three of these tests are discussed here.

13.9.1 Taber

The Taber Abraser has been in use in the coatings industry for a long time. It is a standard ASTM test, D4060-01, "Standard Test Method for Abrasion Resistance of Organic Coatings by the Taber

Abraser." This has been historically a favorite of the coatings industry, in general, because it is inexpensive and easy to run. A test panel is coated that has a hole in the center. It is mounted in the Taber abrader that is shown in Fig.13.19.

A weight is selected and added, as are the types of abrasive wheels. The wheel and weight assembly is lowered onto the test panels, which then rotates, allowing the panel to be abraded by the wheels. Vacuum removes the abraded debris. The panels are rotated for a given number of cycles or until the coating is worn away down to the substrate. By measuring the dry film thickness (DFT), or by weighing before and after the test and knowing the number of cycles, the wear rate can be calculated in terms of DFT loss or weight loss per 1000 cycles.

There are problems with this test method. First, as the fluoropolymer is abraded, it tends to fill in the porosity of the abrading wheels. This makes them less efficient at abrading. Therefore, when studying fluoropolymer-containing coatings, the abrading wheels need to be cleaned or redressed every 100 to 200 cycles. Secondly, the test has poor reproducibility. Comparisons of coatings should be restricted to testing in only one laboratory when numerical wear rates are to be used. Interlaboratory comparisons should use rankings of coatings in place of numerical values. The substrate disk must be completely flat. Aluminum softens and warps sufficiently

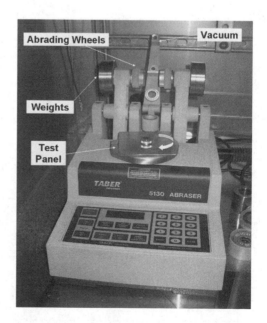

Figure 13.19 Taber Abraser.

during high temperature bakes to cause the abrasive wear on the panel to be inconsistent. If aluminum must be used, as it might because of test coating adhesion problems to the steel, then it must be about 6 mm thick.

13.9.2 Falling Abrasive Test

A simple, inexpensive, reproducible abrasion test is the falling abrasive test described in ASTM D968-93 (2001), "Standard Test Methods for Abrasion Resistance of Organic Coatings by Falling Abrasive." Known weights of sand, gravel, or aluminum oxide are poured onto a panel from a given height through a funnel and tube as shown in Fig. 13.20. The panel

is positioned at a 45° angle to the falling abrasive, which is collected for reuse. The panel DFT and weight is recorded prior to testing. Some laboratories prefer to use 20 grit aluminum oxide.

The author's preferred procedure for comparative evaluation is to pour 4 kg of abrasive onto the panel through the machine and then examine the panel for substrate exposure. If no exposure of substrate is apparent, then the cycle is repeated over and over again until substrate is exposed. From the weight loss and the DFT loss, two wear rates can be calculated as shown in Table 13.3

Abrasive material such as aluminum oxide can be reused many times. When desired, after many tests, the fines can be removed by sieving.

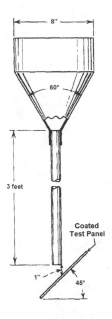

Figure 13.19 Falling Abrasive Abrasion Tester. *(Left)* Schematic showing the funnel and test panel position. *(Right)* Photograph of tester.

Table 13.3 Falling Abrasive Abrasion Tester Worksheet

Panel #	DFT, μm	kg Abrasive/ # passes	Wt Loss, g	Wear Rate Wt loss/ kg Abrasive	Wear Rate DFT loss/ kg Abrasive
1	49	60/15	0.0274	4.57×10^{-4}	0.82
2	45	52/13	0.0239	4.60×10^{-4}	0.87
3	69	80/20	0.0353	4.41×10^{-4}	0.86
4	58	60/15	0.0290	4.83×10^{-4}	0.97
5	54	68/17	0.0310	4.56×10^{-4}	0.79
6	53	64/16	0.0324	5.06×10^{-4}	0.83

13.9.3 Thrust Washer Abrasion Testing

The most sophisticated abrasion test is called the Thrust Washer Abrasion Test and is described by ASTM procedure D3702-94 (2004), "Standard Test Method for Wear Rate and Coefficient of Friction of Materials in Self-Lubricated Rubbing Contact Using a Thrust Washer Testing Machine." The machine can provide a large amount of quality and detailed information about the wear of coatings. However, it is very expensive to own and expensive and time consuming to operate. Some contract laboratories can run these tests. It is the best overall test available. Two manufacturers of the machine are the Falex Corporation and Plint Tribology Products.

The machine tests a coating that is applied to a precision-machined washer shown in Fig. 13.21. The opposing surface is an uncoated ring made of the material by which the coating will be abraded. It is typically steel but could be made of any other metal or even a polymer.

The test specimens are loaded into the test machine as shown in Fig. 13.22.

There are a number of selectable variables for this test:

1. The load pressing the washer and ring together (lbs/in^2 or kg/cm^2).

2. The speed or RPM from which can be calculated the velocity (ft/min or m/min).

3. The ring holder temperature.

4. The time the experiment runs.

An individual experiment can be run in two basic ways:

1. The machine can be set to run a specific number of revolutions or for a specified period of time.

2. The machine monitors the torque applied to the ring by friction; when the torque rises above a threshold (i.e., the coating has worn through), the machine shuts down and the revolutions are recorded.

After the experiment, the film thickness change and weight loss can be measured. From this data, an array of wear measures can be calculated:

1. Wear rate in terms of film thickness lost per unit time (typically μ"/hr, μm/hr)

2. Wear rate in terms of film weight lost per unit time (g/hr)

3. The wear factor, K, based on thickness change (in·min/ft·lb·hr or cm·min/m·kg·hr):

$$\text{Eq. (13.11)} \quad K = \frac{W_{(t)}}{PVT}$$

4. The wear factor, K, based on volume change (in^3·min/ft·lb·hr or cm^3·min/m·kg·hr):

$$\text{Eq. (13.12)} \quad K = \frac{W_{(v)}}{PVT}$$

where:

$K =$ wear factor (see above discussion)

$W_{(t)} =$ thickness lost (in or cm)

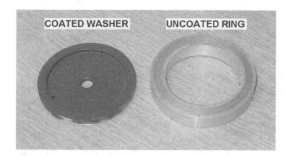

Figure 13.21 Thrust washer test specimens.

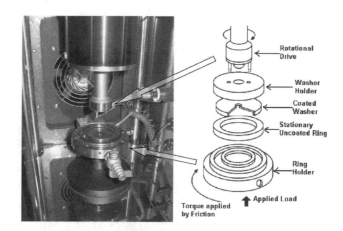

Figure 13.22 Thrust washer test equipment.

$$W_{(v)} = \text{volume lost (in}^3 \text{ or cm}^3)$$
$$P = \text{pressure (ft·lb or m·kg)}$$
$$V = \text{velocity (ft/min or m/min)}$$
$$T = \text{time (min)}$$

Of course, the wear factor varies with pressure or force and velocity. Often the pressure and velocity test parameters are multiplied together giving a PV parameter against which wear rates or coefficient of friction can be plotted.

Occasionally, the PV rating of a coating or material is given. This number is usually the maximum abrasion condition a coating can withstand for a short time. Exceeding this rating leads to almost immediate breakdown of the coating. PV ratings are unusual and have limited applicability in coating evaluation. The PV rating has historically been used by engineers to described roller bearing performance.

13.10 Hardness

There are a large number of hardness tests available, which include:

- Rockwell
- Brinell
- Vickers
- Knoop
- Shore

Most of these are aimed at measuring the hardness of metal alloys or other engineering materials. They are not designed for thin coatings and, though measurements on coatings have been done, they have little actual significance since the hardness of the substrate affects the measurement. The most common hardness test used for fluorinated coatings is the ASTM standard D3363-00, "Standard Test Method for Film Hardness by Pencil Test." It is a simple technique to evaluate the hardness of fluorinated coatings.

13.10.1 Pencil Hardness

Pencils come in a range of hardness from soft to hard:

6B 5B 4B 3B B HB F H 2H 3H 4H 5H 6H
Softer ⟵――――――――――⟶ *Harder*

Pencils from different manufacturers, or even pencils of different lots from the same manufacturer, will not necessarily have exactly equal hardness of the lead even though they have the same hardness rating. Pencil sets should be purchased as a set and replaced as a set.

- Pencils for use in this test should be sharpened with a knife.

- Starting with the softest pencil, the pencil point is sanded flat with 400 grit sand paper so that an even cutting surface is provided around the circumference of the lead. Special devices are available to do this reproducibly.

- The pencil is moved forward on the coating surface at an angle of 45°. A tool is available to hold the pencil at 45° consistently.

- The mark is examined with a magnifier or microscope to see if the lead has cut into the film.

- This procedure is followed with pencils of increasing hardness until the first pencil that cuts into the film is identified.

- The hardness rating of the previous pencil is the rated hardness of the film.

13.11 Cure

Generally there are no cure tests for fluoropolymer coatings that do not contain non-fluoropolymer binders. The main reason is that the cure is a melting process, not a crosslinking process. For the resin-bonded fluorinated coatings, a solvent rub test might be done to measure cure. Coatings based on epoxy have been tested this way. The standard procedure used is the "Solvent Rub Test for Cure – NCCA Technical Bulletin II-18." The NCCA is the National Coil Coaters Association.

13.12 Cookware Testing

Besides some of the tests described earlier in this chapter, there are other special tests for cookware. The testing of cookware performance is the subject of endless disagreements between different pan manufacturers and different coating manufacturers. Most of these disagreements focus on the accelerated tests rather than actual in-home testing. While the accelerated tests can often show wide differences between pans and between coatings, it all comes down to this question, "Does this test really relate to how a coated pan is used in a home kitchen?" The following section summarizes some of the tests and how they are performed.

13.12.1 In-Home Testing

In-home testing is exactly what the words mean. The cookware is distributed to a large number of homes, where the pans are used however the homeowner sees fit, and they are returned periodically for a visual evaluation. What follows is a basic process for in-home testing.

Participants of in-home testing are selected on the basis of questionnaires, which ask for information about family size and cooking habits. A family is given a piece of unidentified non-stick coated cookware and asked to use it as they normally would for three years.

The in-home cookware sample is evaluated visually every six months over the course of the next three years. Researchers evaluate how the sample has withstood real-life abuse, checking for telltale signs of weakness such as scratching, corrosion, and staining of the coating. In-home test results are also compared with the results of accelerated lab tests (discussed next) to ensure those laboratory predictions match real-world experience.

Usually a number rating system is developed for scratch, wear, and corrosion that makes it easier to analyze the results of hundreds of pans that typically would be part of a test. Sometimes, the homeowner also fills out a questionnaire about what they like and dislike about the pan.

This is considered the best overall test, but it takes a very long time and the cost is very high.

13.12.2 Accelerated Cooking Test

The accelerated cooking test further challenges cookware, subjecting test pans to a barrage of ever-changing conditions: temperature, diverse food items, metal utensils, and length of cooking time. The test is designed to simulate, in a greatly accelerated time frame, the wide range of treatment and abuse cookware receives in actual use.

The test (Fig. 13.23) uses a scratching device called a *tiger paw*, a disk with 3 ball-point pens forming a triangle, weighing 1 lb (0.45 kg).

Here is a "recipe" for an accelerated cooking test:

- Heat dry pan to 380°F–400°F (193°C–204°C) (medium high).
- Cook hamburger, heavily salted; 5 minutes.
- Cover; cook additional 5 minutes.
- Flip burger and return cover; cook 5 minutes.
- Add onions around burger and water, if necessary, to prevent burning.
- Cover and cook 5 minutes.
- Add tomato sauce, salt and water; remove burger.
- Paw (stir in a circular pattern) the mixture 25 times clockwise, 25 times counterclockwise.

Figure 13.23 Accelerated cooking test in progress in a test kitchen. *(Photo courtesy of DuPont.)*

- Cover; cook 15 minutes.
- Remove tiger paw and mixture; wash and dry pan.
- Reheat dry pan to 380°F–400°F (193°C–204°C); cook one pancake.
- Plunge pan into warm soapy water; wash and dry pan.
- Reheat dry pan to 380°F–400°F (193°C–204°C); cook egg; using the tiger paw, stir 25 times clockwise, 25 times counterclockwise; empty, wash and dry pan.
- Evaluate the pan for scratches, stains, and defects. Repeat until the pan is no longer usable because of finish failures.

Another coatings manufacturer has a similar test, which they call a Manual Tiger Paw Test, but no food is used. The cookware is filled with a thin layer of cooking oil, and heated to 400°F (204°C). The tiger paw is rotated over the non-stick surface in a circular fashion 2000 times, changing direction every 100 rotations. The coating is then examined for any fraying, blistering, or penetration to bare metal.

13.12.3 Mechanical Tiger Paw (MTP)

Whenever the manual tiger paw is used, there is variability from cooker to cooker. When there are two or more cookers, that variability must be considered in the statistical design of the test. A device that attempts to standardize the pressure and motion of the tiger paw is the Mechanical Tiger Paw or MTP. The MTP attaches the tiger paw to a rotating disk. The tiger paw is free to rotate on its own. A selected weight is placed on the tiger paw assembly. The test pan is heated on a laboratory hot plate. The hot plate is placed on a shaker table that moves back and forth. The temperature, RPM of the tiger paw assembly, the weight, and the hot plate motion are all carefully controlled. MTP measures the overall durability of a coating in resisting point abrasion and scratching. No food is used in this test, it is a dry heat test. A photograph of a typical tiger paw machine is shown in Fig. 13.24.

- A pan is placed on a hot plate of a MTP machine and heated to 400°F (204°C).
- The tiger paw device, consisting of three weighted prongs, is run over the interior surface of the cookware until the surface is abraded to metal.
- Movement of the tiger paw is reversed every 15 minutes.
- The pan is rated for its performance based on the appearance of the scratches the tiger paw puts in the surface. The rating is somewhat subjective, but is usually compared to lab-generated standards.
- The test result is the time it takes for the substrate to be exposed or the time to reach a particular pan rating.

There is another version of the test that varies slightly. The pan is heated to 400°F (204°C), and is filled with a thin layer of cooking oil. To gauge the coating performance, researchers measure the time (in minutes) that it takes for the tiger paw to penetrate the coating film to bare metal.

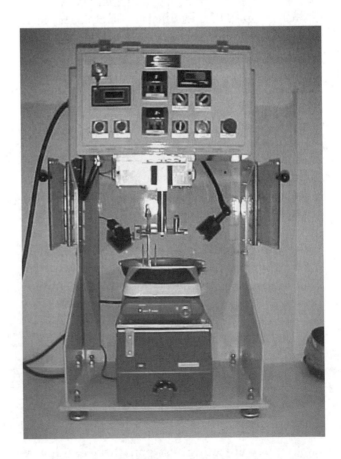

Figure 13.24 Mechanical Tiger Paw (MTP). *(Photo courtesy of DuPont.)*

13.12.4 Steel Wool Abrasion Test (SWAT), Sand Paper Abrasion Test (SPAT)

Several abrasion tests may be run with slight modification on the equipment shown in Fig. 13.25. This machine moves an abrasive pad, steel wool, a Scotch-Brite® pad (a common kitchen cleaning sponge), or sand paper back and forth across the surface of the pan. A weight applies a controlled constant pressure to the surface and the pans can be heated with the hot plate.

The SWAT procedure, the steel wool version of the test, is used to test how well a nonstick coating performs under constant abrasion:

- Initial film build is marked at several spots on a test pan.

- The pan is placed on the hot plate of a SWAT machine and heated to 400°F (204°C) (simulates cooking temperature).

- Steel wool is mechanically rubbed back and forth over the interior surface of the cookware, exerting 9 lb (4.1 kg) of pressure. Every 500 cycles (a cycle is once back and forth by the steel wool), the film build is checked again at each spot to determine the coating loss.

Pans are put through cycles until the coating is worn down to the metal, where food release from the pan will be affected. Film thickness readings taken throughout the test accurately determine re-sistance to abrasion. The number of cycles is counted.

An alternate version of the test uses a Scotch-Brite® pad. A weight placed on the scouring pad is 10 lb (4.54 kg), and the scouring pad is changed every 10,000 strokes. The technicians document the number of cycles that are required to scrape the coating down to bare metal in order to gauge the abrasion resistance of the non-stick system.

Some companies test the abraded pans for release in what they call the Scotch-Brite® Egg Release Test. Every 10,000 strokes, an egg is fried on the "wear track" created by the scouring pad. The number of cycles that a coating can endure before the egg sticks to the wear track is noted.

13.12.5 Accelerated In-Home Abuse Test (AIHAT)

As the name implies, the AIHAT procedure is a cooking test meant to measure the harshest of in-home abuse in a greatly accelerated time frame.

The test measures even the toughest coating's resistance to marring, scratching, cutting, and staining according to a procedure similar to the following:

- Pans are heated to 500°F (260°C); an egg is poured into the center of the pan and fried for three minutes.

- The egg is flipped with a metal spatula six times, cut into nine equal pieces with a knife and fork, and the egg is removed.

- An egg mixture is poured into a pan and scrambled with fork tines thirty times in each direction; the mixture is then removed with high-pressure hot water.

- A hamburger is fried and flipped with a metal spatula and moved to the side of the pan, while a fork is used to make a "Z" motion ten times each at two different angles.

- Tomato sauce is added after the burger is removed and stiffed with a metal whisk in a zigzag motion fifty times.

- The above procedure is repeated ten times to complete one AIHAT cycle.

- The pans are cleaned in a dishwasher once during the ten cycles and again at the end of ten cycles.

Figure 13.25 Oscillating abrasion tester. *(Photo courtesy of DuPont.)*

In this procedure, cookware is exposed to the equivalent of five to seven years of harsh cooking conditions. It is then evaluated for scratches, peeling, and staining. The cookware is put through cycles approximately an hour and a half long (with five "cooks" in each cycle) until the cookware has reached a scratch rating of five. The scratch rating runs from absolutely no damage (a rating of one) to a rating of five, which indicates severe damage. Most consumers would reject a five rating as no longer usable.

13.12.6 Blister Test

One mode of failure of home cookware is the formation of blisters in the coating that can be easily abraded off, leaving an apparent hole in the coating. A blister test is often done to compare the tendency of a coating to fail by this mechanism. The procedure is:

1. Combine tomato sauce, salt, and water

2. Simmer two hours

3. Wash and dry pan

4. Cook one box of oatmeal per package instructions

5. Wash and dry pan

6. Boil salt and water for one hour

7. Wash and dry pan

8. Cook two cups of rice per package directions

9. Wash and dry pan

10. Fill pan with water and 2%–3% liquid detergent, bring to a boil

11. Turn off the heat and cover; leave overnight

12. The following morning, dump the mixture, and evaluate the pan for any blistering of the finish

13. Repeat for two more days

At the end of the three days of testing, the pan is examined for blistering and given a rating between five and zero. Five indicates no blistering, while a zero means that 90%–100% of the pan surface shows corrosion.

13.12.7 Salt Corrosion Test

Even though a majority of cookware is made from aluminum, like steel it is subject to corrosion. The Salt Corrosion Test is a method to evaluate this type of failure.

1. The sample is filled with a salt solution to a level more than half-way up the side (10% NaCl in water).

2. The solution in the sample is boiled for 24 hours continuously or in split cycles of 3 or 4 periods of 6 hours.

3. Water should be added if required to maintain the liquid level.

4. After boiling, any adhering salt is washed from the surface and the surface is immediately examined visually for any film defects or corrosion.

13.13 Summary

There are dozens of other accelerated tests. Many users of coated products develop their own screening tests because they believe they can best simulate their own end-use performance requirements. While useful in screening potential coating systems, the tests can lead one in the wrong direction unless the correlation between the accelerated test and real work performance has been proven. Ultimately, the best test is an actual in-use test.

REFERENCES

1. Paul N. Gardner Company, Inc. (catalog)

2. Resnick, R., and Halliday, D., *Physics,* John Wiley & Sons, Inc., New York, NY (1966)

14 Recognizing, Understanding, and Dealing with Coating Defects

14.1 Introduction

Frequently, an applicator follows the coatings manufacturer's instructions but the applied coating does not look right. Part of solving or fixing the problem is identifying it properly. Having identified the problem, work can then begin to change the coating process to eliminate the problem. Occasionally, the user can not correct the problem. The coating manufacturer should be able to help if the problem is correctly described. Sometimes the "fix" is in a product change, a thinner change, process change, and occasionally, a formula change. Formula change is not common unless the coating is a new product, being applied by a new application technique, or is being used for a new end-use.

14.2 Surface Tension and Shear

Paint defects can form during application, shortly after application but before bake, and during the baking/curing cycle. A large majority of defects form as a result of paint flowing in ways that were not intended. Factors that affect paint flow are surface tension, shear rate, and gravity.[1]

The subject of surface tension is discussed in Ch. 7 on solvents. That discussion focuses on surface tension on a macroscopic scale. The fact that surface tension can vary along the surface of a liquid paint film even on a microscopic scale is not part of that discussion. That microscopic scale variation of surface tension is a very important part of identifying the causes of many coating defects and finding cures for them. Surface tension can vary significantly on a microscopic level due to compositional changes caused by contamination, evaporation rate non-uniformity, temperature differences, or gradients. If surface tension is different at one point on a liquid film relative to a nearby point, then the material will flow from low surface tension to high surface tension as depicted in Fig. 14.1

The second factor is the effect of shear force on the paint flow. Gravity can be the force that causes the shear. Figure 14.2 shows that gravity will induce flow in a vertical paint film leading to defects called runs, drips, or sags.

Shear and surface tension are discussed as this chapter deals with common paint defects.

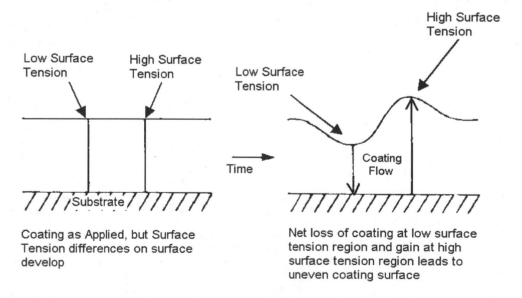

Figure 14.1 Flow of liquid due to a surface tension difference.[2]

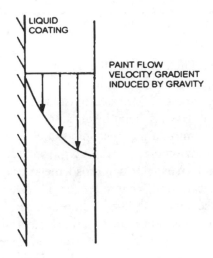

Figure 14.2 Flow of liquid due to gravity.

14.3 Common Coating Defects

Many coating defects have been identified by the paint industry, and the general causes of many are understood. However, for high-temperature baked coatings based on fluoropolymers, the defect often changes after it has formed, making it harder to recognize. Cracks can melt and heal, and solvent pops can change to look like pinholes or craters. Whatever the defect is called, it is important to gain some understanding of how and when the defect formed. Once understood, elimination or minimization of the defect can be attempted.

14.3.1 Air Entrapment

High viscosity and high surface tension coatings exhibit a defect identified as air entrapment, an example of which is shown in Fig. 14.3. This defect occurs during one of three processes: manufacturing, coating preparation, or spraying.

The manufacturing processes frequently include high shear mixing, pumping, and filtration. Air can be incorporated into the coating and if the viscosity is high and surface tension is high, these small air bubbles can remain in the packaged coating indefinitely. Similarly, coating manufacturers usually in-

struct users to mix the product before using it. Mixing instructions may include rolling or stirring, or, rarely, shaking. Care must be taken to avoid air incorporation during these preparative steps. Often, air bubbles introduced can be seen in the coating or can be inferred by measuring the paint density. Measuring density or gallon weight is a simple process (described in ASTM D1475-98, 2003 "Standard Test Method for Density of Liquid Coatings, Inks, and Related Products"). If the density is lower than the specified value by five percent or more, then air is almost certainly entrapped.

Air entrapment can also occur during the application process. It may dissolve in the coating when it is under pressure, such as in a pressure pot. When released to atmospheric pressure, the dissolved air can form bubbles. In spray applications, air bubbles can be introduced during atomization. Droplets can entrap air as they splash onto the substrate. Air can be entrapped in a rough blast profile while paint is being applied. Dip, dip-spin, and dip-and-spin processes have also been known to entrap air.

Solutions to air entrapment problems are many, given that the sources of bubbles are many. First, one should avoid introducing the bubbles in the first place. This applies to the manufacturer and to the user during the preparation process. Mixing instructions before use must be followed precisely. Rolling too fast or mixing too rapidly must be avoided. When

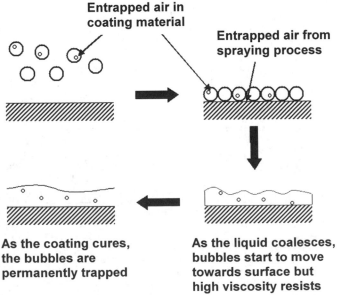

Figure 14.3 Air entrapment.

a mixing blade is used, it must be properly positioned in the coating container.

Removal of air already entrapped in liquid coating material by the preparation process is feasible. One can draw a vacuum on the coating and pull the bubbles out. Sometimes ultrasonic mixers will remove bubbles. For solvent-based coatings that experience high shear mixing without degradation, an enclosed-shaft rotor-stator mixer can be used.[3]

Often the air bubbles will break and leave the applied film. The user can promote this by adding small amount of slow-evaporating solvent. It can also be promoted by gradually warming the coating after application. Gradual warming slows evaporation while surface tension is reduced with a temperature rise. This can be done by the formulator in anticipation of the problem by using a slower evaporating solvent that allows more time for bubbles to break. Lower viscosity will also promote bubble breaking. There are also a very large number of defoamers and bubble-breaking additives that destabilize the bubbles by reducing surface tension. See Ch. 6 for more details on additives.

14.3.2 Decomposition Bubbles or Foam

Not all bubbles are due to air entrapment. When many thick fluoropolymers are baked at high temperature, they start to degrade. If they are perfluorinated, they do not usually turn brown as expected of normal organic polymers. They tend to bubble because degradation products are volatile. An example is shown in Fig. 14.4. The bubbling may not occur uniformly over the entire part. It might occur only in thin areas of a part (substrate) that heat up faster than the thicker areas. It might also occur when the film build is higher. The best solution for this issue is to reduce the baking temperature, or control the film build carefully. Some thermal stabilizers may also work. Fine tin and zinc powders at one percent by weight will raise the bubbling onset temperature to some extent.

14.3.3 Blisters

Blisters are usually formed upon exposure of the coating. They form when liquids diffuse through the coating and reach the substrate or a different layer of the coating such as the primer. They accumulate where there is a substrate or an intercoat adhesion weakness. As the pressure of accumulating liquid increases, adhesion fails and the blister gets larger. When a blister is cut open it is frequently filled with liquid.

Where blistering is a problem, reformulation is usually required. Sometimes, though, blisters form in coating systems because the application process was not correct.

14.3.4 Pinholing, Popping, or Solvent Popping

Popping is a common problem that usually occurs when a slow-boiling solvent is trapped under the coating surface. It is trapped when the coating above it starts to coalesce or skin over. This can occur when the paint layer above it starts to cure or melt. The paint might also skin over due to the drying effects of too much air movement over the surface of the coating. The skinned-over surface traps the solvent below the surface. When the solvent goes from liquid to vapor phase, its volume increases dramatically. It forms a vapor pocket that erupts through the surface, leaving a hole in the coating. This is shown in the drawing in Fig. 14.5.

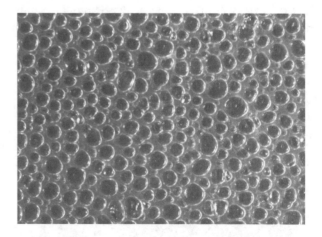

Figure 14.4 Bubbles: thermal decomposition of PFA.

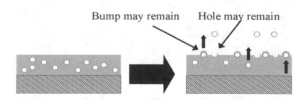

Figure 14.5 Solvent popping or pinholing.

Solvent pops in a fluoropolymer coating are sometimes harder to recognize. That is due to the coating melting and flowing after the pop forms, partially healing the defect. Figure 14.6 shows a fluoropolymer coating that is loaded with pops. These are the dark spots in the photograph. Sometimes this occurs when a coating is ramped up to temperature too quickly. If that is the cause, slowing the ramp or staging the bake at a lower temperature for a time before ramping up to the maximum bake temperature can help.

14.3.5 Mud Cracking, Stress Cracking, and Benard Cells

Cracks are often seen in coatings and many are caused by stresses that form in the coating during drying, curing, with aging, or in-use.

Mud cracking occurs during the drying step. As the solvent leaves the film, shrinkage occurs. Preferably, the shrinkage takes place only in the direction normal to the surface (perpendicular), but often that is not the case. When shrinkage occurs parallel to the surface, mud cracks can form as shown in Fig. 14.7.

Mud cracks in commercial coatings usually form when the coating is applied too thickly or if the drying procedure is not correct. If it occurs during use, then the film thickness must be reduced or a different drying procedure tried. The formulator can affect the tendency to mud-crack. Larger particles, such as fluoropolymer powder particles, start to mud-crack at higher thickness. Often, this phenomenon can not be controlled efficiently, therefore requiring addition of film-forming additives.

The additives can be sacrificial polymers such as acrylics or high boiling solvents such as glycerol or plasticizers.

Cracking is sometimes called checking, crazing, splitting, or alligatoring. The cracks are caused by stresses induced in the coating parallel to the surface. The stresses can be due to chemical crosslinking or shrinkage upon cooling. One such example is shown in Fig.14.8.

Slightly different cracks are sometime formed by another process. They are formed by surface tension gradients that occur during the earliest part of baking. As the coating warms, surface tension and viscosity drop as shown on the right side of Fig.14.9. As the solvent evaporates, it cools the surface, and the surface tension rises as the coating flows away from the center of the cell, called a Benard Cell. The solids also rise on a microscopic scale due to solvent evaporation. This flow generates a "bee-hive" looking structure, as shown, that can crack on the boundaries.

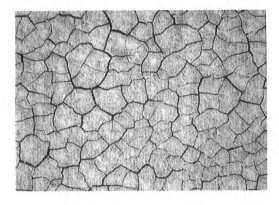

Figure 14.7 Mud cracking in fluoropolymer coating.

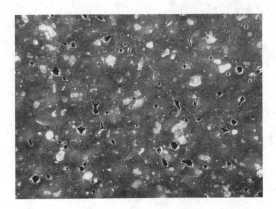

Figure 14.6 Micrograph of solvent pops in fluoropolymer coating.

Figure 14.8 Micrograph of stress cracks in a fluoropolymer coating.

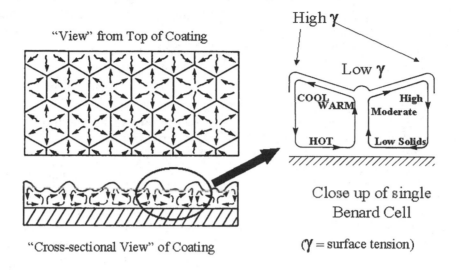

"View" from Top of Coating

"Cross-sectional View" of Coating

High γ

Low γ

COOL WARM — High Moderate

HOT — Low Solids

Close up of single Benard Cell

(γ = surface tension)

Figure 14.9 Formation of Benard Cells.[2]

14.3.6 Cratering

A *crater* is a distinctive, small, bowl-like depression in a paint film. Craters are caused by a particulate contaminant that can usually be seen in the center of the defect with a microscope. As shown in Fig.14.10, a contaminant particle causes a region of low surface tension. The surface tension differential drives coating material flow as shown, forming a circular ridge raised above the surface of the coating around the crater. The center is depressed. The particulate could be nearly anything and could come from nearly anywhere. The particulate could be contamination in the liquid coating, a polymer gel particle, a pigment agglomerate, dirt, oil, etc. The reason particulates depress surface tension is that they usually absorb low surface tension materials such as solvent or oils on their surfaces. Figure 14.11 is a micrograph of a crater that shows the formation well.

Removing the contaminants can eliminate craters. This can be done by filtration to remove particles from the paint, or by cleaning up the air and area where the spraying is done. When using spray guns, the cleanliness of the compressed air must be assured. Anti-cratering additives are also available. Most are surface tension reducers. If the surface tension of the bulk paint is equal to the surface tension of the contaminant particle, then there is no driving force to form the crater. Crater formation has also been mathematically modeled.[4]

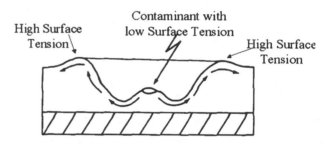

High Surface Tension

Contaminant with low Surface Tension

High Surface Tension

Figure 14.10 Crater formation.[2]

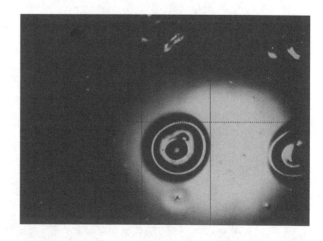

Figure 14.11 Micrograph of a "classic" crater formation.

14.3.7 Fisheyes

Fisheyes are called that because of what they look like. They are also frequently called craters. They are depressions in the coating surfaces that look like little craters of the moon. They are most often caused by contaminants on the surface. The contaminants prevent the coating from wetting that area and flowing out. Fisheyes differ slightly from craters in that a particle is not usually the cause of the surface tension reduction in the coating. Grease or oil contamination, particularly silicone oil, of the substrate is a common cause. The source can be machining oils, fingerprints, or oil in compressed air. Figures 14.12 and 14.13 show a diagram and photograph of fisheyes.

Solving the problem generally involves carefully reviewing the handling and substrate preparation, making sure the compressed air used in the grit-blast-ing equipment and spray equipment is clean. One should check the condition of the oil or water trap on the compressed air source. Additives that reduce surface tension are also used. Occasionally, increasing the pigment level of a coating (either the volume-based Pigment Volume Concentration or the weight-based Pigment/Binder) will eliminate crater formation. Also, spraying the coating with excessive atomization air can minimize fisheye problems. This approach is called "dry spraying." Since the fisheye forms due to surface tension differences, dry spraying causes most of the volatile solvents to evaporate before the atomized paint droplet strike the substrate. This increases the viscosity of the applied film and eliminates surface tension differences from developing upon evaporation.

14.3.8 Crawling and Dewetting

Crawling and dewetting are related to fisheye formation. They are basically the same problems except that very large areas of substrate are involved.

14.3.9 Wrinkling

Wrinkling is not a common defect of fluorinated finishes, but an example is shown in Fig. 14.14. It is almost always caused by application of too much paint at one time. As the surface dries, it can reabsorb solvent from the underlying wet layer, leading to uneven swelling that causes wrinkles to form.

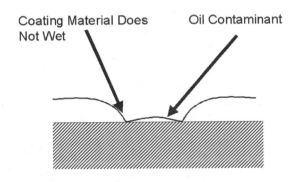

Figure 14.12 Diagram of fisheye formation.

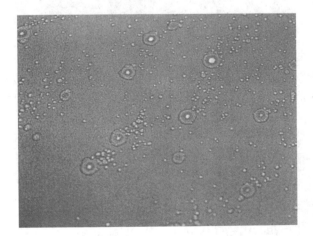

Figure 14.13 Photograph of fisheyes on coating surface.

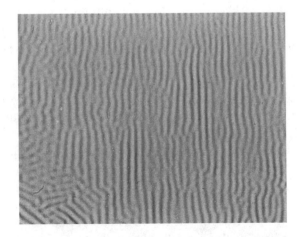

Figure 14.14 Photograph of wrinkled coating surface.

14.4 Summary

The important defects that are commonly encountered in fluorinated coatings have been described in this chapter. Many types of defects have not been discussed in this section but they are compiled elsewhere.[5] The causes of the defects are not always obvious, but it is crucial to identify the causes if they are to be corrected.

REFERENCES

1. Pierce, P. E., and Schoff, C. K., Coating Film Defects, Federation of Societies for Coatings Technology: Philadelphia (1988)

2. Patton, T. C., *Paint Flow and Pigment Dispersion*: *A Rheological Approach to Coating and Ink Technology*, John Wiley & Sons, New York (Apr 1979)

3. LeClair, Mark L., Deaerating and Defoaming Batches Using Enclosed-Shaft Rotor-Stator Mills, *Paint and Coatings Industry* (Nov 1997)

4. Evans, P. L., Schwartz, L. W. and Roy, R. V., A Mathematical Model for Crater Defect Formation in a Drying Paint Layer, *Journal of Colloid and Interface Science,* 227:191–205 (2000)

5. Fitzsimons, B., Weatherhead, R., and Morgan, P., *Fitz's Atlas of Coating Defects,* MPI Group, Hampshire, UK

15 Commercial Applications and Uses

15.1 Introduction

This chapter will first provide an historical perspective to fluoropolymer finishes technology and market development. Some guidance on food contact end-use requirements follow. The rest of this chapter shows a number of sample end-uses. Each end-use description summarizes the type of coatings used and the required properties of the fluorinated finishes.

15.2 A Historical Chronology of Fluoropolymer Finishes Technology

The following discussion contains a chronological perspective of the history of fluoropolymer finishes. Cookware, being the largest single market, dominates the discussion and perhaps should be separated from the industrial applications, but that distinction has not been made here. The discussions of both markets are interspersed in approximate chronological order. The discussion is heavily focused on DuPont for two reasons: the author's experience and DuPont's major role in developing the coating applications and markets.

DuPont's Teflon® finishes business first began in 1948 when 126 gallons of coating was sold for $4,441. Teflon® finishes was just another product line of the Industrial Finishes Department. The first products were 850-200 Single Package Primer and 851-200 Low Build Topcoat. These finishes found many applications where non-stick or dry lubricity was needed. Analogs of these coatings are still manufactured and used today. These first products were manufactured at the Philadelphia paint plant, which stood on the grounds adjacent to the DuPont's Marshall Laboratory in South Philadelphia, Pennsylvania. Buyers of these first coatings were companies such as General Plastics and American Durafilm, which were already users of Teflon® resins. Sales grew to about $450,000 to $550,000 in four to five years and then flattened at that level for the next eight years. While the specific uses for the products during these years are not known, it is generally believed that most of them were used for a variety of industrial applications and some found their way onto cookware.

Dupont chemist Verne Osdal at the Marshall Laboratory made the key technical accomplishment that started the Teflon® Finishes business. His discovery, a method of getting the non-stick Teflon® to adhere to metal, came in 1951.[1] He discovered coating compositions of PTFE and mixed acids that imparted good adhesion to substrates. The first public disclosure of the use of Teflon® coated pans came in 1953.

However, in Paris, France, in the mid 1950s, Marc Grégoire saw the possibilities of using a fluoropolymer coating on cookware and began to make coated pans in a very small operation in Paris. His wife, Colette, was the first sales person and she set up a sales operation on the streets of Paris. Her efforts produced an immediate success and soon the coated cookware found its way into Paris retail shops. Grégoire was granted a patent in 1954 and formed a company to produce en masse in 1956. Grégoire named his company Tefal®. Within several years, pan sales were numbered in the millions.

In 1958, Thomas Hardie received a Tefal® skillet from a friend he had known in Paris during his days as a foreign correspondent for the United Press and International News Service. Hardie became excited about the prospects for the Tefal® pan in the United States and he spent much of the next two years trying to convince American retailers and manufacturers of its bright future but without success. American retailers and manufacturers were uninterested in marketing this type of cookware unless sanctioned by a government agency.

In 1960, after numerous discussions with DuPont and others, US Food and Drug Administration (FDA) disclaimed jurisdiction over home cookware. This meant that no FDA approval was needed to sell Teflon® coated cookware in the United States. Within a few months after that announcement, the FDA stated that "FDA scientists believe pans coated with Teflon® are safe for conventional kitchen use." Macy's Department stores' entrance into selling non-stick pans sparked public interest and brought Hardie's quest to fruition.

The Tefal® process for getting the Teflon® coating to adhere to metal was mechanical rather than chemical in nature. The metal was chemically etched to provide a mechanical bond of the non-stick coating to the metal. This was a patented process, so other interested cookware manufacturers had to look for other technology. This logically led them to DuPont because they had an alternative technology in the Osdal primer patent. American cookware manufacturers rushed to get into the marketplace with their non-stick cookware. In 1961, DuPont's finish sales grew to $1.2 million. However, this surge of new business was short-lived. The inexperience of American cookware manufacturers with the coating process and their lack of quality control led to a flood of poor quality cookware being sold to consumers. This in turn quickly led to consumer complaints, retailer disenchantment, and a drop in DuPont sales of 19% in 1962 rather than the rapid growth that had been expected.

DuPont put together a team including manufacturers and advertisers that conducted extensive research with six thousand consumers that revealed no problem with the concept of non-stick cookware, just a problem with the poor quality of non-stick cookware that was then available. Coatings tended to peel because of poor substrate adhesion, caused by inadequate roughening of the metal. Cookware made of thin metal to reduce costs contributed to the problems due to hot spots and thermal degradation. In some cases, too much Teflon® was applied and this led to cracking of the coated surface.

The DuPont team concluded that the business could be revived and had a bright future but fundamental changes in strategy had to be made and DuPont's role in the marketplace changed dramatically. Instead of merely selling coatings and supplying direct customers with technical support, DuPont recognized an obligation to ensure that their products were used correctly and met the expectations of retailers and consumers. The acceptance of this obligation opened up the opportunity for quality control.

These are some of the new strategies for cookware revival:

- Employ a license agreement with all housewares manufacturers who wished to use DuPont technology and products.

- Set and monitor metal thickness and coating standards for all licensed manufacturers.

- Provide licensees with DuPont certification marks to identify their non-stick cookware after their samples have been submitted, tested, and approved at the Marshall Laboratory in Philadelphia, PA.

- Use the convenience of non-stick cooking with easy clean-up as the unique selling proposition in addition to fat free cooking.

- Test the use of television advertising to build the non-stick cookware category and the awareness of the Teflon® brand.

Around 1960, clear coatings were introduced in order to broaden the thrust into markets that required purer Teflon® topcoats; that also opened up certain electrical applications that could not tolerate pigments in the finish.

DuPont introduced two new coatings in 1961. A topcoat based on FEP was developed for application into non-stick/mold-release areas such as tire molds. This was the first commercial coating based on a melt-processible fluoropolymer. This coating is still sold today under the same product code number.

While single package, premixed primers had been successfully used for ten years, a need for more versatility in the primers led to development of the two-package primers. The two packages included an enamel package containing the fluoropolymer and additives and an accelerator package that included chromic acid. Some end-uses required different accelerator-to-enamel ratios to obtain optimum adhesion, so the two-package approach allowed for that flexibility. The two-package system also extended storage life and eliminated the need for refrigerated storage that was required for the premixed primers.

In 1965, cookware manufacturers began to spray a material harder than aluminum between the aluminum and the Teflon® coating. These materials—aluminum oxide, stainless steel, or ceramic frit—were applied in a discontinuous coating that created a peak and valley profile onto which the Teflon® coating was applied. The idea was that cooking utensils like spatulas would ride on the peaks of the profile and thus would not scratch the coating. While hard bases were not on the vast majority of coated cookware, they were used to some degree by many manufacturers. Many of the performance claims went well beyond what the product could deliver.

At this time, DuPont introduced darker color coatings to mask the staining caused by grease permeation into the relatively porous coating and then carbonizing during continuous cooking, particularly at high cooking temperatures.

The consumer's desire for more durable non-stick coatings drove the technical efforts in the late 1960s. While it was hoped that more durable coatings would be developed for use on cookware, it was also believed that improved durability would open up the possibility of new consumer and industrial applications, thus more opportunities for growth of fluoropolymer coatings.

In the mid 1960s, DuPont's research team discovered a new concept in non-stick coatings. It was found that hard, adhesion-promoting binder resins could be incorporated with FEP to provide one-coat products that had greater abrasive wear-resistance than prior Teflon® coatings while maintaining an acceptable level of dry lubricity and non-stick. These products stratified when baked, leaving the surface mostly comprised of fluoropolymer, and the interface with the metal mostly the tougher resin. DuPont trademarked this new generation of coatings Teflon-S®. Additionally, many one-coat finishes could be cured at lower temperatures. The first such coating was commercialized in 1966. It had a low bake requirement (232°C/450°F). The Teflon-S® coating had very good release and durability. It was a blend of an epoxy resin and a proprietary low melting FEP resin. Several Teflon-S® products were introduced in 1967 and hand-tool manufacturers showed a good deal of interest in their use. DuPont developed a certification program for tools.

In 1967, a new Teflon-S® was developed. This coating was based on a blend of polyamide-imide resin and FEP. It was the hardest and most durable non-stick coating developed until that time. Further, the adhesion of this new coating over smooth metals was excellent. These characteristics coupled with good dry lubricity and fair non-stick properties made this second Teflon-S® coating very marketable. End-uses such as bearings, lawn and garden tools, bakeware, and many others covered the markets for this technology.

The fluoropolymer coatings industry prior to 1969 was almost entirely about Tefal® and DuPont's Teflon®. ICI and Hoechst had a few products using fluoropolymers they manufactured. A new coatings company, one that did not make any fluoropolymer resins, was founded in 1969: Whitford Corporation.

Whitford's first product, Xylan® 1010, was developed in March of 1969. Xylan® 1010 is a *matrix coating,* as Whitford calls it. It was designed specifically to solve two problems:

1. Provide a tough, very low-friction film that could withstand the constant wiping of a rubber seal.

2. Capable of being cured at temperatures sufficiently low to avoid blistering or distortion of the casting of the actuator housing.

Xylan® 1010 remains one of Whitford's most popular products today. Similarly, Weilburger was an old German specialty paint manufacturer who also saw the potential for developing business using the combination of fluoropolymers and other resins. They started manufacturing fluoropolymer-based coatings around 1975. Weilburger Coatings has manufactured non-stick coatings in Germany for over thirty years. These GREBE GROUP companies manufacture and supply a wide comprehensive range of GREBLON® non-stick and Senotherm® high temperature decorative coatings to many major cookware, bakeware, and household electrical appliance manufacturers. The products offer flexibility in design, color coordination, performance, and competitive cost.

In 1973, a great improvement took place in DuPont's fry pan coating technology. A new air-dry primer, and an improved Teflon® enamel topcoat, proved to be significantly superior to Teflon II® in terms of adhesion and stain resistance. The improvement in stain resistance utilized new chemistry that reduced the porosity of the topcoat by 90%.[2] So great was the improvement that for the first time DuPont offered a white Teflon® to the cookware market. A photo of a Teflon II Classic white pan is shown in Fig. 15.1.

In 1973, DuPont also introduced a new resin-bonded coating that could be cured at 350°F (177°C). This low-bake analog, which was a catalyzed epoxy FEP blend, was a relatively inexpensive dry lube coating. These characteristics made it especially attractive for the hardware market.

Figure 15.1 DuPont Teflon® II classic white fry pan.

In 1974, DuPont's first powder coat Teflon® finish was introduced; it was an FEP powder coating. Not only was it pollution free, but it enabled the applicators to get thicker FEP coatings than with liquid FEP coatings.

In the mid 1970s, IBM developed a new generation of high speed printers to go with their mainframe computers and DuPont's technology brought about their use of Teflon-S® technology (low-melt FEP plus epoxy) as a coating for the toner beads. One usual property of fluoropolymers was advantageous in this application. That was the ability to tribocharge (develop a static charge by rubbing against another material, see Ch. 11).

In 1976, SilverStone® three-coat system for top-of-the-range cookware was introduced. A key technical achievement was made by the addition of an acrylic resin to the formula which burned off during the curing bake, but created a film that was less porous and thus less susceptible to staining from carbonized fat during the cooking process.[3] Additionally, it prevented film cracking, which enabled the thickness of the coating system to be increased by 30% to 1.3 mils and higher and thus provided greater durability. All the pigment was included in the midcoat and the topcoat was clear, thus making it very rich in fluoropolymer and enhancing its non-stick characteristic. It was intended for use on premium quality cookware only, and a new set of quality tests built around it had to be met in order to gain the SilverStone® Seal.

In 1976, the system called Spectragraphics® was developed. It employed a technique that enabled putting images in a Teflon® coating (Fig. 15.2). A new series of high performance Teflon® resin-bonded, dry lubrication coatings coded 958-300 was developed. They were specifically designed to meet the needs of a market where high load-bearing conditions are present. Automotive use of this coating was common.

In 1979, Teflon-P® PFA Powder Coating 532-5010 was introduced. This coating has the high temperature properties of PTFE and the additional advantage of being thermoplastic.

During mid to late 1970s, non-stick bakeware coated in flat coils and then stamped into shape began to emerge. This coating process required different finish formulations and the market was supplied by both Whitford and Weilburger.

In 1983, SilverStone® coatings were introduced on glass ovenware, heavy gauge gourmet cookware, and a line with non-stick outside as well as inside. Another new use of SilverStone® that required new technology was launched in 1984 with its use on plastic ovenware for microwave, convection, and conventional oven use. Regal and Northland launched lines that were moderately successful in mid 1980s.

SilverStone SUPRA® was introduced in 1986. This coating technology was based on blends of PTFE and PFA.[4] Laboratory and in-home testing showed this new system to be 50% tougher than SilverStone®, which was based on PTFE only.

Figure 15.2 DuPont Spectrographic® fry pan.

In 1983, fluorinated coatings companies focused more upon specific industrial market segments. DuPont's key market segments that accounted for most of their growth were office automation for copier rollers, toner beads, and computer printers; automobile and truck fasteners; food processing; and shoe molds. Significant technical and market development work was carried out to penetrate other targeted segments including commercial bakeware and chemical processing equipment. Keys to the companies' varying degrees of success were their ability to formulate both liquid and powder systems that met specific requirements of the end-use, and it is expected that this ability will continue to be key to future growth.

The dirty and time-consuming process of grit blasting the metal surface to ensure coating adhesion was always needed for the best performing housewares systems until DuPont developed grit-blast-free primer technology.[5] Mastering control of stratification was the key to this technology. Stratification was discussed in Sec.4.2.

The utilization of PFA in varying amounts in the primer, midcoat, and topcoat also made possible durability improvements beyond SilverStone® and SilverStone SUPRA®. Simultaneously with the introduction of the grit-blast-free systems, DuPont also introduced under the AUTOGRAPH® trademark a ceramic filled system. In essence, this system had greater durability built into the coating as a result of tough ceramic fillers being included in the three coats at varying ratios.[6] A patent on this technology was granted in 1992.

The developments in fluorinated coatings during the late 1990s and early 2000s were formulations for new applications, for new application methods, and for powder coatings.

There were a lot of formulations developed for high efficiency application. For cookware, the application techniques were shifting towards roller coating and curtain coating. Powder coatings for cookware have yet to become a commercial success, though powder coatings in industrial applications have expanded.

15.3 Food Contact

Questions about coatings for food contact are common. The first step towards general understand-ing is determining what the regulations are in the country where the products will be used; that means the end-user, not where they are applied or produced. This information is best obtained from experts in the field, and this section only provides an overview that applies to the United States. In all cases, one should obtain written documentation from the manufacturers of products and materials being used that state compliance under the appropriate regulations.

There are several regulations that govern the use of fluoropolymers as articles or components of articles intended for use in contact with food. These are published in the Code of Federal Regulations commonly called the CFR. The CFR covers just about everything but Section 21 covers food and drugs.

For Articles Intended to Contact Food, one of the important sections is 21 CFR 177.1550, "Perfluorocarbon Resins." Many fluoropolymer resins may be used as articles or components of articles intended to contact food in compliance with this regulation. One should get documentation from the fluoropolymer material manufacturer certifying compliance before use, however.

Some fluoropolymer resins are irradiated to facilitate grinding into fine powders for applications needing a very small particle size as is typical for many coatings applications. Paragraph (c) of this regulation specifies the allowable dose of radiation and maximum particle size for PTFE resins so processed and restricts their use to components of articles intended for repeated use in contact with food.

For coatings, the most important regulation is 21 CFR 175.300, "Resinous and Polymeric Coatings." Most fluoropolymer resins and fluoroadditives may be used as release agents in compliance with this regulation as long as the finished coating meets the extractives' limitations of the regulation.

Product formulators must take care in ingredient selection, not only for the fluoropolymers used, but for all the ingredients including pigments, additives, and other polymers. Written documentation from each manufacturer needs to be kept on file.

Other regulatory agencies may have additional regulations. The United States Department of Agriculture (USDA) has accepted fluoropolymer resins that comply with 21 CFR 177.1550 as components of materials in direct contact with meat or poultry food products prepared under federal inspection. The

Dairy and Food Industries Supply Association, Inc., has its "3-A Sanitary Standards for Multiple-Use Plastic Materials Used as Product Contact Surfaces for Dairy Equipment, Number 20-17," published by the 3-A Secretary, Dairy and Food Industries Supply Association, Inc. US Pharmacopoeia Class V1 (USP) has additional regulations.

Representative samples of fluoropolymers have been tested in accordance with USP protocol, and many meet the requirements of a USP Class VI plastic. These tests on representative samples may not reflect results on articles made from these fluoropolymers, especially if other substances are added during fabrication. Testing of the finished article is the responsibility of the manufacturer or seller of the finished product if certification that it meets USP standards is required. USP testing was done to support use of these fluoropolymers in pharmaceutical processing and food processing applications. While USP Class VI certification is not required for pharmaceutical processing, many pharmaceutical customers seeking ISO-9000 certification have requested it.

15.4 Commercial Applications of Fluorocoatings

The remainder of this chapter reviews just some of the thousands of uses of fluorinated coatings. For each application the discussion focuses on the properties of the coating and how it improves those products, or the problems the coating solved are summarized.

15.4.1 Housewares: Cookware, Bakeware, Small Electrical Appliances (SEA)

Until recently, most coatings were applied by spraying. A pan can be coated on smooth metal, over grit-blasted metal, or over a flame-sprayed, hard, rough surface. Cast or rolled aluminum is the most common metal, though stainless steel is also common. There have even been coated glass and ceramic cookware. The first coat is a primer that is almost always based on polyamide-imide and fluoropolymer. That is typically dried, then coated with a midcoat consisting of mostly fluoropolymer or a blend of fluoropolymers. The blend is usually PTFE

and PFA. The midcoat generally has plenty of pigment to hide the primer/substrate and is usually dark colored to hide the staining of the coating during use. A third coat, a topcoat, is finally applied. It is usually applied right over the wet midcoat. This coating consists of PTFE or a blend, but often has mica in it to provide the common sparkle look. The coating is typically baked, depending on the system, between 700°F–820°F (371°C–438°C).

Recently, more cookware is roller coated or curtain coated. This needs to be done on flat disks that are formed into a pan after completing the coating and curing process. The coating needs to be post-formable as the pans are pressed into shape from a coated disk.

Small electric appliances (SEA) include kitchen devices such as breadmakers, waffle irons, and rice cookers.

15.4.2 Commercial or Industrial Bakeware

Commercial bread and bun bakers have used release coatings for a very long time. Those coatings historically were not fluoropolymer but were polysiloxane, commonly called silicones or silicone glazes. A bread or bun pan would be coated with this clear or yellow material and it would function with the help of sprayed-on oil for 300–600 baking cycles. At this point, the release would become so poor that the pans were removed, coating stripped, and new coating was applied. Generally, this could be done quickly and relatively cheaply, though there were environmental concerns about emissions with this process. A typical bakery needs about 2000 pans to fill a line and might have many different sets. A different pan is required for different bun sizes, shapes, and configurations. Typical pans are shown in Fig. 15.3. A bun pan is shown on the left and a "strap" of bread pans is shown on the right.

Fluoropolymer coatings are more expensive than the silicone glazes, so the coatings needed to last much longer. The goal would be the life of the pan, but 5,000 cycles usually provides economic incentive to the bakers. The pans are usually made from aluminized steel. Aluminized steel is used to minimize corrosion of the bottom of the pans, which see 95% humidity at 105°F (41°C) for up to an hour during the rising of the dough. The coating system is

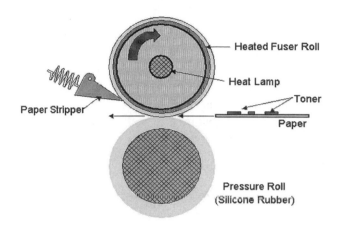

Figure 15.3 Commercial or industrial bakeware coated with fluoropolymer coating: (*left*) bun pan; (*right*) "strap" of bread pans.

Figure 15.4 Schematic of a typical fuser unit of a laser beam printer or photocopier.

a primer with one or more topcoats. One of the preferred systems is a powdered PFA topcoat. FEP liquid coatings and powder coatings are also used. Extra benefits for the baker include reduced oil consumption which is a cost saving, but also leads to a cleaner and safer bakery.

15.4.3   Fuser Rolls

A crucial part of copiers and laser beam printers in a fusing mechanism. This part melts dry powdered toner that has been applied to the paper, melting the toner onto the paper. This process makes the image permanent. The fusing temperature for black and white machines is typically 400°F (200°C). Heat is transferred by a hot metal roller or belt. The fuser must release the molten toner. Fluoropolymer coatings are crucial to this step and have been used since 1966 or earlier. Figure 15.4 shows a schematic of the fusing unit. A photoimaging drum puts the dry powdered toner on the paper in the proper areas.

The paper is fed into the fuser unit. The heated fuser roll and the soft silicone rubber backup roll press together forming an area where the paper is pressed between the two rollers. The toner melts and sticks to the paper. If it stuck to the drum, image quality would be poor. The next time the drum makes a rotation a shadow of the wrong image might print on the paper. Proper release by the fuser drum coating is very important.

As mentioned in Ch. 11, almost everything that rubs against a fluoropolymer picks up a triboelectric charge. To maintain charge balance, the coated fuser drum picks up an opposite charge. Often, the toner has a residual charge, and if the fuser drum has a charge, the toner may jump off the paper onto the drum. When this happens, the sharpness of the image is lost. This is called electrostatic offset. For this reason many fuser drum coatings are made conductive so that any charge that forms is lost to ground.

The coating systems for these rollers, shown in Fig. 15.5, are complex and many are optimized for a particular machine. Most are applied as liquids. Many rollers of laser beam printers consist of sleeves of

Figure 15.5 Typical fluoropolymer coated aluminum photocopier fuser rollers.

PFA that are heat-shrunk onto the metal roller. Those liquid systems usually are two- or three-coat systems. The primers are generally PAI based, but the topcoats can range from PTFE to blends of PTFE and PFA to pure PFA. FEP and other fluoropolymers are rarely used on fuser rolls.

Paper dust is quite abrasive, so the coatings also need abrasion resistance, which is often improved with additives such as silicon carbide or aluminum oxide. A delicate balance is struck between the release and abrasion resistance properties. There is also a thermistor in contact with the rollers that controls temperature. This is the cause of many roller failures because it can cut through the coating system. Picker fingers make sure the paper comes off the fuser roll. These can also abrade the coating, particularly if they pick up toner and paper dust. They are usually made from engineering plastics such as PES or Vespel®. They are frequently coated with a resin-bonded fluoropolymer coating.

15.4.4 Light Bulbs

Coated light bulbs are available in many hardware stores. Bare bulbs with no containing fixtures are often used in factories and in homes. Containment of the glass in a broken bulb is the aim of this fluoropolymer coating end-use. The bulbs can get very hot so thermal resistance and good mechanical properties at high temperature are required.

Fluoropolymer coatings are ideal for glass containment, shown in Fig. 15.6. The coating is usually PFA that is applied as a powder coating. An interesting note is that the PFA does not adhere well to the glass bulb. If a primer is used to improve the adhesion, when the glass breaks the PFA no longer contains the broken glass. Therefore strong adhesion of the coating to the glass is avoided.

15.4.5 Automotive

There are dozens if not hundreds of applications of fluorinated coatings in the automotive industry. They are used for many purposes. Some are important for production processes, some are intended for car performance. The properties of these finishes include combinations of the following:

1. Dry lubrication with a low coefficient of friction
2. Wear resistance
3. Chemical protection—fuels
4. Corrosion protection
5. Non-stick
6. Electrical insulation

A few applications are discussed.

Automotive air conditioner pistons (Fig. 15.7) are often coated with a resin-bonded fluoropolymer coating. The coating must adhere to the aluminum piston. A low baking temperature is required so that the part retains its hardness and precisely machined dimensions. The coating supplies dry lubrication and abrasion resistance. This is particularly important during the first few hours of use. A small percentage of uncoated pistons fail due to galling, which means metal sticking to metal. The warranty repair is very expensive, and the coated piston reduces the failure percentage significantly. The coating also provides some increase in efficiency after break-in.

Figure 15.6 PFA-coated light bulbs.

Figure 15.7 Automotive air conditioner piston.

Coated fuel pumps (Fig. 15.8) are common. The coating provides a low coefficient of friction, wear resistance, and chemical resistance to the fuel. This leads to pumps that last significantly longer, with the ultimate goal of lasting the life of the car. Today's fuels often contain alcohols and other additives that are more corrosive than hydrocarbons. A resin-bonded coating based on PPS and PTFE micropowder is usually used, though the coating systems for fry-pans have also been used in this application.

Seat belt D-rings (Fig. 15.9) are usually visible in a car over the left shoulder of the driver or right shoulder of the passenger. These rings are often coated with a resin-bonded coating of either epoxy/PTFE or PAI/PTFE. Liquid and powder coatings have also been used. The coating lets the seat belt glide more smoothly over the ring, primarily decreasing the abrasion of the belt, leading to longer life of seat belts.

Lots of fasteners are used to assemble an automobile. Many are coated with a fluoropolymer coating. There are two basic reasons to coat the fasteners. One type of fastener is coated to offer improved assembly and corrosion resistance. The improved assembly comes from robotic assembly. Screws are driven by torque sensing wrenches or screwdrivers. A fluoropolymer dry lubrication coating allows the fastener to slip more consistently into or through the materials being fastened. The fasteners are turned to a given torque. Without the dry lube coating, the fastener could snag or gall and might not be properly attached. The fluoropolymer in these resin-bonded coatings allows them to penetrate the materials better, but is designed to still hold the applied torque, keeping the materials attached. The coatings used for this type of fastener are typically resin-bonded PTFE micropowder. They are frequently applied by dip spin.

The second type of fastener is called a weld-nut or stud (Fig. 15.10). They are welded to the chassis. The weld-nut accepts a bolt and the stud accepts a nut. Both are welded in place. Often, weld splatter hits the threads of the fasteners and sticks there. When this happens the fasteners become unusable. Before the use of a fluoropolymer coating that resisted the weld splatter, these fasteners had to be masked. The coating on the fastener does more than prevent weld splatter from sticking, however.

The chassis is primed in its entirety in the first step of vehicle assembly. The chassis in submerged in a primer paint bath called an electrocoat or electroprime bath. Electrical charge is applied to the chassis and the primer is attracted to it and deposits

Figure 15.8 Automotive fuel pump.

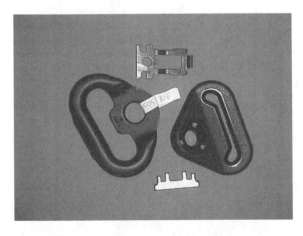

Figure 15.9 Automotive seat belt D-rings.

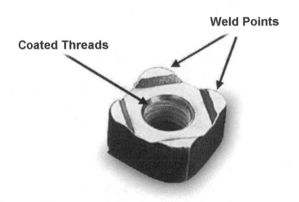

Figure 15.10 Automotive weld-nut with fluoropolymer coating on threads.

uniformly. The problem with this process is that the threads of the nut and studs on the chassis that are used later in the assembly are coated with primer. This makes assembly involving these fasteners very difficult. The fluoropolymer coating that prevents the weld splatter problem is not conductive. Therefore, the electrocoated primer is not attracted to it and does not coat it, leaving the threads clear. Even if a drop or two of primer happens to remain in the threads, it is easily pushed out because of the non-stick character of the coating.

15.4.6 Chemical Processing Industry (CPI)

The chemical processing industry (CPI) has many interesting applications for fluoropolymer coatings. These include such items, many of which are very large, as:

- Ducts for corrosive fumes, fire resistance
- Chemical reactors
- Impellers
- Tanks
- Pipes
- Fasteners
- Ducts

15.4.6.1 Chemical Reactors

Impreglon Canada, a DuPont licensed industrial applicator in Edmonton, Alberta, Canada, did the largest items ever coated with a fluoropolymer coating. These vessels, shown in Fig. 15.11, were large polymer reactors being installed at the Nova Chemicals complex in Joffre, Alberta. The largest of three vessels, which were eventually connected end-to-end, is shown in this figure. The vessel was fabricated by Dacro Industries in Edmonton. It is 23 feet (7 meters) in diameter and 50 feet (15 meters) long and weighs in at 126 tons. The DuPont coating system was applied in three coats at a total film build of only 2 mils. The properties that the coating system addressed were thermal resistance, non-stick of molten polymer, and chemical resistance. The coating is expected to last the life of the reactor.

Chemical reactions are usually run in a reactor tank such as that shown in Fig. 15.12. The reactors often have pipes (Fig. 15.13) and mixing blades. They are frequently run at high temperatures. These vessels, mixers, and pipes are frequently coated with thick fluoropolymer films. The most chemically resistant and highest temperature rated material is PFA. These films are very thick, typically 40 mils (1 mm) or more. They can be applied from liquid coatings or powder coatings. The hot flocking process described in Ch. 11 is often used. Because the topcoats need to be applied in multiple coats with multiple bakes, it

Figure 15.11 Polymer reactor coated by Impreglon Canada. *(Photos courtesy of Impreglon Canada.)*

can take days or even weeks to coat a reactor. Extra care must be taken to avoid any defects, especially bubbles from resin decomposition. While sometimes these defects can be repaired, often the entire coating must be stripped and the process started over.

15.4.6.2 Ducts for Corrosive Fumes, Fire Resistance

The semiconductor industry uses all sorts of aggressive chemicals in the production of chips.

These are produced in a manufacturing site that is called a "fab." The ductworks in the fab carry corrosive and flammable materials. A resin-impregnated fiberglass material called FRP has historically been used for the ductwork. However, this ductwork is not sufficiently fire resistant. Fluoropolymer-coated metal has replaced FRP in many of these applications (Fig. 15.14). The coatings are ETFE or E-CTFE. Both are more chemically resistant and certainly more fire resistant than FRP.

Figure 15.12 Polymer reactor tank coated with a thick film of PFA.

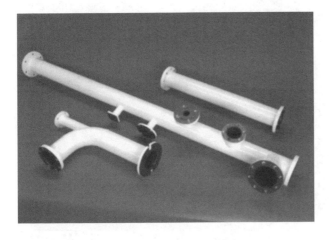

Figure 15.13 Chemical processing pipe coated with a thick film of PFA.

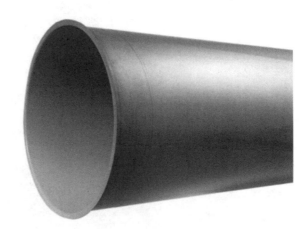

Figure 15.14 Fab fume-ducts coated with ETFE or E-CTFE.

15.4.7 Commercial Dryer Drums

Commercial dryers like those used in hospitals often end up with plastic materials being unintentionally dried. Plastic bags are particularly a problem. These can melt and adhere to the walls of the dryer basket or drum, and are very difficult to remove. Many of these dryer baskets are coated with a fluoropolymer. In Fig. 15.15, only one panel was coated with a fluoropolymer. As can be seen, the fluoropolymer-coated panel in the middle is essentially melt-free. The non-stick and heat resistant properties are used in this application.

15.4.8 Industrial Rollers

Large rollers are used in many industries such as paper making or fabric printing. The rollers are heated to promote drying. These rollers are coated for improved release, easy cleanability, and thermal resistance. One such roller from the paper making industry is shown in Fig. 15.16. The coatings used here can range from chromic acid primer, to PFA, to electroless nickel/PTFE composites.

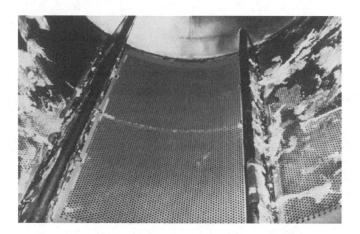

Figure 15.15 Dry drum panel coated with PFA.

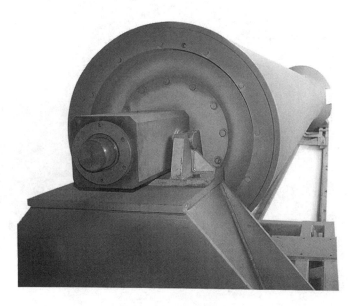

Figure 15.16 Roller used in the paper or fabric industry.

15.4.9 Medical Devices

Fluorinated coatings are used to coat some medical devices. The metered dose inhaler (MDI) is an example a drug delivery device. It consists of a small spray device that delivers a carefully measured volume of propellant and drug into the lungs. The device, shown in Fig. 15.17, is commonly used by asthmatics, but other drugs can also be delivered in this manner. The device has been marketed for decades. Many drugs do not disperse well in the propellant and adhere to wall of the aluminum container. This causes variation in drug dose delivery and a premature drop off in the dose/actuation profile.

A fluoropolymer coating is applied to the inside of the cans to eliminate the adhesion of the drug to the wall permitting the device to deliver the expected number of doses and more complete use of the contents.

The summarized needs for the coating are:

1. Adhesion to smooth aluminum substrate

2. Release characteristics and chemical inertness

3. Optimized application characteristics such as wetting, flow and leveling, and viscosity

4. Regulatory compliance, formulations with FDA listed ingredients and those that use as few ingredients as possible

5. Minimize extractables

Glaxo Smith Kline holds a number of patents[8] on the devices in which they claim a large array of fluorinated coating compositions, most of which are resin bonded fluoropolymers.

15.5 Summary

Fluorinated coatings are often a small part of the final product, but are crucial to their function and commercial success. They frequently take advantage of several of the unique properties of fluoropolymers.

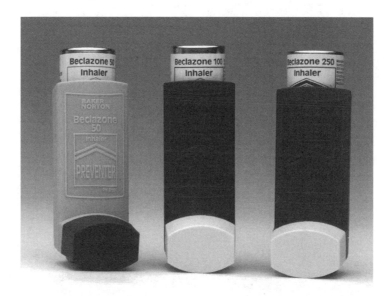

Figure 15.17 Metered dose inhaler (MDI).

REFERENCES

1. US Patent 2,562,118, Osdal Le Verne, assigned to DuPont (Jul 24, 1951)

2. US Patent 4,070,525, Vassiliou, Eustathios and Concannon, Thomas P., assigned to DuPont (Nov 1, 1976)

3. US Patent 4,118,537, Vary; Eva M. and Vassiliou; Eustathios., assigned to DuPont (Oct 3, 1978)

4. US Patent 4,070,525 , Vassiliou, Eustathios and Concannon, Thomas P., assigned to DuPont (Oct 31, 1978)

5. US Patent 5,230,961, Tannenbaum; Harvey P., assigned to DuPont (Jul 27, 1993)

6. US Patent 4,070,525, Vassiliou, Eustathios and Concannon, Thomas P., assigned to DuPont (Oct 31, 1978)

7. Air Flow Products, Ltd., http://www.air-flow.co.nz

8. US Patent 6,131,566, Ashurst; Ian Carl, Herman, Craig Steven, Li-Bovet, Li, Riebe, Michael Thomas, assigned to Glaxo Wellcome Inc. and Glaxo Group Limited (Oct 17, 2000)

16.1 Introduction

This chapter contains information about safe handling and processing of fluoropolymers and coatings. It is based on coatings and resins manufacturers' extensive experience and history. The material in this chapter is not intended as a replacement for the specific information and data supplied by the manufacturers of fluoropolymers and coatings. A handbook entitled *Guide to the Safe Handling of Fluoropolymer Resins,* latest edition, is available from The Society of the Plastics Industry (SPI), 1275 K Street, N.W., Suite 400, Washington D.C. 20005 (202) 371-5233. This book is recommended for information on safe handling resin and cured coatings. Liquid fluoropolymer coatings require supplemental consideration.

This chapter gives an overview of:

- The toxicology of cured fluoropolymer coatings
- Safety guidelines for applying and handling liquid fluoropolymer coatings
- Safety guidelines for applying and handling fluoropolymer powder coatings
- Safety guidelines for baking fluoropolymer coatings
- Safe removal of fluoropolymer coatings
- Food contact and other regulatory considerations
- Disposal of fluoropolymer coating materials

Since their discovery, hundreds of millions of pounds of fluoropolymer resins have been processed at temperatures in excess of 350°C (662°F), and subsequently placed in end-use applications, some of which may have exceeded the rated use temperatures. During this period, spanning more than fifty years, there have been no reported cases of serious injury, prolonged illness, or death resulting from the handling of the resins. This record includes the experience of fluoropolymer manufacturers personnel, thousands of processors, and millions of end-users who handle these products in some form every day.

16.2 Toxicology of Fluoropolymers

Fluoropolymers are chemically stable, inert, and essentially unreactive. Reactivity decreases as fluorine content of the polymer increases. Fluorine induces thermal and chemical stability in the polymers in contrast to chlorine (thermally stability: PVC < PE < PVF). Fluoropolymers have low toxicity and almost no toxicological activity. No neat fluoropolymers or cured coatings have been reported to cause skin sensitivity and irritation in humans. Even polyvinyl fluoride, which contains one fluorine atom and three hydrogen atoms per monomer unit, has been shown to cause no skin reaction in human beings.[1] Excessive human exposure to fluoropolymer resin dust resulted in no toxic effects, although urinary fluoride content increased.[2]

Coatings can contain pigments, surfactants, and other additives to modify the polymer coating properties. These additives are likely to present risks and hazards in the fluoropolymer composition. For example, aqueous fluoropolymer coatings contain surfactants that may produce adverse physiological symptoms. The hazards of using these additives should be considered by themselves and in conjunction with fluoropolymers. Safety information provided by manufacturers of the additives and the compounds should be consulted.

16.3 Safe Handling and Application of Liquid Fluoropolymer Coatings

Aqueous dispersions of fluoropolymers are applied to substrates by coating techniques, described in Ch. 10. Water and surfactants are normally removed in a heating step prior to sintering or curing during which surfactants may decompose. The degradation fragments and the surfactant may be flammable. They may also have adverse health effects. Forced ventilation of the drying oven is necessary to remove the surfactant vapors and minimize build-up of degradation products. Some coating formulations contain organic solvents. Combustion hazards and health effects of these substances should be considered during the handling and processing of the coating.

A number of measures can be taken to reduce and control exposure to monomers and decomposition products during the processing of fluoropolymers. It is important to monitor processing plants and take measures where necessary according to legal requirements established under the Occupational Safety and Health Act (OSHA). The customary precautionary actions for safe fluoropolymer processing are described in this chapter. They include ventilation, processing measures, spillage clean-up, equipment cleaning, and maintenance procedures. A number of general measures should be taken while handling fluoropolymers including protective clothing, personal hygiene, fire hazard, and material incompatibility.

Removal of the decomposition products from the work environment is one of the most important actions taken to reduce and control human exposure. Even at room temperature, small amounts of trapped monomers or other gases could diffuse out of the resin particles. It is a good practice to open the fluoropolymer container in a well-ventilated area. All processing equipment should be ventilated by local exhaust ventilation.

The most effective method of controlling emissions is to capture them close to the source before they are dispersed in the workspace. A fairly small volume of air has to be removed by local exhaust compared to the substantially larger volume of air that must be removed from the entire building. Correct design and operation of local exhaust systems can minimize human exposure.

Exhaust air must enter the exhaust hood or booth to carry the contaminants with it and convey them to the exhaust point. The required air velocity at the point the contaminants are given off to force these contaminants into an exhaust hood is called *capture velocity* and should be at least 200 ft/min (1.0 m/sec).[3] An airflow meter can be used to measure the air velocity. A static pressure gauge can be installed to continuously monitor the air velocity in the hood by pressure drop.

Three publications by The Society of Plastics Industry,[4] American Conference of Governmental Industrial Hygienists,[5] and Canadian Center for Occupational Health and Safety[6] provide detailed information on various aspects of industrial ventilation.

Even with good engineering controls, however, it is recommended that operators use a respirator (see Sec. 16.8 on protective clothing), especially those operators involved in hand-spraying or spray-gun/spray-booth maintenance procedures.

Overspray escape conditions can be minimized by establishing a spray booth maintenance schedule that includes the routine cleaning or replacement of filters and periodic monitoring of air flow face velocities. Automatic spray guns should be properly oriented so that the spray pattern is not directed toward any booth openings. Adjust spray guns to use the minimum atomization pressure, fluid pressure, and spray pattern to suit the job. Avoid strong drafts from any nearby open windows or other building ventilation systems that might adversely affect the operation of the spray booth. If it is necessary to enter the spray booth for any reason, a respirator should be worn and any overspray mist should first be allowed to fully exhaust.

Industrial experience has clearly shown that fluoropolymer coatings can be handled, applied, and cured at elevated temperatures without hazard to personnel provided that good industrial practices are observed. The following is a list of general precautions that should be taken when processing fluoropolymer coatings, but fact sheets and Material Safety Data Sheets issued by the coating manufacturer should be consulted for special precautions.

Keep away from heat and open flame and consider the factors listed below:

- Some liquid fluoropolymer coatings are flammable, many are combustible.

- Keep the lid on the container when not in use.

- Smoking should be prohibited in areas where fluoropolymer products are handled, applied, or baked, and "No Smoking" signs should be prominently displayed.

- Use only with adequate ventilation.

- Avoid breathing vapor, spray mist, and fumes from baking equipment.

- Open containers in well-ventilated areas.

- Some products may irritate skin and eyes.

- Some products may liberate vapors that may be harmful.

- Avoid contact with the skin and eyes.

- Ground containers, funnels, etc., during transfer of the liquids to minimize the chance of static discharge possibly causing a fire.

When agitating, filtering, or transferring these materials, wearing neoprene gloves, a face shield, and protective clothing are strongly recommended. A NIOSH-approved paint-spray respirator should be worn when spraying these products. Under federal regulation (General Industry OSHA Safety & Health Standards, 29CFR.1910 Series, 1910.134), employers who provide respirators are required to train and fit employees in their use.

- Observe good personal hygiene practices. Wash hands thoroughly after handling, especially before eating or smoking.

- Carrying exposed cigarettes or other tobacco products should be prohibited in the areas where fluoropolymer products are handled, applied, or baked: eating, drinking, or carrying food or cosmetic products.

Do not breathe vapors or mists. Wear a properly fitted vapor/particulate respirator approved by NIOSH/OSHA (TC-23C) for use with paints during application and until all vapors and spray mist are exhausted. Follow the respirator manufacturer's directions for respirator use. For prolonged hand-spraying operations, an air-supplied respirator or hood may be preferable. Safety glasses, coveralls, and neoprene gloves are also recommended.

Respirators are not usually required in the baking area where properly operated and ventilated ovens are used. It is still a good practice to wear the respirator when opening oven doors where trapped vapors could be released directly to the face. If ventilation conditions are suspect, a respirator should be required. An air-supplied respirator or hood or a NIOSH/OSHA-approved chemical cartridge respirator with organic vapor cartridges in combination with a high-efficiency particulate filter is recommended.

For general handling (pouring, reducing, filtering, mixing) the following should be used:

- Safety glasses

- Full face shield when handling acids

- Neoprene gloves
- Coveralls
- Neoprene apron (optional)

16.4 Thermal Properties of Fluoropolymers

It is important when using fluoropolymer parts to follow the recommendations and specifications of the resin and part suppliers. From a thermal exposure standpoint, the maximum continuous-use temperature should comply with the values specified by *The Guide to Safe Handling of Fluoropolymers Resins,* published by The Society of Plastics Industry, Inc., summarized in Table 16.1. Polymer manufacturers may specify slightly different values.

The curing, baking, or sintering operation requires heating the polymer in ovens at high temperatures where decomposition products are formed to different extents. Ovens must be equipped with sufficiently strong ventilation to remove the gaseous products and prevent them from entering the work area. It is important that ventilation prevent entrance of the contaminants into the plant area during the operation of the oven and when the door is open.

Table 16.1 Continuous-Use Temperatures of Some Common Fluoropolymers

Fluoropolymer Resin	Maximum Continuous Use Temperature
PCTFE	250°F (120°C)
PVDF	300°F (150°C)
THV	300°F (150°C)
ECTFE	300°F (150°C)
ETFE	300°F (150°C)
FEP	400°F (205°C)
MFA	480°F (250°C)
PFA	500°F (260°C)
PTFE	500°F (260°C)

Typically, ovens operate at high temperatures approaching 820°F (438°C). Overheating should be prevented by installing limit switches to avoid oven runoff that can result in high temperatures at which accelerated decomposition may occur. It is a good practice to operate the sintering and baking ovens at the lowest possible temperature that is adequate for the completion of the part fabrication or coating cure. An overheated oven must be cooled before opening the doors. Proper personal protective equipment, including a self-contained breathing apparatus, must be donned prior to opening the oven doors when overheating has occurred.

Compounds containing fillers are usually more sensitive to thermal decomposition due to the acceleration of thermo-oxidative reactions by a number of additives at elevated temperatures. It may also be possible to sinter compounds at lower temperatures and shorter oven hold-up times due to changes in conductivity of the part. For example, metal-filled fluoropolymer coatings have a significantly higher thermal conductivity than unfilled coatings, which leads to more rapid heating of the part.

16.4.1 Off-Gases During Baking and Curing

While most inorganic engineering materials, such as metals or ceramics, simply soften and lose strength when overheated, organic polymeric materials also undergo some decomposition or actual breakdown in chemical structure. The products thus formed are usually given off in the form of gases or fumes. Fumes from the pyrolysis of many resins and elastomers, as well as those from naturally occurring polymers like cotton, rubber, coal, silk, and wood, may be toxic.

Fluoropolymers and fluoropolymer coatings are used in thousands of vastly different applications. Thermal stability is a major feature of these polymers, spurring their applications where high temperature exposures are encountered. Fluoropolymers, however, can produce toxic products if they are overheated. Precautions should be taken to remove any degradation fragments produced during the baking of the fluoropolymer coatings. Therefore, the ventilation precautions to be observed when heating fluoropolymer resins are similar to those that should be observed when heating such conventional materials.

The high temperature ratings of fluoropolymer resins result from their extremely low rate of thermal decomposition. Even in cases of relatively severe overheating, the quantity of fumes evolved is minute in comparison with that from most organic materials. Even though the degradation products are small in quantity, adequate ventilation should be provided in situations where exposure of personnel can occur.

Thermal decomposition of fluoropolymers has been discussed in previous books in this series.[7][8] The reader should refer to those for a review of this topic. Fluorinated coatings are heated to high temperatures during processing and degrade to some extent. It is important to remember that the type of degradation products and the extent of decomposition depend on several factors. One must consider the following variables during processing:

- Temperature
- Presence of oxygen
- Physical form of the article
- Residence time at temperature
- Presence of additives

The products of decomposition of fluoropolymers fall in three categories: fluoroalkenes, oxidation products, and particulates of low molecular weight fluoropolymers. These products must be removed by adequate ventilation from the work environment to prevent human exposure. A major oxidation product of PTFE is carbonyl fluoride, which is highly toxic and hydrolyzes to yield hydrofluoric acid and carbon dioxide. At 842°F (450°C) in the presence of oxygen, PTFE degrades into carbonyl fluoride and hydrofluoric acid. In the absence of oxygen, at 1472°F (800°C), tetrafluoromethane is formed. It has been suggested that tetrafluoroethylene (TFE) is the only product that is produced when PTFE is heated to melt stage.[9]

16.4.2 Polymer Fume Fever

Fluoropolymers degrade during processing and generate effluents with an increasing rate with temperature. The operation of process equipment at high temperatures may result in generation of toxic gases and particulate fume. Human exposure to heated fluoropolymer resins may cause a temporary flu-like condition similar to the metal fume fever (or Foundryman's fever) known for many years. These symptoms, called "polymer fume fever," are the only adverse effects observed in humans to date. The symptoms do not ordinarily occur until about two or more hours after exposure, and pass within 36 to 48 hours, even in the absence of treatment. Observations indicate that these attacks have no lasting effect and that the effects are not cumulative. When such an attack occurs, it usually follows exposure to vapors evolved from the polymer at high temperatures used in the baking operation, or from smoking cigarettes or tobacco contaminated with the polymer. If such attacks occur, it is recommended that the patient be removed immediately to fresh air and that a physician be called. It is prudent to ban tobacco products from fluoropolymer work areas. It has been suggested that no health hazards exist unless the fluoropolymer is heated above 572°F (300°C).[10]

Johnston and his coworkers[11] have proposed that heating PTFE gives rise to fumes which contain very fine particulates. The exposure of lung tissues to these particulates can result in a toxic reaction causing pulmonary edema or excessive fluid build-up in the lung cells. Severe irritation of the tissues along with the release of blood from small vessels is another reaction to exposure. In controlled experiments, animals were exposed to filtered air from which fumes had been removed and to unfiltered air. Unfiltered air produced the expected fume fever response. Animals exposed to the filtered air did not develop any of the symptoms of polymer fume fever.

The products of fluoropolymer decomposition produce certain health effects upon exposure, summarized in Table.16.2. Resin manufacturers can supply available exposure information.

16.5 Removal of Fluoropolymer Films and Coatings

Grit blasting is the only recommended method of removing fully cured fluoropolymer.

CAUTION: Although this method has been used extensively and safely for many years, a potential fire/explosion hazard exists if the fluoropolymer and metal dusts generated from this operation are heated to 796°F (424°C).

Although solid films of fluoropolymer coatings do not present any particular fire hazard, small particles (fines) of these films can become extremely combustible in the presence of various metal fines when exposed to temperatures above 796°F (424°C). An intimate mixture of such finely divided fluorocarbon and metal powder (e.g., aluminum, magnesium) particles will react violently when subjected to high temperatures and the resulting fire can only be extinguished by radically lowering the

Table 16.2 Health Effects of Some Decomposition Products

Decomposition Species	Health Effects
Hydrogen Fluoride (HF)	Initial symptoms: choking, coughing, severe eye irritation, nose and throat irritation. Followed by: fever, chills, breathing difficulty, cyanosis, pulmonary edema. Can absorb through the skin, acute or chronic over exposure can damage liver and kidneys.
Carbonyl Fluoride (COF$_2$)	Skin irritation with discomfort and rash, eye corrosion with ulceration, respiratory irritation with cough, discomfort, difficulty breathing and shortness of breath.
Tetrafluoroethylene (TFE)	Acute respiratory and eye irritation, mild central nervous system depression, nausea, vomiting, and dry cough.
Perfluoroisobutylene (PFIB)	Pulmonary edema

temperature of the material. It may be possible to generate such mixtures from operations such as belt sanding, grit blasting with abrasives for reclaiming previously coated parts, or from grinding or buffing operations to remove cured coatings.

Good housekeeping and compliance to governmental laws and fire/insurance codes should prevent the accumulation of these fines mixtures, but when their collection is absolutely necessary, they should be maintained in a wet state. If ventilation is used as the primary method of dust control, a maintenance schedule should be established to keep the system in efficient working order. In all cases, adequate means of quenching heat quickly (e.g., sprinkler systems, water-based portable fire extinguishers) should be accessible in all vulnerable locations. In addition, the fluorocarbon dust can contaminate cigarettes or other tobacco products, which subsequently may lead to polymer fume fever. It is important, therefore, that good housekeeping practices be observed in the area where grit-blasting operations are performed.

Some cured resin bonded fluoropolymer coatings can sometimes be removed by immersion in commercially available alkaline paint strippers. High-pressure water can also be used.

Under no circumstances is high-temperature bum-off recommended, even though it is commonly done.

16.6 Fire Hazard

Fluoropolymers do not ignite easily and do not sustain flame. However, they can decompose in a flame and evolve toxic gases. For example, PTFE will sustain flame in an ambient of >95% oxygen (Limiting Oxygen Index by ASTM D2863). In less oxygen rich environments, burning stops when the flame is removed. The Underwriters Laboratory rating of perfluoropolymers is 94–V0. Self-ignition temperature of PTFE is 932°F–1040°F (500°C–560°C) according to ASTM D1929, far above most other organic materials.[12] PTFE does not form flammable dust clouds under normal conditions as determined in the Godwert Greenwald test at 1832°F (1000°C). Polytetrafluoroethylene falls in the explosion class ST1.[11]

Cured fluoropolymer films have a comparatively low fuel value and, in a fire situation, resist ignition and do not themselves promote flame spread. When ignited by sustained flame from other sources, their contribution of heat to the fire is exceptionally low and at a relatively slow rate.

In the liquid state, solvent-based fluoropolymer coatings are flammable or combustible and should be used in well-ventilated areas where smoking is prohibited. Although not as obvious, many water-based fluoropolymer products, even though they may not support combustion, exhibit a flash point and must be treated as combustible liquids. A good reference for flammability for solvent based coatings is "Working with Modern Hydrocarbon and Oxygenated Solvents: A Guide to Flammability," published by the American Solvent Council[13]

In the event of a fire, personnel entering the area should have full protection, including acid-resistant clothing and self-contained breathing apparatus with full face piece operated in the pressure demand or other positive pressure mode. In case of direct exposure to combustion products, the affected areas should be washed promptly with copious amounts of water.

In the case of exposure of the eyes, medical attention should be provided as soon as they have been thoroughly flushed with water.

16.7 Spillage Cleanup

Fluoropolymers can create a slippery surface when they are rubbed against a hard surface because they are soft and easily abrade away and coat the surface. Any spills during handling should be cleaned up immediately. It is helpful to cover the floors of the processing area with anti-slip coatings.

16.8 Personal Protective Equipment

Appropriate personal protective equipment should be worn to avoid hazards during the processing of fluoropolymers. They include safety glasses, gloves, and gauntlets (arm protection). Dust masks or respirators should be worn to prevent inhalation of dust and particulates of fluoropolymers during grinding and machining. Additional protection may be required when working with filled compounds.

Skin contact with fluoropolymer dispersions should be avoided by wearing gloves, overalls, and safety glasses due to their surfactant or solvent content. Fluoropolymer coatings must be sprayed in a properly equipped spray booth. Overspray should be captured in a water bath. The spray operator should wear a disposable Tyvek® suit or Nomex® coveralls, goggles, gloves, and a respirator or self-contained breathing apparatus.

16.9 Personal Hygiene

Tobacco products should be banned from the work areas to prevent polymer fume fever. Street clothing should be stored separately from work clothing. Thorough washing after removal of work clothing will remove powder residues from the body.

16.10 Food Contact and Medical Applications

Fluoropolymer resins are covered by Federal Food, Drug and Cosmetic Act, 21 CFR & 177.1380 & 177.1550 in the United States and EC Directive 90/128 in the European Union. The U.S. Food and Drug Administration has approved many fluoropolymers (e.g., PTFE, PFA, and FEP) for food contact. Additives such as pigments, stabilizers, antioxidants, and others must be approved under a food additive regulation if they do not have prior clearance.

Some fluoropolymers have been used in the construction of FDA regulated medical devices. FDA only grants approval for a complete device, not components such as resin. Resin suppliers usually have specific policies regarding the use of their products in medical devices. Thorough review of these policies and regulatory counsel advice would be prudent before initiating any activity in this area.

16.11 Fluoropolymer Scrap and Recycling

Fluoroplastics described in this book are thermoplastics and can be reused under the right circumstances. There are a few sources of waste fluoropolymer. Various processing steps of fluo-

ropolymers such as preforming, molding, machining, grinding, and cutting create debris and scrap. Some of the scrap material is generated prior to sintering, but the majority is produced after sintering. A third category of scrap is polymer that does not meet specifications and cannot be used in its intended applications. Efforts have been made to recycle PTFE soon after its discovery. The incentive in the early days to recycle scrap was economic due to the high cost of polytetrafluoroethylene. Today, a small industry has evolved around recycling fluoropolymers.

Scrap PTFE has to be processed for conversion to usable feedstock. The extent of processing depends on the amount of contamination in the debris. The less contaminant in the scrap, the higher the value of the material will be. Machine cuttings and debris usually contain organic solvents, metals, moisture, and other contaminants. Conversion of this material to useful feedstock requires chemical and thermal treatment. The clean PTFE feedstock can be converted to a number of powders.

A large quantity of scrap PTFE is converted into micropowder (fluoroadditive) by methods discussed in Ch. 2. Fluoroadditives are added to plastics, inks, oils, lubricants, and coatings to impart fluoropolymer-like properties such as reduced wear rate and friction. Part of the PTFE is converted back into molding powders, which are referred to as "repro," short for reprocessed. Unmelted new polymer is, by contrast, called "virgin."

16.12 Environmental Protection and Disposal Methods

None of fluoropolymers or their decomposition products poses any threats to the ozone layer. None are subject to any restrictive regulations under the Montreal Protocol and the US Clean Air Act. Reacting HF with chloroform produces the main fluorinated ingredient of tetrafluoroethylene synthesis: $CHClF_2$. It has a small ozone depleting potential but is excluded from the Montreal Protocol regulation due to its intermediate role and destruction from the environment.

The preferred methods of disposing fluoropolymers are recycling and landfilling according to the various regulations. In the case of suspensions and dispersions, solids should be removed from the liquid and disposed. Liquid discharge to waste water systems should be according to the permits. None of the polymers should be incinerated unless the incinerator is equipped to scrub out hydrogen fluoride, hydrogen chloride, and other acidic products of combustion.

In the disposal of fluoropolymer scrap containing pigments, additives, or solvents, additional consideration must be given to the regulation for the disposal of the non-fluoropolymer ingredients. Some of the compounds and mixtures may require compliance with the Hazardous Material Acts.

REFERENCES

1. Harris, L. R. and Savadi, D. G., Synthetic Polymers, *Patty's Industrial Hygiene and Toxicology*, 4th ed., Vol. 2, Part E (George D. Clayton and Florence E. Clayton, Eds.), John Wiley & Sons, New York (1994)

2. *Guide for the Safe Handling of Fluoropolymer Resins*, Association of Plastics Manufacturers in Europe, Brussels, Belgium (1995)

3. ANSI/AIHA Z9.3-1994 Standard for Spray Finishing Operations - Safety Code for Design, Construction, and Ventilation

4. *The Guide to Safe Handling of Fluoropolymers Resins*, published by The Society of Plastics Industry, Inc. (1998)

5. Industrial Ventilation: A Manual of Recommended Practice, published by American Conference of Governmental Industrial Hygienists

6. A Basic Guide to Industrial Ventilation, published by Canadian Center for Occupational Health and Safety, Hamilton, Ontario, Canada L8N 1H6, Pub. No. 88-7E.

7. Ebnesajjad, S., *Fluoroplastics*, Vol. 1: *Non-Melt Processible Fluoroplastics, The Definitive User's Guide and Databook*, William Andrew, Inc., Norwich, NY (2000)

8. Ebnesajjad, S., *Fluoroplastics*, Vol. 2: *Melt Processible Fluoroplastics, The Definitive User's Guide and Databook*, William Andrew, Inc., Norwich, NY (2003)

9. *The Guide to Safe Handling of Fluoropolymers Resins*, published by The Society of Plastics Industry, Inc. (1998)

10. Rose, C. A., *Inhalation Fevers*, in: *Environmental and Occupational Medicine*, 2nd ed. (Rom, W. N., ed.), pp. 373–380, Little, Brown and Company, Boston (1992)

11. Johnston, C. J., Finkelstein, J. N., Gelein, R., Baggs, R., and Obrduster, G., Characterization of Early Pulmonary Inflammatory Response Associated with PTFE Fume Exposure, *Toxicology and Applied Pharmacology*, Article No. 0208, Academic Press (May, 1996)

12. *Guide for the Safe Handling of Fluoropolymer Resins*, Association of Plastics Manufacturers in Europe, Brussels, Belgium (1995)

13. Working with Modern Hydrocarbon and Oxygenated Solvents: A Guide to Flammability, published by the American Solvent Council.

Appendix I: Chemical Resistance of Fluoropolymers

I.1 PDL Chemical Resistance Guidelines

This appendix contains extensive chemical resistance data for a number of commercial fluoropolymers. The data in these tables come entirely from the *PDL Handbook Series, Chemical Resistance,* Volume 1. These appendices are edited versions from the PDL work.[1] Data not related to coatings has been removed to produce these tables. Most of the chemicals are frequently encountered in processing operations. The data for each fluoropolymer are organized alphabetically, using the common name of each chemical. The reader should review the next section (Sec. I.2) to understand the basis for the PDL Rating. Exposure conditions for each chemical have been listed because the same chemical could behave in a different way if the conditions of exposure (such as temperature or concentration) are altered. Where data have been available, the effect of exposure on the physical properties such as weight change and tensile properties have been listed.

I.2 PDL Resistance Rating

The PDL Resistance Rating is determined using a weighted value scale developed by PDL and reviewed by experts. Each of the ratings is calculated from test results provided for a material after exposure to a specific environment. It gives a general indication of a material's resistance to a specific environment. In addition, it allows users to search for materials most likely to be resistant to a specific exposure medium.

After assigning the weighted value to each field for which information is available, the PDL Resistance Rating is determined by adding together all weighted values and dividing this number by the number of values added together. All numbers to the right of the decimal are truncated to give the final result. If the result is equal to 10, a resistance rating of 9 is assigned. Each reported field is given equal importance in assigning the resistance rating since, depending on the end use, different factors play a role in the suitability for use of material in a specific environment. Statistically, it is necessary to consider all available information in assigning the rating. Supplier resistance ratings are also figured into the calculation of the PDL Resistance Rating. Weighted values assigned depend on the scale used by the supplier.

Table I.1 gives the values and guidelines used in assigning the PDL Resistance Rating. The guidelines—especially in the case of visual observations—are sometimes subject to an educated judgement. An effort is made to maintain consistency and accuracy.

Table I.1 PDL Chemical Resistance Ratings

Weighted Value	Weight Change	Diameter Length Change	Thickness Change	Volume*1 Change	Mechanical*2 Property Retained	Visual*3 Observed Change	BTT*4 (min)	Permeation Rate (µg/cm²/min)	Hardness Change (Units)
10	0-0.25	>0-0.1	0-0.25	0-2.5	>97	No change	≤51	≤0.9	0-2
9	>0.25-0.5	>0.1-0.2	>0.25-0.5	>2.5-5.0	94 to <97		>1 to ≤2		>2-4
8	>0.5-0.75	>0.2-0.3	>0.5-0.75	>5.0-10.0	90 to <94		>2 to ≤5	>0.9-9	>4-6
7	>0.75-1.0	>0.3-0.4	>0.75-1.0	>10.0-20.0	85 to <90	Slightly discolored, slightly bleached	>5 to ≤10		>6-9
6	>1.0-1.5	>0.4-0.5	>1.0-1.5	>20.0-30.0	80 to <85	Discolored yellows, slightly flexible	>10 to ≤30	>9-90	>9-12
5	>1.5-2.0	>0.5-0.75	>1.5-2.0	>30.0-40.0	75 to <80	Possible stress crack agent, flexible, possible oxidizing agent, slightly crazed	>30 to ≤120		>12-15
4	>2.0-3.0	>0.75-1.0	>2.0-3.0	>40.0-50.0	70 to <75	Distorted, warped, softened, slight swelling, blistered, known stress crack agent	>120 to ≤240	>90-900	>15-18
3	>3.0-4.0	>1.0-1.5	>3.0-4.0	>50.0-70.0	60 to <70	Cracking, crazing, brittle, plasticizer, oxidizer, softened swelling, surface hardness	>240 to ≤480		>18-21
2	>4.0-6.0	>1.5-2.0	>4.0-6.0	>60.9-90.0	50 to <60	Severe distortion, oxidizer and plasticizer deteriorated	>480 to ≤960	>900-9000	>21-25
1	>6.0	>2.0	>6.0	>90.0	>0 to <50	decomposed	>960		>25
					0	Solvent dissolved, disintegrated		>9000	

*1 All values are given as percent change from original.

*2 Percent mechanical properties retained include tensile strength, elongation, modulus, flexural strength, and impact strength. If the % retention is greater than 100%, a value of 200 minus the %property retained is used in the calculation.

*3 Due to the variety of information of this type reported, this table can be used only as a guideline.

*4 Breakthrough time: time from initial chemical contact to detection.

I.3   Chemical Resistance Tables

Tables I.2 through I.8 contain chemical resistance data for commercial fluoropolymers including PTFE, ECTFE, ETFE, FEP, PCTFE, PFA, and PVDF. The data in these tables come entirely from the *PDL Handbook Series, Chemical Resistance,* Volume 1. These tables are edited versions from the PDL work.[1]

Table I.2 Chemical Resistance of Polytetrafluoroethylene (PTFE)

Reagent	Reagent Note	Conc.	Temp (°C)	Time (days) Load	PDL Rating	% Change Weight	% Retained Tensile Strength	Elongation	Resistance Note	Material Note
Abietic Acid	up to boiling point				8				Compatible	DuPont Teflon® PTFE
Acetic Acid		10	23		8				no effect	Saint Gobain Rulon J
	up to boiling point				8				Compatible	DuPont Teflon® PTFE
	concentrated		23		8				no effect	Saint Gobain Rulon J
Acetic Anhydride	up to boiling point				8				Compatible	DuPont Teflon® PTFE
Acetone	up to boiling point				8				Compatible	DuPont Teflon® PTFE
			25	365	9	0.3			No Significant Change	DuPont Teflon® PTFE
			50	365	9	0.4			No Significant Change	DuPont Teflon® PTFE
			70	14	9	0			No Significant Change	DuPont Teflon® PTFE
Acetophenone	up to boiling point				8				Compatible	DuPont Teflon® PTFE
Acrylic Anhydride	up to boiling point				8				Compatible	DuPont Teflon® PTFE
Acrylonitrile	up to boiling point				8				Compatible	DuPont Teflon® PTFE
Allyl Acetate	up to boiling point				8				Compatible	DuPont Teflon® PTFE
Aluminum Chloride	up to boiling point				8				Compatible	DuPont Teflon® PTFE
Ammonia	liquid to boiling point				8				Compatible	DuPont Teflon® PTFE
Ammonium Chloride	up to boiling point				8				Compatible	DuPont Teflon® PTFE
Ammonium Hydroxide	concentrated	10	25	365	9	0			No Significant Change	DuPont Teflon® PTFE
		10	70	365	9	0.1			No Significant Change	DuPont Teflon® PTFE
		30	23		2				Vigorous Attack	Saint Gobain Rulon J
			23		2				Vigorous Attack	Saint Gobain Rulon J
Aniline	up to boiling point				8				Compatible	DuPont Teflon® PTFE
Animal Oils	up to boiling point				8				Compatible	DuPont Teflon® PTFE
Benzene			78	4	9	0.5			Compatible	DuPont Teflon® PTFE
			100	0.33	8	0.6			Compatible	DuPont Teflon® PTFE
			200	0.33	7	1			Compatible	DuPont Teflon® PTFE
Benzonitrile	up to boiling point				8				Compatible	DuPont Teflon® PTFE
Benzoyl Chloride	up to boiling point				8				Compatible	DuPont Teflon® PTFE
Benzyl Alcohol	up to boiling point				8				Compatible	DuPont Teflon® PTFE
Borax	up to boiling point				8				Compatible	DuPont Teflon® PTFE
Boric Acid	up to boiling point				8				Compatible	DuPont Teflon® PTFE
Bromine	up to boiling point				8				Compatible	DuPont Teflon® PTFE
Butyl Acetate	up to boiling point				8				Compatible	DuPont Teflon® PTFE
Butyl Methacrylate	up to boiling point				8				Compatible	DuPont Teflon® PTFE
Butylamine	up to boiling point				8				Compatible	DuPont Teflon® PTFE
Calcium Chloride	up to boiling point				8				Compatible	DuPont Teflon® PTFE
Carbon Tetrachloride			25	365	8	0.6			No Significant Change	DuPont Teflon® PTFE
			50	365	5	1.6			No Significant Change	DuPont Teflon® PTFE
			70	14	5	1.99			No Significant Change	DuPont Teflon® PTFE
			100	0.33	4	2.5			No Significant Change	DuPont Teflon® PTFE
			200	0.33	3	3.7			No Significant Change	DuPont Teflon® PTFE
Carbon Disulfide	Up to boiling point				8				Compatible	DuPont Teflon® PTFE
Cetane	Up to boiling point				8				Compatible	DuPont Teflon® PTFE

(Cont'd.)

Table I.2 Chemical Resistance of Polytetrafluoroethylene (PTFE) (Cont'd.)

Reagent	Reagent Note	Conc.	Temp (°C)	Time (days) Load	PDL Rating	% Change Weight	% Retained Tensile Strength	% Retained Elongation	Resistance Note	Material Note
Chlorine	up to boiling point				8				Compatible	DuPont Teflon® PTFE
Chloroform	up to boiling point				8				Compatible	DuPont Teflon® PTFE
Chlorosulfonic Acid	up to boiling point				8				Compatible	DuPont Teflon® PTFE
Chromic Acid	up to boiling point				8				Compatible	DuPont Teflon® PTFE
Cyclohexane	up to boiling point				8				Compatible	DuPont Teflon® PTFE
Detergents	up to boiling point				8				Compatible	DuPont Teflon® PTFE
Dibutyl Phthalate	up to boiling point				8				Compatible	DuPont Teflon® PTFE
Dibutyl Sebacate	up to boiling point				8				Compatible	DuPont Teflon® PTFE
Diethyl Carbonate	up to boiling point				8				Compatible	DuPont Teflon® PTFE
Diisobutyl Adipate	up to boiling point				8				Compatible	DuPont Teflon® PTFE
Dimethyl Ether	up to boiling point				8				Compatible	DuPont Teflon® PTFE
Dimethyl formamide	up to boiling point				8				Compatible	DuPont Teflon® PTFE
Dimethylhydrazine	up to boiling point				8				Compatible	DuPont Teflon® PTFE
Dioxane	up to boiling point				8				Compatible	DuPont Teflon® PTFE
Ethyl Acetate	up to boiling point				8				Compatible	DuPont Teflon® PTFE
			25	365	9	0.5			No Significant Change	DuPont Teflon® PTFE
			50	365	8	0.7			No Significant Change	DuPont Teflon® PTFE
			70	14	8	0.7			No Significant Change	DuPont Teflon® PTFE
Ethyl Alcohol		95	25	365	9	0			No Significant Change	DuPont Teflon® PTFE
		95	50	365	9	0			No Significant Change	DuPont Teflon® PTFE
		95	70	14	9	0			No Significant Change	DuPont Teflon® PTFE
		95	100	0.33	9	0.1			No Significant Change	DuPont Teflon® PTFE
		95	200	0.33	9	0.3			No Significant Change	DuPont Teflon® PTFE
	up to boiling point				8				Compatible	DuPont Teflon® PTFE
Ethyl Ether	up to boiling point				8				Compatible	DuPont Teflon® PTFE
Ethyl Hexoate	up to boiling point				8				Compatible	DuPont Teflon® PTFE
Ethylene Bromide	up to boiling point				8				Compatible	DuPont Teflon® PTFE
Ethylene Glycol	up to boiling point				8				Compatible	DuPont Teflon® PTFE
Ferric Chloride	up to boiling point				8				Compatible	DuPont Teflon® PTFE
Ferric Phosphate	up to boiling point				8				Compatible	DuPont Teflon® PTFE
Fluoronaphthalene	up to boiling point				8				Compatible	DuPont Teflon® PTFE
Fluoronitrobenzene	up to boiling point				8				Compatible	DuPont Teflon® PTFE
Formaldehyde	up to boiling point				8				Compatible	DuPont Teflon® PTFE
Formic Acid	up to boiling point				8				Compatible	DuPont Teflon® PTFE
Furan	up to boiling point				8				Compatible	DuPont Teflon® PTFE
Gasoline	up to boiling point				8				Compatible	DuPont Teflon® PTFE
Hexachloroethane	up to boiling point				8				Compatible	DuPont Teflon® PTFE
Hexane	up to boiling point				8				Compatible	DuPont Teflon® PTFE
Hydrazine	up to boiling point				8				Compatible	DuPont Teflon® PTFE
Hydrobromic Acid		48			8				No effect	Saint Gobain Rulon J

(Cont'd.)

Table I.2 Chemical Resistance of Polytetrafluoroethylene (PTFE) *(Cont'd.)*

Reagent	Reagent Note	Conc.	Temp (°C)	Time (days) Load	PDL Rating	% Change Weight	% Retained Tensile Strength	% Retained Elongation	Resistance Note	Material Note
Hydrochloric Acid		10	23		8	0			No effect	DuPont Teflon® PTFE
		10	25	365	9	0			No Significant Change	DuPont Teflon® PTFE
		10	50	365	9	0			No Significant Change	DuPont Teflon® PTFE
		10	70	0.33	9	0			No Significant Change	DuPont Teflon® PTFE
		20	100	0.33	9	0			No Significant Change	DuPont Teflon® PTFE
		20	200		9	0			No Significant Change	DuPont Teflon® PTFE
	up to boiling point				8				Compatible	Saint Gobain Rulon J
Hydrofluoric Acid	concentrated solution	60	23		7				no effect	Saint Gobain Rulon J
	up to boiling point		23		8				dielectric reduced	DuPont Teflon® PTFE
Hydrogen Peroxide	up to boiling point				8				Compatible	DuPont Teflon® PTFE
Lead	up to boiling point				8				Compatible	DuPont Teflon® PTFE
Magnesium Chloride	up to boiling point				8				Compatible	DuPont Teflon® PTFE
Mercury	up to boiling point				8				Compatible	DuPont Teflon® PTFE
Methacrylic Acid	up to boiling point				8				Compatible	DuPont Teflon® PTFE
Methyl Alcohol	up to boiling point				8				Compatible	DuPont Teflon® PTFE
Methyl Ethyl Ketone	up to boiling point				8				Compatible	DuPont Teflon® PTFE
Methyl Methacrylate	up to boiling point				8				Compatible	DuPont Teflon® PTFE
Naphthalene	up to boiling point				8				Compatible	DuPont Teflon® PTFE
Naphthols	up to boiling point				8				Compatible	DuPont Teflon® PTFE
Nitric Acid		10	23		3				Strongly attacks	Furon Rulon J
		10	25	365	9	0			No Significant Change	DuPont Teflon® PTFE
		10	70	365	9	0.1			No Significant Change	DuPont Teflon® PTFE
		40	23		3				Strongly attacks	Furon Rulon J
	up to boiling point		23		8				Compatible	DuPont Teflon® PTFE
	concentrated				3				strongly attacks	Furon Rulon J
Nitro-2-Methylpropanol (2-)	up to boiling point				8				Compatible	DuPont Teflon® PTFE
Nitrobenzene	up to boiling point				8				Compatible	DuPont Teflon® PTFE
Nitrobutanol (2-)	up to boiling point				8				Compatible	DuPont Teflon® PTFE
Nitrogen Tetraoxide	up to boiling point				8				Compatible	DuPont Teflon® PTFE
Nitromethane	up to boiling point				8				Compatible	DuPont Teflon® PTFE
Octadecyl Alcohol	up to boiling point				8				Compatible	DuPont Teflon® PTFE
Ozone	up to boiling point				8				Compatible	DuPont Teflon® PTFE
Pentachlorobenzamide	up to boiling point				8				Compatible	DuPont Teflon® PTFE
Perchloroethylene	up to boiling point				8				Compatible	DuPont Teflon® PTFE
Perfluoroxylene	up to boiling point				8				Compatible	DuPont Teflon® PTFE
Phenol					8				Compatible	DuPont Teflon® PTFE
Phosphoric Acid		10	23		8				No effect	Saint Gobain Rulon J
		30	23		8				No effect	Saint Gobain Rulon J
	up to boiling point				8				Compatible	DuPont Teflon® PTFE
	concentrated		23		8				No effect	Saint Gobain Rulon J

(Cont'd.)

Table I.2 Chemical Resistance of Polytetrafluoroethylene (PTFE) (Cont'd.)

Reagent	Reagent Note	Conc.	Temp (°C)	Time (days) Load	PDL Rating	% Change Weight	% Retained		Resistance Note	Material Note
							Tensile Strength	Elongation		
Phosphorous Pentachloride	up to boiling point				8				Compatible	DuPont Teflon® PTFE
Phthalic Acid	up to boiling point				8				Compatible	DuPont Teflon® PTFE
Pinene	up to boiling point				8				Compatible	DuPont Teflon® PTFE
Piperidine	up to boiling point				8				Compatible	DuPont Teflon® PTFE
Potassium Acetate	up to boiling point				8				Compatible	DuPont Teflon® PTFE
Potassium Hydroxide		10	50		5				Attacked	Saint Gobain Rulon J
		10	100		7				slight effects on dielectric	Saint Gobain Rulon J
Toluene	up to boiling point		25	365	9	0.3			No Significant Change	Dupont Teflon® PTFE
			50	365	8	0.6			No Significant Change	DuPont Teflon® PTFE
			70	14	8	0.6			No Significant Change	DuPont Teflon® PTFE
Trichloroacetic Acid	up to boiling point				8				Compatible	DuPont Teflon® PTFE
Trichloroethylene	up to boiling point				8				Compatible	DuPont Teflon® PTFE
Tricresyl Phosphate	up to boiling point				8				Compatible	DuPont Teflon® PTFE
Triethanolamine	up to boiling point				8				Compatible	DuPont Teflon® PTFE
Vegetable Oils	up to boiling point				8				Compatible	DuPont Teflon® PTFE
Vinyl Methacrylate	up to boiling point				8				Compatible	DuPont Teflon® PTFE
Water	up to boiling point				8				Compatible	DuPont Teflon® PTFE
Xylene	up to boiling point				8				Compatible	DuPont Teflon® PTFE
Zinc Chloride	up to boiling point				8				Compatible	DuPont Teflon® PTFE

Table I.3 Chemical Resistance of Ethylene Chlorotrifluoroethylene Copolymer (ECTFE)

Reagent	Reagent Note	Conc.	Temp (°C)	Time (days)	PDL Rating	% Change Weight	% Retained Tensile Strength	% Retained Modulus	% Retained Elongation	Resistance Note	Material Note
Acetic Acid		10	23		8					Recommended for use	Halar®
		10	121		8					Recommended for use	Halar®
		20	23		8					Recommended for use	Halar®
		20	121		8					Recommended for use	Halar®
		50	23		8					Recommended for use	Halar®
		50	121		8					Recommended for use	Halar®
		80	23		8					Recommended for use	Halar®
		80	66		8					Recommended for use	Halar®
	Glacial		23		8					No cracking observed	Halar® 2.3mm thick
	Glacial		66	11	8					Recommended for use	Halar®
	Glacial		230	11	4					No cracking observed	Halar® 2.3mm thick
Acetic Anhydride			23		8					Recommended for use	Halar®
Acetone			23		8					Recommended for use	Halar®
			23	11	8	0.1		80-100	80-100	No cracking observed	Halar® 2.3mm thick
			66		8					Recommended for use	Halar®
			121		2					Not recommended for use	Halar®
			125	11	4	4		25-50	80-100	No cracking observed	Halar® 2.3mm thick
Aluminum Nitrate			23		8					Recommended	Halar®
			149		8					Recommended	Halar®
Aluminum Oxychloride			23		8					Recommended	Halar®
			66		8					Recommended	Halar®
Aluminum Sulfate			23		8					Recommended	Halar®
			149		8					Recommended	Halar®
Ammonia	aqueous solution	10	23		8					Recommended	Halar®
	gas	10	121		8					Recommended	Halar®
			23		8					Recommended	Halar®
			149		8					Recommended	Halar®
Ammonium Acetate			23		8					Recommended	Halar®
			66		8					Recommended	Halar®
Ammonium Alum			28		8					Recommended	Halar®
			149		8					Recommended	Halar®
Ammonium Bifluoride			23		8					Recommended	Halar®
			149		8					Recommended	Halar®
Ammonium Bisulfide			23		8					Recommended	Halar®
			149		8					Recommended	Halar®
Ammonium Carbonate			23		8					Recommended	Halar®
			149		8					Recommended	Halar®
Ammonium Chloride			23		8					Recommended	Halar®
			149		8					Recommended	Halar®
Ammonium Dichromate			23		8					Recommended	Halar®

(Cont'd.)

Table I.3 Chemical Resistance of Ethylene Chlorotrifluoroethylene Copolymer (ECTFE) *(Cont'd.)*

Reagent	Reagent Note	Conc.	Temp (°C)	Time (days)	PDL Rating	% Change Weight	% Retained Tensile Strength	% Retained Modulus	% Retained Elongation	Resistance Note	Material Note
Ammonium Fluoride		10	23		8					Recommended	Halar®
		10	149		8					Recommended	Halar®
		25	23		8					Recommended	Halar®
		25	149		8					Recommended	Halar®
Ammonium Hydroxide		28	23	11	8	<0.1	59.1	80-100	80-100	no stress cracking	Halar® 2.3 mm thick
		28	66	11	8	0.2	89.7	80-100	80-100	no stress cracking	Halar®
			149		8					Recommended	Halar®
Ammonium Metaphosphate			23		8					Recommended	Halar®
			149		8					Recommended	Halar®
Ammonium Nitrate			23		8					Recommended	Halar®
			149		8					Recommended	Halar®
Ammonium Persulfate			23		8					Recommended	Halar®
			149		8					Recommended	Halar®
Ammonium Phosphate			23		8					Recommended	Halar®
			149		8					Recommended	Halar®
Ammonium Sulfate			23		8					Recommended	Halar®
			149		8					Recommended	Halar®
Ammonium Sulfide			23		8					Recommended	Halar®
			149		8					Recommended	Halar®
Amyl Acetate			23		8					Recommended	Halar®
			66		8					Recommended	Halar®
			121		2					Not Recommended	Halar®
Amyl Alcohol			23		8					Recommended	Halar®
			149		8					Recommended	Halar®
Amyl Chloride			23		8					Recommended	Halar®
			149		8					Recommended	Halar®
Aniline			23		8					Recommended	Halar®
			23	11	8	<0.1		80-100	80-100	No stress cracking	Halar® 2.3 mm thick
			66		2					Not Recommended	Halar®
			121	11	1					Attacked	Halar® 2.3 mm thick
Animal Oils	lard oil		23		8					Recommended	Halar®
			149		8					Recommended	Halar®
Anthraquinone			23		8					Recommended	Halar®
			66		8					Recommended	Halar®
Anthraquinonesulfonic Acid			23		8					Recommended	Halar®
			66		8					Recommended	Halar®
Antimony Trichloride			23		8					Recommended	Halar®
Aqua Regia			23		8					Recommended	Halar®
			23	11	8	0.1		80-100	80-100	No stress cracking, may darken	Halar® 2.3 mm thick
			75-105	11	7	0.5		80-100	80-100	No stress cracking, may darken	Halar®
			121		8					Recommended	Halar®
Arsenic Acid			23		8					Recommended	Halar®
			149		8					Recommended	Halar®

(Cont'd.)

Table I.3 Chemical Resistance of Ethylene Chlorotrifluoroethylene Copolymer (ECTFE) *(Cont'd.)*

Reagent	Reagent Note	Conc.	Temp (°C)	Time (days)	PDL Rating	% Change Weight	Tensile Strength	Modulus	Elongation	Resistance Note	Material Note
Barium Carbonate			P3		8					Recommended	Halar®
			149		8					Recommended	Halar®
Barium Chloride			23		8					Recommended	Halar®
			149		8					Recommended	Halar®
Barium Hydroxide			23		8					Recommended	Halar®
			149		8					Recommended	Halar®
Barium Nitrate			23		8					Recommended	Halar®
Barium Sulfate			23		8					Recommended	Halar®
			149		8					Recommended	Halar®
Barium Sulfide			23		8					Recommended	Halar®
			149		8					Recommended	Halar®
Beer			23		8					Recommended	Halar®
Beet Sugar Liquors			23		8					Recommended	Halar®
			149		8					Recommended	Halar®
Benzaldehyde		10	23		8					Recommended	Halar®
		10	66		8					Recommended	Halar®
		10	121		2					Not Recommended	Halar®
		>10	23		8					recommended	Halar®
		>10	66		2					not recommended	Halar®
			23	11	8	0.2		80-100	80-100	no stress cracking	Halar® 2.3 mm thick
			121	11	3	10.5		<25	80-100	no stress cracking	Halar® 2.3 mm thick
Benzene	benzole		23		8					recommended	Halar®
	benzole		23	11	8	0.6		80-100	80-100	no stress cracking	Halar® 2.3 mm thick
	benzole		66		8					recommended	Halar®
	benzole		74	11	3	7		<25	80-100	no stress cracking	Halar® 2.3 mm thick
	benzole		121		2					not recommended	Halar®
Benzenesulfonic Acid		10	23		8					recommended	Halar®
		10	66		8					recommended	Halar®
		10	121		2					not recommended	Halar®
Benzoic Acid			23		8					recommended	Halar®
			121		8					recommended	Halar®
Benzyl Alcohol			23		8					recommended	Halar®
			149		8					recommended	Halar®
Bismuth Carbonate			23		8					recommended	Halar®
Black Liquor			23		8					recommended	Halar®
			149		8					recommended	Halar®
Bleach	12.5% chlorine		23		8					recommended	Halar®
	5.5% chlorine		23		8					recommended	Halar®
	12.5% chlorine		149		8					recommended	Halar®
	5.5% chlorine		149		8					recommended	Halar®

(Cont'd.)

Table I.3 Chemical Resistance of Ethylene Chlorotrifluoroethylene Copolymer (ECTFE) (*Cont'd.*)

Reagent	Reagent Note	Conc.	Temp (°C)	Time (days)	PDL Rating	% Change Weight	Tensile Strength	Modulus	Elongation	Resistance Note	Material Note
Borax			23		8					recommended	Halar®
			149		8					recommended	Halar®
Boric Acid			23		8					recommended	Halar®
			149		8					recommended	Halar®
Brines	brine acid		23		8					recommended	Halar®
	brine acid		121		8					recommended	Halar®
Bromine	bromine vapor	25	23		8					recommended	Halar®
	bromine vapor	25	66		8					recommended	Halar®
	bromine vapor	25	121		2					not recommended	Halar®
	bromine liquid		23		8					recommended	Halar®
	bromine water		23		8					recommended	Halar®
			23	11	7	1.4		80-100	80-100	no stress cracking	Halar® 2.3 mm thick
			23	180	4	10.4	86	79	75	no stress cracking	Halar® 2.3 mm thick
	bromine liquid		66		8					recommended	Halar®
	bromine water		121		8					recommended	Halar®
Bromobenzene			23		8					recommended	Halar®
			66		2					not recommended	Halar®
Bromotoluene			23		8					recommended	Halar®
			58		8					recommended	Halar®
			121		2					not recommended	Halar®
Butadiene			23		8					recommended	Halar®
			121		8					recommended	Halar®
Butane			23		8					recommended	Halar®
					8					recommended	Halar®
Butyl Acetate			23		8					recommended	Halar®
			23	11	8	0.7		80-100	80-100	no stress cracking	Halar® 2.3 mm thick
Butyl Acetate			66		8					recommended	Halar®
			121		2					not recommended	Halar®
			121	11	3	10.5		<25	80-100	no stress cracking	Halar® 2.3 mm thick
Butyl Alcohol	butanol		23		8					recommended	Halar®
	1-butanol		23		8					recommended	Halar®
	2-butanol		23		8					recommended	Halar®
	butanol		23	11	8	<0.1		80-100	80-100	no stress cracking	Halar® 2.3 mm thick
	butanol		118	11	5	2		50-75	80-100	no stress cracking	Halar® 2.3 mm thick
	butanol		149		8					recommended	Halar®
	1-butanol		149		8					recommended	Halar®
	2-butanol		149		8					recommended	Halar®
Butyl Alcohol (sec-)			23		8					recommended	Halar®
			149		8					recommended	Halar®
Butyl Cellosolve			23		8					recommended	Halar®
Butyl Phenol			23		8					recommended	Halar®
			121		8					recommended	Halar®

(*Cont'd.*)

Table I.3 Chemical Resistance of Ethylene Chlorotrifluoroethylene Copolymer (ECTFE) *(Cont'd.)*

Reagent	Reagent Note	Conc.	Temp (°C)	Time (days)	PDL Rating	% Change Weight	Tensile Strength	Modulus	Elongation	Resistance Note	Material Note
								% Retained			
Butyl Stearate			23		8					recommended	Halar®
Butylene			23		8					recommended	Halar®
			149		8					recommended	Halar®
Butyric Acid			23		8					recommended	Halar®
			121		8					recommended	Halar®
Cadmium Cyanide			23		8					recommended	Halar®
			66		8					recommended	Halar®
Calcium Bisulfide			23		8					recommended	Halar®
			149		8					recommended	Halar®
Calcium Bisulfite			23		8					recommended	Halar®
Calcium Carbonate			23		8					recommended	Halar®
			149		8					recommended	Halar®
Calcium Chlorate			23		8					recommended	Halar®
			149		8					recommended	Halar®
Calcium Chloride			23		8					recommended	Halar®
			149		8					recommended	Halar®
Calcium Hydroxide			23		8					recommended	Halar®
			149		8					recommended	Halar®
Calcium Hypochlorite			23		8					recommended	Halar®
			149		8					recommended	Halar®
Calcium Nitrate			23		8					recommended	Halar®
			149		8					recommended	Halar®
Calcium Oxide			23		8					recommended	Halar®
			149		8					recommended	Halar®
Calcium Sulfate			23		8					recommended	Halar®
			149		8					recommended	Halar®
Cane Sugar	can sugar liquors		23		8					recommended	Halar®
	can sugar liquors		66		8					recommended	Halar®
Caprylic Acid			23		8					recommended	Halar®
			66		8					recommended	Halar®
Carbon Dioxide	dry		23		8					recommended	Halar®
	wet		23		8					recommended	Halar®
Carbon Dioxide	dry		149		8					recommended	Halar®
	wet		149		8					recommended	Halar®
Carbon Disulfide			23		8					recommended	Halar®
Carbon Monoxide			23		8					recommended	Halar®
			66		8					recommended	Halar®
Carbon Tetrachloride			23		8					recommended	Halar®
			149		8					recommended	Halar®
Carbonic Acid			23		8					recommended	Halar®
			149		8					recommended	Halar®

(Cont'd.)

Table I.3 Chemical Resistance of Ethylene Chlorotrifluoroethylene Copolymer (ECTFE) *(Cont'd.)*

Reagent	Reagent Note	Conc.	Temp (°C)	Time (days)	PDL Rating	% Change Weight	Tensile Strength	% Retained Modulus	% Retained Elongation	Resistance Note	Material Note
Castor Oil			23		8					recommended	Halar®
			149		8					recommended	Halar®
Caustic Potash			23		8					recommended	Halar®
			149		8					recommended	Halar®
Cellosolve	2-ethoxyethanol		23		8					recommended	Halar®
			149		8					recommended	Halar®
Cellosolve Acetate			23		8					recommended	Halar®
Chloral Hydrate			23		8					recommended	Halar®
			66		8					recommended	Halar®
Chloramines			23		8					recommended	Halar®
Chlorine	chlorine water		23		8					recommended	Halar®
	dry gas		21		8					recommended	Halar®
	liquid		23		8					recommended	Halar®
	moist gas		23		8					recommended	Halar®
	dry gas		66		8					recommended	Halar®
	chlorine water		121		8					recommended	Halar®
	dry gas		121		2					not recommended	Halar®
	liquid		121		8					recommended	Halar®
	moist gas		121		8					recommended	Halar®
Chloroacetic Acid			23		8					recommended	Halar®
			121		8					recommended	Halar®
Chlorobenzene			23		8					recommended	Halar®
			23	11	6	0.9		50-75	80-100	no stress cracking	Halar® 2.3 mm thick
			66		8					recommended	Halar®
			121		2					not recommended	Halar®
			121	11	3	19.5		<25	80-100	no stress cracking	Halar® 2.3 mm thick
Chlorobenzyl Chloride			243		8					Recommended	Halar®
			66		2					not recommended	Halar®
Chloroform			23		8					recommended	Halar®
			23	11	4	4.5		50-75	80-100	no stress cracking	Halar® 2.3 mm thick
			121		8					recommended	Halar®
Chlorosulfonic Acid		60	23	11	8	0.1		80-100	80-100	no stress cracks, may darken	Halar® 2.3 mm thick
			23		8					recommended	Halar®
Chrome Alum			23		8					recommended	Halar®
			149		8					recommended	Halar®
Chromic Acid		10	23		8					recommended	Halar®
		10	121		8					recommended	Halar®
		30	23		8					recommended	Halar®
		30	121		8					recommended	Halar®
		40	23		8					recommended	Halar®

(Cont'd.)

Table I.3 Chemical Resistance of Ethylene Chlorotrifluoroethylene Copolymer (ECTFE) *(Cont'd.)*

Reagent	Reagent Note	Conc.	Temp (°C)	Time (days)	PDL Rating	% Change Weight	Tensile Strength	Modulus	Elongation	Resistance Note	Material Note
Chromic Acid		40	121		8					recommended	Halar®
		50	23		8					recommended	Halar®
		50	23	11	8	<0.1		80-100	80-100	no stress cracking; may darken	Halar® 2.3 mm thick
		50	111	11	7	0.4		80-100	80-100	no stress cracking; may darken	Halar® 2.3 mm thick
		50	121		8					recommended	Halar®
Citric Acid			23		8					recommended	Halar®
			149		8					recommended	Halar®
Coconut Oil			23		8					recommended	Halar®
			149		8					recommended	Halar®
Coke Oven Gas			23		8					recommended	Halar®
			121		8					recommended	Halar®
Copper Carbonate			23		8					recommended	Halar®
			66		8					recommended	Halar®
Copper Chloride			23		8					recommended	Halar®
			121		8					recommended	Halar®
Copper Cyanide			23		8					recommended	Halar®
			121		8					recommended	Halar®
Copper Fluoride			23		8					recommended	Halar®
			121		8					recommended	Halar®
Copper Nitrate			23		8					recommended	Halar®
			121		8					recommended	Halar®
Copper Sulfate			23		8					recommended	Halar®
			121		8					recommended	Halar®
Corn Syrup			23		8					recommended	Halar®
			121		8					recommended	Halar®
Cottonseed Oil			23		8					recommended	Halar®
			121		8					recommended	Halar®
Cresol			23		8					recommended	Halar®
			66		8					recommended	Halar®
			121		2					not recommended	Halar®
Cresylic Acid		50	23		8					recommended	Halar®
		50	66		8					recommended	Halar®
		50	121		2					not recommended	Halar®
Crotonaldehyde			23		8					recommended	Halar®
			66		2					not recommended	Halar®
Crude Oils	sour crude		23		8					recommended	Halar®
	sour crude		23		8					recommended	Halar®
			149		8					recommended	Halar®
			149		8					recommended	Halar®
Cupric Chloride		25	23	11	8	<0.1		80-100	80-100	no stress cracking no stress	Halar® 2.3 mm thick
		25	103	11	8	<0.1		80-100	80-100	cracking	Halar® 2.3 mm thick

(Cont'd.)

Table I.3 Chemical Resistance of Ethylene Chlorotrifluoroethylene Copolymer (ECTFE) *(Cont'd.)*

Reagent	Reagent Note	Conc.	Temp (°C)	Time (days)	PDL Rating	% Change Weight	Tensile Strength	% Retained Modulus	Elongation	Resistance Note	Material Note
Cupric Fluoride			23		8					recommended	Halar®
			149		8					recommended	Halar®
Cupric Sulfate			23		8					recommended	Halar®
			149		8					recommended	Halar®
Cuprous Chloride			23		8					recommended	Halar®
			121		8					recommended	Halar®
Cutting Fluids	thread cutting oils		23		8					recommended	Halar®
	thread cutting oils		149		8					recommended	Halar®
Cyclohexane			23		8					recommended	Halar®
			149		8					recommended	Halar®
Cyclohexanone			23		8					recommended	Halar®
			66		2					not recommended	Halar®
Cyclohexyl Alcohol			23		8					recommended	Halar®
			66		8					recommended	Halar®
			121		2					not recommended	Halar®
Detergents	heavy duty sol.		23		8					recommended	Halar®
	heavy duty sol.		23		8					recommended	Halar®
			149		8					recommended	Halar®
			149		8					recommended	Halar®
Dextrin			23		8					recommended	Halar®
			121		8					recommended	Halar®
Dextrose			23		8					recommended	Halar®
			121		8					recommended	Halar®
Diacetone Alcohol			23		8					recommended	Halar®
			66		8					recommended	Halar®
			121		2					not recommended	Halar®
Dichlorobenzene			23		8					recommended	Halar®
			66		2					recommended	Halar®
Dichloroethylene			23		8					recommended	Halar®
			66		2					recommended	Halar®
Diesel Fuels			23		8					recommended	Halar®
			149		8					recommended	Halar®
Diethyl Cellosolve			23		8					recommended	Halar®
			149		8					recommended	Halar®
Diethyl Ether			23		8					recommended	Halar®
Diethylamine			23		8					recommended	Halar®
			66		2					not recommended	Halar®
Diglycolic Acid			23		8					recommended for use	Halar®
Dimethyl Phthalate			23	11	8	.0.1		80-100	80-100	no stress cracking	Halar® 2.3 mm thick
			121	11	4	3.5		50-75	80-100	no stress cracking	Halar® 2.3 mm thick
Dimethyl Sulfoxide					8	0.1		80-100	80-100	no stress cracking	Halar® 2.3 mm thick
					8	3		80-100	80-100	no stress cracking	Halar® 2.3 mm thick

(Cont'd.)

Table I.3 Chemical Resistance of Ethylene Chlorotrifluoroethylene Copolymer (ECTFE) *(Cont'd.)*

Reagent	Reagent Note	Conc.	Temp (°C)	Time (days)	PDL Rating	% Change Weight	Tensile Strength	% Retained Modulus	% Retained Elongation	Resistance Note	Material Note
Dimethylamine			23		8					recommended	Halar®
			66		2					not recommended	Halar®
Dimethyl formamide			23	11	5	2		50-75	80-100	no stress cracking	Halar® 2.3 mm thick
			121	11	3	7.5		<25	80-100	no stress cracking	Halar® 2.3 mm thick
Dimethylhydrazine			23		8					recommended	Halar®
			66		2					not recommended	Halar®
Dioctyl Phthalate			23		8					recommended	Halar®
			66		2					not recommended	Halar®
Dioxane			23		8					recommended	Halar®
			23	11	7	0.9		80-100	80-100	no stress cracking	Halar® 2.3 mm thick
			66		8					recommended	Halar®
			102	11	5	16		80-100	80-100	no stress cracking	Halar® 2.3 mm thick
			121		2					not recommended	Halar®
Dioxane (1,4-)			23		8					recommended	Halar®
			66		8					recommended	Halar®
			121		2					not recommended	Halar®
Disodium Phosphate			23		8					recommended	Halar®
			149		8					recommended	Halar®
Divinylbenzene			23		8					recommended	Halar®
			66		2					not recommended	Halar®
Epsom Salts			23		8					recommended	Halar®
			140		8					recommended	Halar®
Ethyl Acetate			23		8					recommended	Halar®
			23	11	7	1.5		80-100	80-100	no stress cracking	Halar® 2.3 mm thick
			60		8					recommended	Halar®
			71	11	3	6.5		25-50	80-100	no stress cracking	Halar® 2.3 mm thick
Ethyl Acetoacetate			23		8					recommended	Halar®
Ethyl Acrylate			23		a					recommended	Halar®
			66		8					recommended	Halar®
			121		2					not recommended	Halar®
Ethyl Alcohol			23		8					recommended	Halar®
			149							recommended	Halar®
Ethyl Chloride			23		8					recommended	Halar®
			149		8					recommended	Halar®
Ethyl Chloroacetate			23		8					recommended	Halar®
Ethyl Ether			23		8					recommended	Halar®
			23	11	7	0.9		80-100	80-100	no stress cracking	Halar® 2.3 mm thick
			66		a					recommended	Halar®
Ethylene Bromide			23		8					recommended	Halar®
			149		8					recommended	Halar®
Ethylene Chloride			23		8					recommended	Halar®
			149		8					recommended	Halar®

(Cont'd.)

Table I.3 Chemical Resistance of Ethylene Chlorotrifluoroethylene Copolymer (ECTFE) *(Cont'd.)*

Reagent	Reagent Note	Conc.	Temp (°C)	Time (days)	PDL Rating	% Change Weight	% Retained Tensile Strength	% Retained Modulus	% Retained Elongation	Resistance Note	Material Note
Ethylene Chlorohydrin			23		8					recommended	Halar®
			66		2					not recommended	Halar®
Ethylene Dichloride			23		8					recommended	Halar®
			23	11	7	1		80-100	80-100	no stress cracking	Halar® 2.3 mm thick
			66		2					recommended	Halar®
			85	11	5	9.5		80-100	80-100	no stress cracking	Halar® 2.3 mm thick
Ethylene Glycol			23		8					recommended	Halar®
			149		8					recommended	Halar®
Ethylene Oxide			23		8					recommended	Halar®
			149		8					recommended	Halar®
Ethylenediamine			23		8					recommended	Halar®
			23	11	8	0.2		80-100	80-100	no Stress Cracking	Halar® 2.3 mm thick
			66		2					not recommended	Halar®
			118	11	1					attacked	Halar® 2.3 mm thick
Fatty Acids			22		8					recommended	Halar®
			149		8					recommended	Halar®
Ferric Chloride		25	23	11	8	0.1		80-100	80-100	no stress cracking	Halar® 2.3 mm thick
		25	103	11	8	0.1		80-100	80-100	no stress cracking	Halar® 2.3 mm thick
			23		8					recommended	Halar®
			149		8					recommended	Halar®
Ferric Nitrate			23		8					recommended	Halar®
			149		8					recommended	Halar®
Ferric Sulfate			23		8					recommended	Halar®
			149		6					recommended	Halar®
Ferrous Chloride			23		8					recommended	Halar®
			149		8					recommended	Halar®
Ferrous Nitrate			23		8					recommended	Halar®
			149		8					recommended	Halar®
Ferrous Sulfate			23		8					recommended	Halar®
			149		8					recommended	Halar®
Fluoboric Acid	wet gas		22		8					recommended	Halar®
Fluorine			23		8					recommended	Halar®
Fluosilicic Acid			23		8					recommended	Halar®
			149		8					recommended	Halar®
Formaldehyde		35	23		8					recommended	Halar®
		35	66		8					recommended	Halar®
		37	23		8					recommended	Halar®
		37	66		8					recommended	Halar®
		50	23		8					recommended	Halar®

(Cont'd.)

Table I.3 Chemical Resistance of Ethylene Chlorotrifluoroethylene Copolymer (ECTFE) *(Cont'd.)*

Reagent	Reagent Note	Conc.	Temp (°C)	Time (days)	PDL Rating	% Change Weight	% Retained Tensile Strength	% Retained Modulus	% Retained Elongation	Resistance Note	Material Note
Formic Acid	anhydrous		23		8					recommended	Halar®
			23		8					recommended	Halar®
	anhydrous		121		8					recommended	Halar®
			121		8					recommended	Halar®
Freon® 11			23		8					recommended	Halar®
			66		8					recommended	Halar®
Freon® 113			23		8					recommended	Halar®
			66		8					recommended	Halar®
Freon® 114			23		8					recommended	Halar®
			66		8					recommended	Halar®
Freon® 12			23		8					recommended	Halar®
			66		8					recommended	Halar®
Freon® 21			23		8					recommended	Halar®
			66		8					recommended	Halar®
Freon® 22			23		8					recommended	Halar®
			66		8					recommended	Halar®
Fruit Juices	and pulp		23		8					recommended	Halar®
	and pulp		66		8					recommended	Halar®
Hydrochloric Acid		37	21		8					recommended	Halar®
		37	23	11	8	<0.1		80-100	80-100	no stress cracks; may darken	Halar® 2.3 mm thick
		37	75-105	11	8	0.1		80-100	80-100	no stress cracks; may darken	Halar® 2.3 mm thick
		37	149		8					recommended	Halar®
		60	23	11	8	<0.1		80-100	80-100	no stress cracks; may darken	Halar® 2.3 mm thick
Hydrocyanic Acid		10	23		8					recommended	recommended
		10	149		8					recommended	recommended
			23		8					recommended	Halar®
			149		8					recommended	Halar®
Hydrofluoric Acid		30	23		8					recommended	Halar®
		30	121		8					recommended	Halar®
		40	23		8					recommended	Halar®
		40	121		8					recommended	Halar®
		50	23		8					recommended	Halar®
		50	121		8					recommended	Halar®
	dilute		23		8					recommended	Halar®
	dilute		149		8					recommended	Halar®
Hydrofluorosilicic Acid			23		8					recommended	Halar®
			149		8					recommended	Halar®
Hydrogen			23		8					recommended	Halar®
			149		8					recommended	Halar®
Hydrogen Cyanide			23		8					recommended	Halar®
			149		8					recommended	Halar®

(Cont'd.)

Table I.3 Chemical Resistance of Ethylene Chlorotrifluoroethylene Copolymer (ECTFE) *(Cont'd.)*

Reagent	Reagent Note	Conc.	Temp (°C)	Time (days)	PDL Rating	% Change Weight	Tensile Strength	Modulus	Elongation	Resistance Note	Material Note
Hydrogen Peroxide		50	23		8					recommended	Halar®
		50	66		8					recommended	Halar®
		90	23		8					recommended	Halar®
		90	66		8					recommended	Halar®
			23		8					recommended	Halar®
Hydrogen Phosphide			23		8					recommended	Halar®
			66		8					recommended	Halar®
Hydrogen Sulfide	aqueous solution		23		8					recommended	Halar®
	dry		23		8					recommended	Halar®
	aqueous solution		66		8					recommended	Halar®
	dry		149		8					recommended	Halar®
Hydroquinone			23		8					recommended	Halar®
			121		8					recommended	Halar®
Hypochlorous Acid			23		8					recommended	Halar®
			149		8					recommended	Halar®
Iodine	solution	10	23		8					recommended	Halar®
	solution	10	121		8					recommended	Halar®
			23		8					recommended	Halar®
			121		8					recommended	Halar®
Isooctane			23		8					recommended	Halar®
			23	11	6	<0.1		80-100	80-100	no stress cracking	Halar® 2.3 mm thick
			115	11	4	3.3		50-75	80-100	no stress cracking	Halar® 2.3 mm thick
Isopropyl Alcohol	isopropanol		23		8					recommended	Halar®
	isopropanol		149		8					recommended	Halar®
Isopropyl Ether			23		8					recommended	Halar®
Gallic Acid			23		8					recommended	Halar®
			66		8					recommended	Halar®
Gasoline	leaded		23		8					recommended	Halar®
	natural		23		8					recommended	Halar®
	sour		23		8					recommended	Halar®
	unleaded		23		8					recommended	Halar®
	leaded		149		8					recommended	Halar®
	natural		149		8					recommended	Halar®
	sour		49		8					recommended	Halar®
	unleaded		149		8					recommended	Halar®
Gelatins			23		8					recommended	Halar®
			121		8					recommended	Halar®
Genetron 113	chlorinated solvent		23	11	8	0.4		80-100	80-100	no stress cracking	Halar® 2.3 mm thick
	chlorinated solvent		49	11	4	2		<25	80-100	no stress cracking	Halar® 2.3 mm thick

(Cont'd.)

Table I.3 Chemical Resistance of Ethylene Chlorotrifluoroethylene Copolymer (ECTFE) *(Cont'd.)*

Reagent	Reagent Note	Conc.	Temp (°C)	Time (days)	PDL Rating	% Change Weight	Tensile Strength	Modulus	Elongation	Resistance Note	Material Note
Gin			23		8					recommended	Halar®
			149		8					recommended	Halar®
Glucose			23		8					recommended	Halar®
			149		8					recommended	Halar®
Glycerin	Glycerol		23		8					recommended	Halar®
	glycerin		23		8					recommended	Halar®
	Glycerol		149		8					recommended	Halar®
	glycerin		149		8					recommended	Halar®
Glycolic Acid			23		8					recommended	Halar®
			66		8					recommended	Halar®
Glycols			23		8					recommended	Halar®
			149		8					recommended	Halar®
Heptane			23		8					recommended	Halar®
			149		8					recommended	Halar®
Hexane			23		8					recommended	Halar®
			23	11	8	0.1		80-100	80-100	no stress cracking	Halar® 2.3 mm thick
			54	11	5	1.4		50-75	80-100	no stress cracking	Halar® 2.3 mm thick
			149		8					recommended	Halar®
Hexyl Alcohol			23		8					recommended	Halar®
			149		8					recommended	Halar®
Hydrobromic Acid		20	23		8					recommended	Halar®
		20	149		8					recommended	Halar®
		50	23		8					recommended	Halar®
		50	149		8					recommended	Halar®
Jet Aircraft Fuels	JP 4		23		8					recommended	Halar®
	JP 5		23		8					recommended	Halar®
	JP 4		149		8					recommended	Halar®
	JP 5		149		8					recommended	Halar®
Kerosene			23		8					recommended	Halar®
			149		8					recommended	Halar®
Lactic Acid		25	23		8					recommended	Halar®
		25	66		8					recommended	Halar®
		80	23		8					recommended	Halar®
Lauric Acid			23		8					recommended	Halar®
			121		8					recommended	Halar®
Lauryl Chloride			23		8					recommended	Halar®
			121		8					recommended	Halar®
Lead Acetate			23		8					recommended	Halar®
			149		8					recommended	Halar®

(Cont'd.)

Table I.3 Chemical Resistance of Ethylene Chlorotrifluoroethylene Copolymer (ECTFE) (Cont'd.)

Reagent	Reagent Note	Conc.	Temp (°C)	Time (days)	PDL Rating	% Change Weight	Tensile Strength	Modulus	Elongation	Resistance Note	Material Note
Lead Chloride			23		8					recommended	Halar®
			149		8					recommended	Halar®
Lead Nitrate			23		8					recommended	Halar®
			149		8					recommended	Halar®
Lead Sulfate			23		8					recommended	Halar®
			149		8					recommended	Halar®
Lemon Oil			23		8					recommended	Halar®
			121		8					recommended	Halar®
Lime Sulfur			23		8					recommended	Halar®
			66		8					recommended	Halar®
Linoleic Acid			23		8					recommended	Halar®
			121		8					recommended	Halar®
Linoleic Oil			23		8					recommended	Halar®
			121		8					recommended	Halar®
Linseed Oil	blue		23		8					recommended	Halar®
			23		8					recommended	Halar®
			121		8					recommended	Halar®
	blue		121		8					recommended	Halar®
Lithium Bromide			23		8					recommended	Halar®
			66		8					recommended	Halar®
Lubricating Oils	ASTM Oil No. 1		23		8					recommended	Halar®
	ASTM Oil No. 2		23		8					recommended	Halar®
	ASTM Oil No. 3		23		8					recommended	Halar®
	ASTM Oil No. 1		149		8					recommended	Halar®
	ASTM Oil No. 2		149		8					recommended	Halar®
	ASTM Oil No. 3		149		8					recommended	Halar®
Magnesium Carbonate			23		8					recommended	Halar®
			149		8					recommended	Halar®
Magnesium Chloride			23		8					recommended	Halar®
			149		8					recommended	Halar®
Magnesium Hydroxide			23		8					recommended	Halar®
			149		8					recommended	Halar®
Magnesium Nitrate			23		8					recommended	Halar®
			149		8					recommended	Halar®
Magnesium Sulfate			23		8					recommended	Halar®
			149		8					recommended	Halar®
Maleic Acid			23		8					recommended	Halar®
			121		8					recommended	Halar®
Malic Acid			23		8					recommended	Halar®
			121		8					recommended	Halar®

(Cont'd.)

Table I.3 Chemical Resistance of Ethylene Chlorotrifluoroethylene Copolymer (ECTFE) *(Cont'd.)*

Reagent	Reagent Note	Conc.	Temp (°C)	Time (days)	PDL Rating	% Change Weight	% Retained Tensile Strength	Modulus	Elongation	Resistance Note	Material Note
Methanol			23	11	8	0.1		80-100	80-100	no stress cracking	Halar® 2.3 mm thick
			60	11	6	0.4		50-75	80-100	no stress cracking	Halar® 2.3 mm thick
Mercuric Chloride			23		8					recommended	Halar®
			121		8					recommended	Halar®
Mercuric Cyanide			23		8					recommended	Halar®
			121		8					recommended	Halar®
Mercuric Sulfate			23		8					recommended	Halar®
			121		8					recommended	Halar®
Mercurous Nitrate			23		8					recommended	Halar®
			121		8					recommended	Halar®
Mercury			23		8					recommended	Halar®
			149		8					recommended	Halar®
Methane			23		8					recommended	Halar®
			121		8					recommended	Halar®
Methoxyethyl Oleate			23		8					recommended	Halar®
Methyl Alcohol			23		8					recommended	Halar®
			149		8					recommended	Halar®
Methyl Bromide			23		8					recommended	Halar®
			149		8					recommended	Halar®
Methyl Cellosolve			23		8					recommended	Halar®
			149		8					recommended	Halar®
Methyl Chloride			23		8					recommended	Halar®
			149		8					recommended	Halar®
Methyl Ethyl Ketone			23		8					recommended	Halar®
			23	11	7	1		80-100	80-100	no stress cracking	Halar® 2.3 mm thick
			66		8					recommended	Halar®
			79	11	3	6		<25	80-100	no stress cracking	Halar® 2.3 mm thick
			121		2					not recommended	Halar®
Methyl Isobutyl Ketone			23		8					recommended	Halar®
			66		8					recommended	Halar®
			121		2					not recommended	Halar®
Methyl Isopropyl Ketone			23	11	8	0.5		50-75	80-100	no stress cracking	Halar® 2.3 mm thick
			116	11	3	9		<25	80-100	no stress cracking	Halar® 2.3 mm thick
Methyl Methacrylate			23		8					recommended	Halar®
Methyl Sulfate			23		8					recommended	Halar®
			149		8					recommended	Halar®
Methyl Sulfuric Acid			23		8					recommended	Halar®
			66		8					recommended	Halar®
Methylamine			23		8					recommended	Halar®
			66		2					not recommended	Halar®

(Cont'd.)

Table I.3 Chemical Resistance of Ethylene Chlorotrifluoroethylene Copolymer (ECTFE) (Cont'd.)

Reagent	Reagent Note	Conc.	Temp (°C)	Time (days)	PDL Rating	% Change Weight	% Retained			Resistance Note	Material Note
							Tensile Strength	Modulus	Elongation		
Methylene Bromide			23		8					recommended	Halar®
			66		2					not recommended	Halar®
Methylene Chloride			23		8					recommended	Halar®
			23	11	3	8		25-50	80-100	no stress cracking	Halar®: 2.3 mm thick
			41	11	3	9		<25	80-100	no stress cracking	Halar®: 2.3 mm thick
			66		2					not recommended	Halar®
Methylene Iodine			23		8					recommended	Halar®
			66		2					not recommended	Halar®
Milk			23		8					recommended	Halar®
			121		8					recommended	Halar®
Mineral Oils			23		8					recommended	Halar®
			149		8					recommended	Halar®
Molasses			23		8					recommended	Halar®
			66		8					recommended	Halar®
Motor Oils			23		8					recommended	Halar®
			149		8					recommended	Halar®
Naphtha			23		8					recommended	Halar®
			149		8					recommended	Halar®
Naphthalene			23		8					recommended	Halar®
			66		8					recommended	Halar®
Natural Gas			23		8					recommended	Halar®
			66		8					recommended	Halar®
Nickel Acetate			23		8					recommended	Halar®
Nickel Chloride			23		8					recommended	Halar®
			149		8					recommended	Halar®
Nickel Nitrate			23		8					recommended	Halar®
			149		8					recommended	Halar®
Nickel Sulfate			23		8					recommended	Halar®
			149		8					recommended	Halar®
Nicotine			23		8					recommended	Halar®
			66		8					recommended	Halar®
Nicotinic Acid			23		8					recommended	Halar®
			121		8					recommended	Halar®

(Cont'd.)

Table I.3 Chemical Resistance of Ethylene Chlorotrifluoroethylene Copolymer (ECTFE) *(Cont'd.)*

Reagent	Reagent Note	Conc.	Temp (°C)	Time (days)	PDL Rating	% Change Weight	Tensile Strength	Modulus	Elongation	Resistance Note	Material Note
							% Retained				
Nitric Acid		10	23		8					recommended	Halar®
		10	149		8					recommended	Halar®
		30	23		8					recommended	Halar®
		30	121		8					recommended	Halar®
		40	23		8					recommended	Halar®
		40	121		8					recommended	Halar®
		50	23		8					recommended	Halar®
		50	66		8					recommended	Halar®
		50	121		2					not recommended	Halar®
		70	23		8					recommended	Halar®
		70	23	11	8	<0.1		80-100	80-100	no stress cracks may darken	Halar® 2.3 mm thick
		70	66		8					recommended	Halar®
		70	100	11	7	0.5		80-100	80-100	no stress cracks may darken	Halar® 2.3 mm thick
		70	121		2					not recommended	Halar®
		70	121	11	4	0.8		<25	50-75	no stress cracks may darken	Halar® 2.3 mm thick
		90	23	180	8	0.7	98.6	97	112	no stress cracks may darken	Halar® 3.2 mm thick
		100	121		2					not recommended	Halar®
	red fuming		23	180	7	1.8	92	87	106	no stress cracks may darken	Halar® 3.2 mm thick
	red fuming		23		8					recommended	Halar®
	red fuming		66		8					recommended	Halar®
Nitrobenzene			23		8					recommended	Halar®
			23	11	8	0.2		80-100	80-100	no stress cracking	Halar® 2.3 mm thick
			66		8					recommended	Halar®
			121		2					not recommended	Halar®
			121	11	3	11.5		<25	80-100	no stress cracking	Halar® 2.3 mm thick
Nitrous Acid		10	23		8					recommended	Halar®
		10	121		8					recommended	Halar®
Nitrous Oxide			23		8					recommended	Halar®
Oleic Acid			23		8					recommended	Halar®
			121		8					recommended	Halar®
Oleum		30	23	180	8	0.3	92.9	100	81.2	no stress cracking	Halar®, 3.2 mm thick
			23		8					recommended	Halar®
			66		2					not recommended	Halar®
Oxalic Acid		50	23		8					recommended	Halar®
		50	121		2					not recommended	Halar®
			23		8					recommended	Halar®
			66		8					recommended	Halar®
Oxygen	gas		23		8					recommended	Halar®
	gas		149		8					recommended	Halar®
Ozone			23		8					recommended	Halar®
			149		8					recommended	Halar®

(Cont'd.)

Table I.3 Chemical Resistance of Ethylene Chlorotrifluoroethylene Copolymer (ECTFE) (*Cont'd.*)

Reagent	Reagent Note	Conc.	Temp (°C)	Time (days)	PDL Rating	% Change Weight	Tensile Strength	Modulus	Elongation	Resistance Note	Material Note
Palmitic Acid		10	23		8					recommended	Halar®
		10	121		8					recommended	Halar®
			23		8					recommended	Halar®
			121		8					recommended	Halar®
Paraffin			23		8					recommended	Halar®
			66		8					recommended	Halar®
Perchloric Acid		10	23		8					recommended	Halar®
		10	66		8					recommended	Halar®
		70	23		8					recommended	Halar®
		70	66		8					recommended	Halar®
Perchloroethylene			23	11	7	1		80-100	80-100	no stress cracking	Halar® 2.3 mm thick
			121	11	3	29		<25	80-100	no stress cracking	Halar® 2.3 mm thick
Perphosphate			23		8					recommended	Halar®
Petroleum Oils	refined		23		8					recommended	Halar®
	sour		23		8					recommended	Halar®
	refined		66		8					recommended	Halar®
	sour		66		8					recommended	Halar®
Phenol			23		8					recommended	Halar®
			66		8					recommended	Halar®
			121		2					not recommended	Halar®
Phenylhydrazine			23		8					Recommended	Halar®
Phosphoric Acid		10	23		8					recommended	Halar®
		10	149		8					recommended	Halar®
		50	23		8					recommended	Halar®
		50	121		8					recommended	Halar®
		85	23		8					recommended	Halar®
		85	121		8					recommended	Halar®
Phosphorous	yellow		23		8					Recommended	Halar®
Phosphorous Pentoxide			28		8					recommended	Halar®
			121		8					recommended	Halar®
Phosphorous Trichloride			23		8					recommended	Halar®
			121		8					recommended	Halar®
Photographic Solutions			23		8					recommended	Halar®
			66		8					recommended	Halar®
Picric Acid			23		8					Recommended	Halar®

(*Cont'd.*)

Table I.3 Chemical Resistance of Ethylene Chlorotrifluoroethylene Copolymer (ECTFE) *(Cont'd.)*

Reagent	Reagent Note	Conc.	Temp (°C)	Time (days)	PDL Rating	% Change Weight	Tensile Strength	Modulus	Elongation	Resistance Note	Material Note
								% Retained			
Plating Solutions	brass		23		8					recommended	Halar®
	cadmium		23		8					recommended	Halar®
	Chrome		23		8					recommended	Halar®
	copper		23		8					recommended	Halar®
	gold		23		8					recommended	Halar®
	lead		23		8					recommended	Halar®
	nickel		23		8					recommended	Halar®
	rhodium		23		8					recommended	Halar®
	silver		23		8					recommended	Halar®
	tin		23		8					recommended	Halar®
	zinc		23		8					recommended	Halar®
	brass		66		8					recommended	Halar®
	cadmium		66		8					recommended	Halar®
	chrome		66		8					recommended	Halar®
	copper		66		8					recommended	Halar®
Plating Solutions	gold		66		8					recommended	Halar®
	lead		66		8					recommended	Halar®
	nickel		66		8					recommended	Halar®
	rhodium		66		8					recommended	Halar®
	silver		66		8					recommended	Halar®
	fin		66		8					recommended	Halar®
	zinc		66		8					recommended	Halar®
Potash			23		8					recommended	Halar®
			149		8					recommended	Halar®
Potassium Alum			23		8					recommended	Halar®
			66		8					recommended	Halar®
			121		8					recommended	Halar®
			149		8					recommended	Halar®
Potassium Aluminum Sulfate			23		8					recommended	Halar®
			149		8					recommended	Halar®
Potassium Bichromate			23		8					recommended	Halar®
			121		8					recommended	Halar®
Potassium Bisulfate			23		8					recommended	Halar®
			121		8					recommended	Halar®
Potassium Borate			23		8					recommended	Halar®
			66		8					recommended	Halar®
Potassium Bromide			23		8					recommended	Halar®
			149		8					recommended	Halar®
Potassium Carbonate			23		8					recommended	Halar®
			149		8					recommended	Halar®

(Cont'd.)

Table I.3 Chemical Resistance of Ethylene Chlorotrifluoroethylene Copolymer (ECTFE) *(Cont'd.)*

Reagent	Reagent Note	Conc.	Temp (°C)	Time (days)	PDL Rating	% Change Weight	% Retained Tensile Strength	% Retained Modulus	% Retained Elongation	Resistance Note	Material Note
Potassium Chlorate	aqueous		23		8					recommended	Halar®
			23		8					recommended	Halar®
	aqueous		149		8					recommended	Halar®
			149		8					recommended	Halar®
Potassium Chloride			23		8					recommended	Halar®
			149		8					recommended	Halar®
Potassium Chromate			23		8					recommended	Halar®
			149		8					recommended	Halar®
Potassium Cyanide			23		8					recommended	Halar®
			149		8					recommended	Halar®
Potassium Dichromate			23		8					recommended	Halar®
			149		8					recommended	Halar®
Potassium Ferricyanide			23		8					recommended	Halar®
			149		8					recommended	Halar®
Potassium Ferrocyanide			23		8					recommended	Halar®
			149		8					recommended	Halar®
Potassium Hydroxide			23		8					recommended	Halar®
			66		8					recommended	Halar®
Potassium iodide			23		8					recommended	Halar®
			121		8					recommended	Halar®
Potassium Nitrate			23		8					recommended	Halar®
			149		8					recommended	Halar®
Potassium Perchlorate			23		8					recommended	Halar®
Potassium Permanganate		10	23		8					recommended	Halar®
		10	149		6					recommended	Halar®
		25	23		8					recommended	Halar®
		25	149		8					recommended	Halar®
Potassium Persulfate			23		8					recommended	Halar®
			121		8					recommended	Halar®
Potassium Sulfate			23		8					recommended	Halar®
			149		8					recommended	Halar®
Propane			23		8					recommended	Halar®
			149		8					recommended	Halar®
Propyl Alcohol	1-propanol		23		8					recommended	Halar®
	1-propanol		149		8					recommended	Halar®
Propylene Oxide			23		2					Not recommended	Halar®
			23	11	3	6		<25	80-100	no stress cracking	Halar® 2.3 mm thick
Pyridine			23		2					not recommended	Halar®
Pyrogallic Acid			23		8					recommended	Halar®
			66		8					recommended	Halar®

(Cont'd.)

Table I.3 Chemical Resistance of Ethylene Chlorotrifluoroethylene Copolymer (ECTFE) (Cont'd.)

Reagent	Reagent Note	Conc.	Temp (°C)	Time (days)	PDL Rating	% Change Weight	% Retained			Resistance Note	Material Note
							Tensile Strength	Modulus	Elongation		
Salicylaldehyde			23		8					recommended	Halar®
			66		8					recommended	Halar®
Salicylic Acid			23		8					recommended	Halar®
			66		8					recommended	Halar®
Sea Water			23		8					recommended	Halar®
			149		8					recommended	Halar®
Sewage	sewage water		23		8					recommended	Halar®
	sewage water		149		8					recommended	Halar®
Silicic Acid			23		8					recommended	Halar®
Silicone Oils			23		8					recommended	Halar®
Silver Cyanide			23		8					recommended	Halar®
			149		8					recommended	Halar®
Silver Nitrate			23		8					recommended	Halar®
			149		8					recommended	Halar®
Silver Sulfate			23		8					recommended	Halar®
			149		8					recommended	Halar®
Soap			23		8					recommended	Halar®
			66		8					recommended	Halar®
Sodium Acetate			23		8					recommended	Halar®
			149		8					recommended	Halar®
Sodium Alum			23		8					recommended	Halar®
			149		8					recommended	Halar®
Sodium Benzoate			23		8					recommended	Halar®
			149		8					recommended	Halar®
Sodium Bicarbonate			23		8					recommended	Halar®
			149		8					recommended	Halar®
Sodium Bichromate			23		8					recommended	Halar®
			66		8					recommended	Halar®
Sodium Bisulfate			23		8					recommended	Halar®
			149		8					recommended	Halar®
Sodium Bisulfite			23		8					recommended	Halar®
			149		8					recommended	Halar®
Sodium Bromide			23		8					recommended	Halar®
			149		8					recommended	Halar®
Sodium Carbonate			23		8					recommended	Halar®
			149		8					recommended	Halar®
Sodium Chlorate			23		8					recommended	Halar®
			149		8					recommended	Halar®
Sodium Chloride			23		8					recommended	Halar®
			149		8					recommended	Halar®

(Cont'd.)

Table I.3 Chemical Resistance of Ethylene Chlorotrifluoroethylene Copolymer (ECTFE) *(Cont'd.)*

Reagent	Reagent Note	Conc.	Temp (°C)	Time (days)	PDL Rating	% Change Weight	Tensile Strength	Modulus	Elongation	Resistance Note	Material Note
Sodium Cyanide			23		8					recommended	Halar®
			149		8					recommended	Halar®
Sodium Dichromate			23		8					recommended	Halar®
			66		8					recommended	Halar®
Sodium Fluoride			23		8					recommended	Halar®
			149		8					recommended	Halar®
Sodium Hydroxide		15	23		8					recommended	Halar®
		15	149		8					recommended	Halar®
		30	23		8					recommended	Halar®
		30	121		8					recommended	Halar®
		so	23		8					recommended	Halar®
		50	23	11	8	<0.1		80-100	80-100	no stress cracking	Halar® 2.3 mm thick
		50	121		8					recommended	Halar®
		so	121	11	8	<0.1		80-100	80-100	no stress cracking	Halar® 2.3 mm thick
		70	23		8					recommended	Halar®
		70	66		8					recommended	Halar®
Sodium Hypochlorite			23		8					recommended	Halar®
			149		8					recommended	Halar®
Sodium Iodide			23		8					recommended	Halar®
			149		8					recommended	Halar®
Sodium Metaphosphate			23		8					recommended	Halar®
			149		8					recommended	Halar®
Sodium Nitrate			23		8					recommended	Halar®
			149		8					recommended	Halar®
Sodium Nitrite			23		8					recommended	Halar®
			149		8					recommended	Halar®
Sodium Perchlorate			23		8					recommended	Halar®
Sodium Peroxide			23		8					recommended	Halar®
			149		8					recommended	Halar®
Sodium Phosphate	acid		23		8					recommended	Halar®
	alkaline		23		8					recommended	Halar®
	acid		149		8					recommended	Halar®
	alkaline		149		8					recommended	Halar®
Sodium Silicate			23		8					recommended	Halar®
			149		8					recommended	Halar®
Sodium Sulfate			23		8					recommended	Halar®
			149		8					recommended	Halar®
Sodium Sulfide			23		8					recommended	Halar®
			149		8					recommended	Halar®

(Cont'd.)

Table I.3 Chemical Resistance of Ethylene Chlorotrifluoroethylene Copolymer (ECTFE) *(Cont'd.)*

Reagent	Reagent Note	Conc.	Temp (°C)	Time (days)	PDL Rating	% Change Weight	Tensile Strength	% Retained Modulus	Elongation	Resistance Note	Material Note
Sodium Sulfite			23		8					recommended	Halar®
			149		8					recommended	Halar®
Sodium Thiosulfate			23		8					recommended	Halar®
			149		8					recommended	Halar®
Stannic Chloride			23		8					recommended	Halar®
			149		8					recommended	Halar®
Stannous Chloride			23		8					recommended	Halar®
			149		8					recommended	Halar®
Starch			23		8					recommended	Halar®
			149		8					recommended	Halar®
Stearic Acid			23		8					recommended	Halar®
			66		8					recommended	Halar®
Stoddard Solvents	white spirits		23		8					recommended	Halar®
	white spirits		149		8					recommended	Halar®
Succinic Acid			23		8					recommended	Halar®
			121		8					recommended	Halar®
Sulfates	sulfate liquors		23		8					recommended	Halar®
Sulfite Liquors			23		8					recommended	Halar®
Sulfur			23		8					recommended	Halar®
			121		8					recommended	Halar®
Sulfur Chloride			23		8					recommended	Halar®
Sulfur Dioxide	dry		23		8					recommended	Halar®
	moist		23		8					recommended	Halar®
	moist		66		8					recommended	Halar®
	dry		121		8					recommended	Halar®
Sulfuric Acid		10	23		8					recommended	Halar®
		10	121		8					recommended	Halar®
		30	23		8					recommended	Halar®
		30	121		8					recommended	Halar®
		50	23		8					recommended	Halar®
		50	121l		8					recommended	Halar®
		60	23		8					recommended	Halar®
		60	121		8					recommended	Halar®
		70	23		8					recommended	Halar®
		70	121		8					recommended	Halar®
	60 deg. Be	78	23	11	8	<0.1		80-100	80-100	no stress cracks; may darken	Halar® 2.3 mm thick
	60 deg. Be	78	121	11	8	<0.1		80-100	80-100	no stress cracks; may darken	Halar® 2.3 mm thick
		80			8					recommended	Halar®
		80			8					recommended	Halar®

(Cont'd.)

Table I.3 Chemical Resistance of Ethylene Chlorotrifluoroethylene Copolymer (ECTFE) *(Cont'd.)*

Reagent	Reagent Note	Conc.	Temp (°C)	Time (days)	PDL Rating	% Change Weight	Tensile Strength	% Retained Modulus	% Retained Elongation	Resistance Note	Material Note
Sulfuric Acid		90	23		8					recommended	Halar®
		90	66		8					recommended	Halar®
		93	23		8					recommended	Halar®
		93	66		8					recommended	Halar®
		94	23		8					recommended	Halar®
		94	66		8					recommended	Halar®
		95	23		8					recommended	Halar®
		95	66		8					recommended	Halar®
		96	23		8					recommended	Halar®
		96	66		8					recommended	Halar®
		98	23		8					recommended	Halar®
		98	66		8					recommended	Halar®
		100	23		8					recommended	Halar®
Sulfurous Acid			23		8					recommended	Halar®
			121		8					recommended	Halar®
Tall Oil			23		8					recommended	Halar®
			149		8					recommended	Halar®
Tannic Acid			23		8					recommended	Halar®
			121		8					recommended	Halar®
Tanning Solutions	tanning liquors		23		8					recommended	Halar®
			121		8					recommended	Halar®
Tar			23		8					recommended	Halar®
			149		8					recommended	Halar®
Tartaric Acid			23		8					recommended	Halar®
			121		8					recommended	Halar®
Tetraethyllead			23		8					recommended	Halar®
			149		8					recommended	Halar®
Tetrahydrofuran			23		2					not recommended	Halar®
			21	11	3	4.5		25-60	80-100	no stress cracking	Halar® 2.3 mm thick
			63	11	3	11		<25	80-100	no stress cracking	Halar® 2.3 mm thick
Thionyl Chloride			23		8					recommended	Halar®
			66		8					recommended	Halar®
Toluene			23		8					recommended	Halar®
	Toluol		23	11	8	0.6		80-100	80-100	no stress cracking	Halar® 2.3 mm thick
	Toluol		66		8					recommended	Halar®
	toluol		110	11	3	8.5		<25	80-100	no stress cracking	Halar® 2.3 mm thick
			121		2					not recommended	Halar®
Tomato Juice			23		8					recommended	Halar®
			121		8					recommended	Halar®

(Cont'd.)

Table I.3 Chemical Resistance of Ethylene Chlorotrifluoroethylene Copolymer (ECTFE) *(Cont'd.)*

Reagent	Reagent Note	Conc.	Temp (°C)	Time (days)	PDL Rating	% Change Weight	Tensile Strength	% Retained Modulus	% Retained Elongation	Resistance Note	Material Note
Transformer Oils	DTE/30		23		8					recommended	Halar®
	DTE/30		23		8					recommended	Halar®
			66		8					recommended	Halar®
			121		8					recommended	Halar®
Tributyl Phosphate			23		8					recommended	Halar®
			66		2					not recommended	Halar®
Trichloroacetic Acid			23		8					recommended	Halar®
			66		8					recommended	Halar®
			121		2					not recommended	Halar®
Trichloroethane	methyl chloroform		23		8					recommended	Halar®
	methyl chloroform		66		8					recommended	Halar®
	methyl chloroform		121		2					not recommended	Halar®
Trichloroethylene			23		8					recommended	Halar®
			23	11	3	9		25-50	80-100	no stress cracking	Halar® 2.3 mm thick
			85	11	3	16.5		<25	80-100	no stress cracking	Halar® 2.3 mm thick
			149		8					recommended	Halar®
Triethanolamine			23		8					recommended	Halar®
			66		2					not recommended	Halar®
Triethylamine			23		8					recommended	Halar®
			66		8					recommended	Halar®
			121		2					not recommended	Halar®
Trisodium Phosphate			23		8					recommended	Halar®
			149		8					recommended	Halar®
Turpentine			23		8					recommended	Halar®
			149		8					recommended	Halar®
Urea			23		8					recommended	Halar®
			121		8					recommended	Halar®
Urine			23		8					recommended	Halar®
			66		8					recommended	Halar®
Vaseline			23		8					recommended	Halar®
			66		8					recommended	Halar®
Vegetable Oils			23		8					recommended	Halar®
			149		8					recommended	Halar®
Vinegar	white		23		8					recommended	Halar®
	white		23		8					recommended	Halar®
			121		8					recommended	Halar®
Vinyl Acetate			23		8					recommended	Halar®
			121		8					recommended	Halar®

(Cont'd.)

Table I.3 Chemical Resistance of Ethylene Chlorotrifluoroethylene Copolymer (ECTFE) (*Cont'd.*)

Reagent	Reagent Note	Conc.	Temp (°C)	Time (days)	PDL Rating	% Change Weight	% Retained Tensile Strength	Modulus	Elongation	Resistance Note	Material Note
Water	acid mine water		23		8					recommended	Halar®
	demineralized		23		8					recommended	Halar®
	distilled or fresh		23		8					recommended	Halar®
	salt water		23		8					recommended	Halar®
	sewage water		23		8					recommended	Halar®
	acid mine water		149		8					recommended	Halar®
	demineralized		149		8					recommended	Halar®
	distilled or fresh		149		8					recommended	Halar®
	sell water		149		8					recommended	Halar®
	sewage water		149		8					recommended	Halar®
			149		8					recommended	Halar®
Whiskey			23		8					recommended	Halar®
			149		8					recommended	Halar®
White Liquor			23		8					recommended	Halar®
			121		8					recommended	Halar®
Wines			23		8					recommended	Halar®
			121		8					recommended	Halar®
xylene	xylol		23		8					recommended	Halar®
	xylol		23		8					recommended	Halar®
	xylol		66		8					recommended	Halar®
	xylol		66		8					recommended	Halar®
Zinc Chloride		25	23	11	8	<0.1		80-100	80-100	no stress cracking	Halar® 2.3 mm thick
		25	104	11	8	<0.1		80-100	80-100	no stress cracking	Halar® 2.3 mm thick
			23		8					recommended	Halar®
			149		8					recommended	Halar®
Zinc Nitrate			23		8					recommended	Halar®
			149		8					recommended	Halar®
Zinc Sulfate			23		8					recommended	Halar®
			149		8					recommended	Halar®

Table I.4 Chemical Resistance of Ethylene Tetrafluoroethylene Copolymer (ETFE)

Reagent	Reagent Note	Conc.	Temp (°C)	Time (days)	PDL Rating	% Change Weight	% Retained Tensile Strength	% Retained Elongation	Resistance Note	Material Note
Acetaldehyde			95		8				Temp. is max. recommended	DuPont Tefzel®
Acetamide			120		8				Temp. is max. recommended	DuPont Tefzel®
Acetic Acid	glacial	50	120		8				Temp. is max. recommended	DuPont Tefzel®
	glacial		110	7	5				Temp. is max. recommended	DuPont Tefzel®
			118	7	5	3.4	82	80		DuPont Tefzel®
Acetic Anhydride			139	7	9	0	100	100		DuPont Tefzel®
			150		8				Temp. is max. recommended	DuPont Tefzel®
Acetone		50	65		8				Temp. is max. recommended	DuPont Tefzel®
			56	7	4	4.1	80	83		DuPont Tefzel®
			65		8				Temp. is max. recommended	DuPont Tefzel®
Acetonitrile			65		8				Temp. is max. recommended	DuPont Tefzel®
Acetophenone			150		8				Temp. is max. recommended	DuPont Tefzel®
			180	7	6	1.5	80	80		DuPont Tefzel®
Acetyl Chloride			65		8				Temp. is max. recommended	DuPont Tefzel®
Acetylene			120		8				Temp. is max. recommended	DuPont Tefzel®
Acetylene Tetrabromide			150		8				Temp. is max. recommended	DuPont Tefzel®
Acetylene Tetrachloride			150		8				Temp. is max. recommended	DuPont Tefzel®
Acrylonitrile			65		8				Temp. is max. recommended	DuPont Tefzel®
Adipic Acid			135		8				Temp. is max. recommended	DuPont Tefzel®
Aerosafe			149	7	6	3.9	92	93		DuPont Tefzel®
Air			150		8				Temp. is max. recommended	DuPont Tefzel®
Allyl Alcohol			100		8				Temp. is max. recommended	DuPont Tefzel®
Allyl Chloride			100		8				Temp. is max. recommended	DuPont Tefzel®
Aluminum Ammonium Sulfate			150		8				Temp. is max. recommended	DuPont Tefzel®
			150		8				Temp. is max. recommended	DuPont Tefzel®
Aluminum Chloride			150		8				Temp. is max. recommended	DuPont Tefzel®
Aluminum Fluoride			150		8				Temp. is max. recommended	DuPont Tefzel®
Aluminum Hydroxide			150		8				Temp. is max. recommended	DuPont Tefzel®
Aluminum Nitrate			150		8				Temp. is max. recommended	DuPont Tefzel®
Aluminum Oxychloride			150		8				Temp. is max. recommended	DuPont Tefzel®
Aluminum Potassium Sulfate			150		8				Temp. is max. recommended	DuPont Tefzel®
Amino Acids			100		8				Temp. is max. recommended	DuPont Tefzel®
Ammonia		30	110		8				Temp. is max. recommended	DuPont Tefzel®
	anhydrous		150		8				Temp. is max. recommended	DuPont Tefzel®
Ammonium Bifluoride		50	150		8				Temp. is max. recommended	DuPont Tefzel®
Ammonium Bromide			135		8				Temp. is max. recommended	DuPont Tefzel®
Ammonium Carbonate			150		8				Temp. is max. recommended	DuPont Tefzel®
Ammonium Chloride			150		8				Temp. is max. recommended	DuPont Tefzel®
Ammonium Dichromate			135		8				Temp. is max. recommended	DuPont Tefzel®
Ammonium Fluoride			150		8				Temp. is max. recommended	DuPont Tefzel®
Ammonium Hydroxide			66	7	9	0	97	97		DuPont Tefzel®
					8				Temp. is max. recommended	DuPont Tefzel®

(Cont'd.)

Table I.4 Chemical Resistance of Ethylene Tetrafluoroethylene Copolymer (ETFE) *(Cont'd.)*

Reagent	Reagent Note	Conc.	Temp (°C)	Time (days)	PDL Rating	% Change Weight	% Retained Tensile Strength	% Retained Elongation	Resistance Note	Material Note
Ammonium Nitrate	concentrated		110		8				Temp. is max. recommended	DuPont Tefzel®
Ammonium Perchlorate			135		8				Temp. is max. recommended	DuPont Tefzel®
Ammonium Persulfate			150		8				Temp. is max. recommended	DuPont Tefzel®
Ammonium Phosphate			150		8				Temp. is max. recommended	DuPont Tefzel®
Ammonium Sulfate			150		8				Temp. is max. recommended	DuPont Tefzel®
Ammonium Sulfide			150		8				Temp. is max. recommended	DuPont Tefzel®
Ammonium Thiocyanate			150		8				Temp. is max. recommended	DuPont Tefzel®
Amyl Acetate			120		8				Temp. is max. recommended	DuPont Tefzel®
Amyl Alcohol			150		8				Temp. is max. recommended	DuPont Tefzel®
Amyl Chloride			150		8				Temp. is max. recommended	DuPont Tefzel®
Aniline			110		8				Temp. is max. recommended	DuPont Tefzel®
			120	7	6	2.7	81	99	Temp. is max. recommended	DuPont Tefzel®
			120	30	7		93	82	Temp. is max. recommended	DuPont Tefzel®
			180	7	8		95	90	Temp. is max. recommended	DuPont Tefzel®
Aniline Hydrochloride		10	65		8				Temp. is max. recommended	DuPont Tefzel®
Animal Oils	lard oil		150		8				Temp. is max. recommended	DuPont Tefzel®
Anthraquinone			135		8				Temp. is max. recommended	DuPont Tefzel®
Anthraquinonesulfonic Acid			135		8				Temp. is max. recommended	DuPont Tefzel®
Antimony Trichloride			100		8				Temp. is max. recommended	DuPont Tefzel®
Aqua Regia			90	0.25	8	0.2	93	89	Temp. is max. recommended;	DuPont Tefzel®
			100		8					DuPont Tefzel®
Arsenic Acid			150		8				Temp. is max. recommended	DuPont Tefzel®
Barium Carbonate			150		8				Temp. is max. recommended	DuPont Tefzel®
Barium Chloride			150		8				Temp. is max. recommended	DuPont Tefzel®
Barium Hydroxide			150		8				Temp. is max. recommended	DuPont Tefzel®
Barium Sulfate			150		8				Temp. is max. recommended	DuPont Tefzel®
Barium Sulfide			150		8				Temp. is max. recommended	DuPont Tefzel®
Battery Acid			120		8				Temp. is max. recommended	DuPont Tefzel®
Benzaldehyde			100		8				Temp. is max. recommended	DuPont Tefzel®
Benzene			80	7	9	0	100	100		DuPont Tefzel®
			100		8				Temp. is max. recommended	DuPont Tefzel®
Benzenesulfonic Acid			100		8				Temp. is max. recommended	DuPont Tefzel®
Benzoic Acid			135		8				Temp. is max. recommended	DuPont Tefzel®
Benzoyl Chloride			65		8				Temp. is max. recommended	DuPont Tefzel®
			120	7	9		94	95		DuPont Tefzel®
			120	30	9		100	100		DuPont Tefzel®
Benzyl Alcohol			120	7	9		97	90	Temp. is max. recommended	DuPont Tefzel®
			150		8				Temp. is max. recommended	DuPont Tefzel®
Benzyl Chloride			150		8				Temp. is max. recommended	DuPont Tefzel®
Bismuth Carbonate			150		8				Temp. is max. recommended	DuPont Tefzel®
Black Liquor			150		8				Temp. is max. recommended	DuPont Tefzel®

(Cont'd.)

Table I.4 Chemical Resistance of Ethylene Tetrafluoroethylene Copolymer (ETFE) *(Cont'd.)*

Reagent	Reagent Note	Conc.	Temp (°C)	Time (days)	PDL Rating	% Change Weight	% Retained Tensile Strength	% Retained Elongation	Resistance Note	Material Note
Bleach	12.5% chlorine		100		8				Temp. is max. recommended	DuPont Tefzel®
	5.5% chlorine		100		8				Temp. is max. recommended	DuPont Tefzel®
Borax			150		8				Temp. is max. recommended	DuPont Tefzel®
Boric Acid			150		8				Temp. is max. recommended	DuPont Tefzel®
Brines	chlorinated		120		8				Temp. is max. recommended	DuPont Tefzel®
			150		8				Temp. is max. recommended	DuPont Tefzel®
Bromic Acid			120		8				Temp. is max. recommended	DuPont Tefzel®
Bromine	bromine water	10	110		8				Temp. is max. recommended	DuPont Tefzel®
	anhydrous		23	7	7	1.2	90	90		DuPont Tefzel®
	anhydrous		57	7	9		99	100		DuPont Tefzel®
	anhydrous		57	30	6	3.4	94	93		DuPont Tefzel®
	dry		65		8				Temp. is max. recommended	DuPont Tefzel®
Bromobenzene			100		8				Temp. is max. recommended	DuPont Tefzel®
Bromoform			100		8				Temp. is max. recommended	DuPont Tefzel®
Butadiene			120		8				Temp. is max. recommended	DuPont Tefzel®
Butane			150		8				Temp. is max. recommended	DuPont Tefzel®
Butanediol			135		8				Temp. is max. recommended	DuPont Tefzel®
Butyl Acetate	n-butyl acetate		110		8				Temp. is max. recommended	DuPont Tefzel®
			127	7	8	0	80	60		DuPont Tefzel®
Butyl Acrylate			110		8				Temp. is max. recommended	DuPont Tefzel®
Butyl Alcohol	n-butanol		150		8				Temp. is max. recommended	DuPont Tefzel®
Butyl Alcohol (sec-)			150		8				Temp. is max. recommended	DuPont Tefzel®
Butyl Alcohol (tert-)			150		8				Temp. is max. recommended	DuPont Tefzel®
Butylamine (tert-)			50		8				Temp. is max. recommended	DuPont Tefzel®
Butyl Bromide			150		8				Temp. is max. recommended	DuPont Tefzel®
Butyl Chloride			150		8				Temp. is max. recommended	DuPont Tefzel®
Butyl Mercaptan			150		8				Temp. is max. recommended	DuPont Tefzel®
Butyl Phenol			110		8				Temp. is max. recommended	DuPont Tefzel®
Butyl Phthalate			65		8				Temp. is max. recommended	DuPont Tefzel®
Butylamine	n-butylamine		50		8				Temp. is max. recommended	DuPont Tefzel®
	n-butylamine		76	7	3	4.4	71	73		DuPont Tefzel®
Butylamine (sec-)			50		8				Temp. is max. recommended	DuPont Tefzel®
Butylene			150		8				Temp. is max. recommended	DuPont Tefzel®
Butyraldehyde			100		8				Temp. is max. recommended	DuPont Tefzel®
Butyric Acid			120		8				Temp. is max. recommended	DuPont Tefzel®
Calcium Bisulfate			150		8				Temp. is max. recommended	DuPont Tefzel®
Calcium Bisulfide			150		8				Temp. is max. recommended	DuPont Tefzel®
Calcium Carbonate			150		8				Temp. is max. recommended	DuPont Tefzel®
Calcium Chlorate			150		8				Temp. is max. recommended	DuPont Tefzel®
Calcium Chloride			150		8				Temp. is max. recommended	DuPont Tefzel®
Calcium Hydroxide			150		8				Temp. is max. recommended	DuPont Tefzel®
Calcium Hypochlorite			150		8				Temp. is max. recommended	DuPont Tefzel®

(Cont'd.)

Table I.4 Chemical Resistance of Ethylene Tetrafluoroethylene Copolymer (ETFE) *(Cont'd.)*

Reagent	Reagent Note	Conc.	Temp (°C)	Time (days)	PDL Rating	% Change Weight	% Retained Tensile Strength	% Retained Elongation	Resistance Note	Material Note
Calcium Nitrate			150		8				Temp. is max recommended	DuPont Tefzel®
Calcium Oxide			135		8				Temp. is max recommended	DuPont Tefzel®
Calcium Sulfate			150		8				Temp. is max recommended	DuPont Tefzel®
Calcium Sulfide			120		8				Temp. is max recommended	DuPont Tefzel®
Caprylic Acid			100		8				Temp. is max recommended	DuPont Tefzel®
Carbon Dioxide	dry		150		8				Temp. is max recommended	DuPont Tefzel®
	wet		150		8				Temp. is max recommended	DuPont Tefzel®
Carbon Disulfide			65		8				Temp. is max recommended	DuPont Tefzel®
Carbon Monoxide			150		8				Temp. is max recommended	DuPont Tefzel®
Carbon Tetrachloride			78	7	5	4.5	90	80		DuPont Tefzel®
			135		8				Temp. is max recommended	DuPont Tefzel®
Carbonic Acid			150		8				Temp. is max recommended	DuPont Tefzel®
Castor Oil			150		8				Temp. is max recommended	DuPont Tefzel®
Caustic Potash		50	100		8				Temp. is max recommended	DuPont Tefzel®
Caustic Soda		50	100		8				Temp. is max recommended	DuPont Tefzel®
Cellosolve	2-ethoxyethanol		150		8				Temp. is max recommended	DuPont Tefzel®
Chloral Hydrate			100		8				Temp. is max recommended	DuPont Tefzel®
Chlorine	dry		100	7	8				Temp. is max recommended	DuPont Tefzel®
	wet		120		8				Temp. is max recommended	DuPont Tefzel®
	anhydrous		120		4	7	85	84		DuPont Tefzel®
Chlorine Dioxide			120		8				Temp. is max recommended	DuPont Tefzel®
Chloroacetic Acid		50	110		8				Temp. is max recommended	DuPont Tefzel®
Chlorobenzene			100		8				Temp. is max recommended	DuPont Tefzel®
			110		8				Temp. is max recommended	DuPont Tefzel®
Chlorobenzyl Chloride			65		8				Temp. is max recommended	DuPont Tefzel®
Chloroform			61	7	6	4	85	100		DuPont Tefzel®
			100		8				Temp. is max recommended	DuPont Tefzel®
Chlorohydrin	liquid		65		8				Temp. is max recommended	DuPont Tefzel®
			25		8				Temp. is max recommended	DuPont Tefzel®
Chlorosulfonic Acid			65		8				Temp. is max recommended	DuPont Tefzel®
Chromic Acid		50	65	7	8				Temp. is max recommended	DuPont Tefzel®
			125		2		66	25		DuPont Tefzel®
Chromic Chloride	5		100	7	8				Temp. is max recommended	DuPont Tefzel®
Chromyl Chloride			100		8				Temp. is max recommended	DuPont Tefzel®
Clorox	.5% chlorine		100		8				Temp. is max recommended	DuPont Tefzel®
Coal Gas			100		8				Temp. is max recommended	DuPont Tefzel®
Copper Chloride			150		8				Temp. is max recommended	DuPont Tefzel®
Copper Cyanide			150		8				Temp. is max recommended	DuPont Tefzel®
Copper Fluoride			150		8				Temp. is max recommended	DuPont Tefzel®
Copper Nitrate			150		8				Temp. is max recommended	DuPont Tefzel®
Copper Sulfate			150		8				Temp. is max recommended	DuPont Tefzel®
Cresol			135		8				Temp. is max recommended	DuPont Tefzel®
Cresol (o-)			180		9	0	100	100		DuPont Tefzel®
Cresylic Acid			135		8				Temp. is max recommended	DuPont Tefzel®
Crotonaldehyde			100		8				Temp. is max recommended	DuPont Tefzel®

(Cont'd.)

Table I.4 Chemical Resistance of Ethylene Tetrafluoroethylene Copolymer (ETFE) (Cont'd.)

Reagent	Reagent Note	Conc.	Temp (°C)	Time (days)	PDL Rating	% Change Weight	% Retained Tensile Strength	% Retained Elongation	Resistance Note	Material Note
Crude Oils	sour crude		150		8				Temp. is max recommended	DuPont Tefzel®
			150		8				Temp. is max recommended	DuPont Tefzel®
Cyclohexane			150		8				Temp. is max recommended	DuPont Tefzel®
Cyclohexanone			150		8				Temp. is max recommended	DuPont Tefzel®
			156	7	8	0	90	85		DuPont Tefzel®
Cyclohexyl Alcohol			120		8				Temp. is max. recommended	DuPont Tefzel®
DDT			100		8				Temp. is max. recommended	DuPont Tefzel®
Decalin			120	7	8				Temp. is max. recommended	DuPont Tefzel®
			120		8		89	95		DuPont Tefzel®
Decane			150		8				Temp. is max. recommended	DuPont Tefzel®
Dextrin			150		8				Temp. is max. recommended	DuPont Tefzel®
Diacetone Alcohol			100		8				Temp. is max. recommended	DuPont Tefzel®
Dibromopropane (1,2-)			95		8				Temp. is max. recommended	DuPont Tefzel®
Dibutyl Phthalate			65		8				Temp. is max. recommended	DuPont Tefzel®
Dibutylamine			110		8				Temp. is max. recommended	DuPont Tefzel®
			120	7	7		81	96		DuPont Tefzel®
			120	30	9		100	100		DuPont Tefzel®
			160	7	3		55	75		DuPont Tefzel®
Dichloroacetic Acid			65		8				Temp. is max. recommended	DuPont Tefzel®
Dichlorobenzene (0-)			65		8				Temp. is max. recommended	DuPont Tefzel®
Dichloroethylene			32	7	7	2.8	95	100		DuPont Tefzel®
			65		8				Temp. is max. recommended	DuPont Tefzel®
Dichloropropionic Acid			65		8				Temp. is max. recommended	DuPont Tefzel®
Diesel Fuels			150		8				Temp. is max. recommended	DuPont Tefzel®
Diethyl Benzene			135		8				Temp. is max. recommended	DuPont Tefzel®
Diethyl Cellosolve			150		8				Temp. is max. recommended	DuPont Tefzel®
Diethyl Ether			100		8				Temp. is max. recommended	DuPont Tefzel®
Diethylamine			110		8				Temp. is max. recommended	DuPont Tefzel®
Diethylenetriamine	DETA		100		8				Temp. is max. recommended	DuPont Tefzel®
Diglycolic Acid			100		8				Temp. is max. recommended	DuPont Tefzel®
Diisobutyl Ketone			100		8				Temp. is max. recommended	DuPont Tefzel®
Diisobutylene			135		8				Temp. is max. recommended	DuPont Tefzel®
Dimethyl Phthalate			100		8				Temp. is max. recommended	DuPont Tefzel®
Dimethyl Sulfate			65		8				Temp. is max. recommended	DuPont Tefzel®
Dimethyl Sulfoxide			90	7	7	1.6	95	90		DuPont Tefzel®
			100		8				Temp. is max. recommended	DuPont Tefzel®
Dimethylamine			50		8				Temp. is max. recommended	DuPont Tefzel®
Dimethylaniline			135		8				Temp. is max. recommended	DuPont Tefzel®
Dimethylaniline (N,N-)			120	7	8		82	97		DuPont Tefzel®
Dimethyl formamide			90	7	8	1.5	100	100		DuPont Tefzel®
			120		8				Temp. is max. recommended	DuPont Tefzel®
			120	7	5	5.5	76	92		DuPont Tefzel®

(Cont'd.)

264

Table I.4 Chemical Resistance of Ethylene Tetrafluoroethylene Copolymer (ETFE) *(Cont'd.)*

Reagent	Reagent Note	Conc.	Temp (°C)	Time (days)	PDL Rating	% Change Weight	% Retained Tensile Strength	% Retained Elongation	Resistance Note	Material Note
Dioctyl Phthalate			65		8				Temp. is max. recommended	DuPont Tefzel®
Dioxane (p-)			65		8				Temp. is max. recommended	DuPont Tefzel®
Diphenyl Oxide			80		8				Temp. is max. recommended	DuPont Tefzel®
Divinylbenzene			80		8				Temp. is max. recommended	DuPont Tefzel®
Epichlorohydrin			65		8				Temp. is max. recommended	DuPont Tefzel®
Esters			65		8				Temp. is max. recommended	DuPont Tefzel®
Ethers			100		8				Temp. is max. recommended	DuPont Tefzel®
Ethyl Acetate			65		8				Temp. is max. recommended	DuPont Tefzel®
			77	7	6	0	85	60		DuPont Tefzel®
Ethyl Acetoacetate			65		8				Temp. is max. recommended	DuPont Tefzel®
Ethyl Acrylate			100		8				Temp. is max. recommended	DuPont Tefzel®
Ethyl Alcohol			65		8				Temp. is max. recommended	DuPont Tefzel®
			150		8				Temp. is max. recommended	DuPont Tefzel®
Ethyl Chloride			150		8				Temp. is max. recommended	DuPont Tefzel®
Ethyl Chloroacetate			100		8				Temp. is max. recommended	DuPont Tefzel®
Ethyl Cyanoacetate			100		8				Temp. is max. recommended	DuPont Tefzel®
Ethylamine			40		8				Temp. is max. recommended	DuPont Tefzel®
Ethylene Bromide			150		8				Temp. is max. recommended	DuPont Tefzel®
Ethylene Chloride			150		8				Temp. is max. recommended	DuPont Tefzel®
Ethylene Chlorohydrin			65		8				Temp. is max. recommended	DuPont Tefzel®
Ethylene Glycol			150		8				Temp. is max. recommended	DuPont Tefzel®
Ethylene Oxide			110		8				Temp. is max. recommended	DuPont Tefzel®
Ethylenediamine			50		8				Temp. is max. recommended	DuPont Tefzel®
Fatty Acids			150		8				Temp. is max. recommended	DuPont Tefzel®
Ferric Chloride		25	100	7	9	0	95	95		DuPont Tefzel®
		50	150		8				Temp. is max. recommended	DuPont Tefzel®
Ferric Hydroxide			150		8				Temp. is max. recommended	DuPont Tefzel®
Ferric Nitrate			150		8				Temp. is max. recommended	DuPont Tefzel®
Ferric Sulfate			150		8				Temp. is max. recommended	DuPont Tefzel®
Ferrous Chloride			150		8				Temp. is max. recommended	DuPont Tefzel®
Ferrous Hydroxide			150		8				Temp. is max. recommended	DuPont Tefzel®
Ferrous Nitrate			150		8				Temp. is max. recommended	DuPont Tefzel®
Ferrous Sulfate			150		8				Temp. is max. recommended	DuPont Tefzel®
Fluoboric Acid			135		8				Temp. is max. recommended	DuPont Tefzel®
Fluorine	gaseous		40		8				Temp. is max. recommended	DuPont Tefzel®
Fluosilicic Acid			135		8				Temp. is max. recommended	DuPont Tefzel®
Formaldehyde		37	110		8				Temp. is max. recommended	DuPont Tefzel®
Formic Acid			135		8				Temp. is max. recommended	DuPont Tefzel®
Freon 11			110		8				Temp. is max. recommended	DuPont Tefzel®
Freon 113			46	7	9	0.8	100	100		DuPont Tefzel®
Freon 12			110		8				Temp. is max. recommended	DuPont Tefzel®

(Cont'd.)

Table I.4 Chemical Resistance of Ethylene Tetrafluoroethylene Copolymer (ETFE) *(Cont'd.)*

Reagent	Reagent Note	Conc.	Temp (°C)	Time (days)	PDL Rating	% Change Weight	% Retained Tensile Strength	% Retained Elongation	Resistance Note	Material Note
Freon 22			110		8				Temp . is max . recommended	DuPont Tefzel®
Fuel Oils			150		8				Temp . is max . recommended	DuPont Tefzel®
Fumaric Acid			95		8				Temp . is max . recommended	DuPont Tefzel®
Furan			65		8				Temp . is max . recommended	DuPont Tefzel®
Furfural			100		8				Temp . is max . recommended	DuPont Tefzel®
Gallic Acid			100		8				Temp . is max . recommended	DuPont Tefzel®
Gases	manufactured		150		8				Temp . is max . recommended	DuPont Tefzel®
	natural		150		8				Temp . is max . recommended	DuPont Tefzel®
Gasoline	leaded		150		8				Temp . is max . recommended	DuPont Tefzel®
	sour		150		8				Temp . is max . recommended	DuPont Tefzel®
	unleaded		150		8				Temp . is max . recommended	DuPont Tefzel®
Glycerin			150		8				Temp . is max . recommended	DuPont Tefzel®
Glycolic Acid			120		8				Temp . is max . recommended	DuPont Tefzel®
Glycols			135		8				Temp . is max . recommended	DuPont Tefzel®
Heptane			140		8				Temp . is max . recommended	DuPont Tefzel®
Hexane			150		8				Temp . is max . recommended	DuPont Tefzel®
Hydrazine	50/50 with UDMH		40		8				Temp . is max . recommended	DuPont Tefzel®
			50		8				Temp . is max . recommended	DuPont Tefzel®
Hydrazine Dihydrochloride			50		8				Temp . is max . recommended	DuPont Tefzel®
Hydriodic Acid			150		8				Temp . is max . recommended	DuPont Tefzel®
Hydrobromic Acid	concentrated	50	150		8				Temp . is max . recommended	DuPont Tefzel®
			125	7	9		100	100		DuPont Tefzel®
Hydrochloric Acid	concentrated	20	150		8				Temp . is max . recommended	DuPont Tefzel®
	concentrated		23	7	9	0	100	90		DuPont Tefzel®
	concentrated		106	7	9	0.1	96	100		DuPont Tefzel®
	gas		150		8				Temp . is max . recommended	DuPont Tefzel®
			150		8				Temp . is max . recommended	DuPont Tefzel®
Hydrocyanic Acid			150		8				Temp . is max . recommended	DuPont Tefzel®
Hydrofluoric Acid		35	135		8				Temp . is max . recommended	DuPont Tefzel®
		70	120		8				Temp . is max . recommended	DuPont Tefzel®
	concentrated		23	7	9	0.1	97	95		DuPont Tefzel®
			110		8				Temp . is max . recommended	DuPont Tefzel®
Hydrofluosilicic Acid			160		8				Temp . is max . recommended	DuPont Tefzel®
Hydrogen			150		8				Temp . is max . recommended	DuPont Tefzel®
Hydrogen Cyanide			150		8				Temp . is max . recommended	DuPont Tefzel®
Hydrogen Peroxide		30	23	7	9	0	99	98		DuPont Tefzel®
		30	120		8				Temp . is max . recommended	DuPont Tefzel®
		90	65		8				Temp . is max . recommended	DuPont Tefzel®
Hydrogen Phosphide			65		8				Temp . is max . recommended	DuPont Tefzel®
Hydrogen Sulfide	dry		150		8				Temp . is max . recommended	DuPont Tefzel®
	moist		150		8				Temp . is max . recommended	DuPont Tefzel®

(Cont'd.)

Table I.4 Chemical Resistance of Ethylene Tetrafluoroethylene Copolymer (ETFE) (Cont'd.)

Reagent	Reagent Note	Conc.	Temp (°C)	Time (days)	PDL Rating	% Change Weight	% Retained Tensile Strength	Elongation	Resistance Note	Material Note
Hydroquinone			120		8				Temp. is max. recommended	DuPont Tefzel®
Hypochlorous Acid			150		8				Temp. is max. recommended	DuPont Tefzel®
Inert Gases			150		8				Temp. is max. recommended	DuPont Tefzel®
Iodine	dry		110		8				Temp. is max. recommended	DuPont Tefzel®
	Moist		110		8				Temp. is max. recommended	DuPont Tefzel®
Iodoform			110		8				Temp. is max. recommended	DuPont Tefzel®
Isobutyl Alcohol			135		8				Temp. is max. recommended	DuPont Tefzel®
Isopropylamine			50		8				Temp. is max. recommended	DuPont Tefzel®
Jet Aircraft Fuels	JP 4		110		8				Temp. is max. recommended	DuPont Tefzel®
	JP 5		110		8				Temp. is max. recommended	DuPont Tefzel®
Lactic Acid			120		8				Temp. is max. recommended	DuPont Tefzel®
Lauric Acid			120		8				Temp. is max. recommended	DuPont Tefzel®
Lauryl Chloride			135		8				Temp. is max. recommended	DuPont Tefzel®
Lauryl Sulfate			120		8				Temp. is max. recommended	DuPont Tefzel®
Lead Acetate			150		8				Temp. is max. recommended	DuPont Tefzel®
Linoleic Acid			135		8				Temp. is max. recommended	DuPont Tefzel®
Linseed Oil			150		8				Temp. is max. recommended	DuPont Tefzel®
Lithium Bromide	saturated		120		8				Temp. is max. recommended	DuPont Tefzel®
Lithium Hydroxide			150		8				Temp. is max. recommended	DuPont Tefzel®
Lubricating Oils			150		8				Temp. is max. recommended	DuPont Tefzel®
Magnesium Carbonate			150		8				Temp. is max. recommended	DuPont Tefzel®
Magnesium Chloride			150		8				Temp. is max. recommended	DuPont Tefzel®
Magnesium Hydroxide			150		8				Temp. is max. recommended	DuPont Tefzel®
Magnesium Nitrate			150		8				Temp. is max. recommended	DuPont Tefzel®
Magnesium Sulfate			150		8				Temp. is max. recommended	DuPont Tefzel®
Maleic Acid			135		8				Temp. is max. recommended	DuPont Tefzel®
Maleic Anhydride			95		8				Temp. is max. recommended	DuPont Tefzel®
Malic Acid			135		8				Temp. is max. recommended	DuPont Tefzel®
Mercuric Chloride			135		8				Temp. is max. recommended	DuPont Tefzel®
Mercuric Cyanide			135		8				Temp. is max. recommended	DuPont Tefzel®
Mercuric Nitrate			135		8				Temp. is max. recommended	DuPont Tefzel®
Mercury			135		8				Temp. is max. recommended	DuPont Tefzel®
Metabromotoluene			100		8				Temp. is max. recommended	DuPont Tefzel®
Methacrylic Acid			95		8				Temp. is max. recommended	DuPont Tefzel®
Methane			120		8				Temp. is max. recommended	DuPont Tefzel®
Methanesulfonic Acid		50	110		8				Temp. is max. recommended	DuPont Tefzel®
Methyl Alcohol			150		8				Temp. is max. recommended	DuPont Tefzel®
Methyl Benzoate			120		8				Temp. is max. recommended	DuPont Tefzel®
Methyl Bromide			150		8				Temp. is max. recommended	DuPont Tefzel®
Methyl Cellosolve			150		8				Temp. is max. recommended	DuPont Tefzel®
Methyl Chloride			150		8				Temp. is max. recommended	DuPont Tefzel®
Methyl Chloromethyl Ether			80		8				Temp. is max. recommended	DuPont Tefzel®
Methyl Cyanoacetate			80		8				Temp. is max. recommended	DuPont Tefzel®

(Cont'd.)

Table I.4 Chemical Resistance of Ethylene Tetrafluoroethylene Copolymer (ETFE) (Cont'd.)

Reagent	Reagent Note	Conc.	Temp (°C)	Time (days)	PDL Rating	% Change Weight	% Retained Tensile Strength	% Retained Elongation	Resistance Note	Material Note
Methyl Ethyl Ketone			80	7	9	0	100	100		DuPont Tefzel®
			110		8				Temp. is max. recommended	DuPont Tefzel®
Methyl Isobutyl Ketone			110		8				Temp. is max. recommended	DuPont Tefzel®
Methyl Methacrylate			80		8				Temp. is max. recommended	DuPont Tefzel®
Methyl Salicylate			95		8				Temp. is max. recommended	DuPont Tefzel®
Methyl Sulfuric Acid			100		8				Temp. is max. recommended	DuPont Tefzel®
Methyl Trichlorosilane			95		8				Temp. is max. recommended	DuPont Tefzel®
Methylaniline (n-)			120		8				Temp. is max. recommended	DuPont Tefzel®
			120	7	8		85	95		DuPont Tefzel®
			120	30	9		100	100		DuPont Tefzel®
Methylene Bromide			100		8				Temp. is max. recommended	DuPont Tefzel®
Methylene Chloride			40	7	8	0	85	85		DuPont Tefzel®
			40	7	8	0	85	85	Temp. is max. recommended	DuPont Tefzel®
			100		8				Temp. is max. recommended	DuPont Tefzel®
Methylene Iodide			100	7	8	0	90	60		DuPont Tefzel®
Mineral Oils			150		8				Temp. is max. recommended	DuPont Tefzel®
			180		7				Temp. is max. recommended	DuPont Tefzel®
Morpholine			65		8				Temp. is max. recommended	DuPont Tefzel®
Naphtha			100	7	9	0.5	100	100		DuPont Tefzel®
			150		8				Temp. is max. recommended	DuPont Tefzel®
Naphthalene			150		8				Temp. is max. recommended	DuPont Tefzel®
Natural Gas			150		8				Temp. is max. recommended	DuPont Tefzel®
Nickel Chloride			150		8				Temp. is max. recommended	DuPont Tefzel®
Nickel Nitrate			150		8				Temp. is max. recommended	DuPont Tefzel®
Nickel Sulfate			150		8				Temp. is max. recommended	DuPont Tefzel®
Nicotine			100		8				Temp. is max. recommended	DuPont Tefzel®
Nicotonic Acid			120		8				Temp. is max. recommended	DuPont Tefzel®
Nitric Acid		25	100	14	9		100	100		DuPont Tefzel®
		50	65		8				Temp. is max. recommended	DuPont Tefzel®
		50	105	14	6		87	81		DuPont Tefzel®
	concentrated	70	23	105	9		100	100		DuPont Tefzel®
		70	25		8	0.6			Temp. is max. recommended	DuPont Tefzel®
		70	60	53	9		100	100		DuPont Tefzel®
		70	120	2	6		72	91		DuPont Tefzel®
Nitric Acid	concentrated	70	120	3	1		58	5		DuPont Tefzel®
	concentrated	70	120	7	0		0	0		DuPont Tefzel®
	50/50 w/ H₂SO₄		100		8				Temp. is max. recommended	DuPont Tefzel®
Nitrobenzene,			150		8					DuPont Tefzel®
Nitrogen	gas		150		8				Temp. is max. recommended	DuPont Tefzel®
Nitrogen Dioxide			100		8				Temp. is max. recommended	DuPont Tefzel®
Nitromethane			100		8				Temp. is max. recommended	DuPont Tefzel®

(Cont'd.)

Table I.4 Chemical Resistance of Ethylene Tetrafluoroethylene Copolymer (ETFE) *(Cont'd.)*

Reagent	Reagent Note	Conc.	Temp (°C)	Time (days)	PDL Rating	% Change Weight	% Retained Tensile Strength	% Retained Elongation	Resistance Note	Material Note
Nitrous Acid	1%		100		8				Temp. is max. recommended	DuPont Tefzel®
Octane			150		8				Temp. is max. recommended	DuPont Tefzel®
Octane			150		8				Temp. is max. recommended	DuPont Tefzel®
Oleic Acid			135		8				Temp. is max. recommended	DuPont Tefzel®
Oleum			50		8				Temp. is max. recommended	DuPont Tefzel®
Orthophenylphenol			100		8				Temp. is max. recommended	DuPont Tefzel®
Oxalic Acid			110		8				Temp. is max. recommended	DuPont Tefzel®
Oxygen			150		8				Temp. is max. recommended	DuPont Tefzel®
Ozone			100		8				Temp. is max. recommended	DuPont Tefzel®
Palmitic Acid			135		8				Temp. is max. recommended	DuPont Tefzel®
Perchloric Acid		10	110		8				Temp. is max. recommended	DuPont Tefzel®
Perchloric Acid		72	65		8				Temp. is max. recommended	DuPont Tefzel®
Perchloroethylene			135		8				Temp. is max. recommended	DuPont Tefzel®
Petrolatum			150		8				Temp. is max. recommended	DuPont Tefzel®
Petroleum			150		8				Temp. is max. recommended	DuPont Tefzel®
Petroleum Ether			100		8				Temp. is max. recommended	DuPont Tefzel®
Phenol		10	110		8				Temp. is max. recommended	DuPont Tefzel®
Phenol	chlorinated phenol		100		8				Temp. is max. recommended	DuPont Tefzel®
Phenol			100		8				Temp. is max. recommended	DuPont Tefzel®
Phenolsulfonic Acid			100		8				Temp. is max. recommended	DuPont Tefzel®
Phenylhydrazine			100		8				Temp. is max. recommended	DuPont Tefzel®
Phenylhydrazine Hydrochloride			100		8				Temp. is max. recommended	DuPont Tefzel®
Phosgene									Temp. is max. recommended	DuPont Tefzel®
Phosphoric Acid	concentrated	30	150		8				Temp. is max. recommended	DuPont Tefzel®
Phosphoric Acid		85	135	7	8	0	94	93	Temp. is max. recommended	DuPont Tefzel®
Phosphoric Acid			120		9					DuPont Tefzel®
Phosphoric Oxychloride			104		9					DuPont Tefzel®
Phosphoric Trichloride			75	7	9		100	100		DuPont Tefzel®
Phosphorous Oxychloride			100	7	8		100	98	Temp. is max. recommended	DuPont Tefzel®
Phosphorous Pentachloride			100		8				Temp. is max. recommended	DuPont Tefzel®
Phosphorous Pentoxide			110		8				Temp. is max. recommended	DuPont Tefzel®
Phosphorous Trichloride			120		8				Temp. is max. recommended	DuPont Tefzel®
Phthalic Acid			100		8				Temp. is max. recommended	DuPont Tefzel®
Phthalic Anhydride			100		8				Temp. is max. recommended	DuPont Tefzel®
Phthaloyl Chloride			120	30	9		100	100		DuPont Tefzel®
Picric Acid			50		8				Temp. is max. recommended	DuPont Tefzel®
Plating Solutions	brass		135		8				Temp. is max. recommended	DuPont Tefzel®
	cadmium		135		8				Temp. is max. recommended	DuPont Tefzel®
	chrome		135		8				Temp. is max. recommended	DuPont Tefzel®
	copper		135		8				Temp. is max. recommended	DuPont Tefzel®
	gold		135		8				Temp. is max. recommended	DuPont Tefzel®

(Cont'd.)

Table I.4 Chemical Resistance of Ethylene Tetrafluoroethylene Copolymer (ETFE) *(Cont'd.)*

Reagent	Reagent Note	Conc.	Temp (°C)	Time (days)	PDL Rating	% Change Weight	% Retained Tensile Strength	% Retained Elongation	Resistance Note	Material Note
Polyvinyl Acetate		50	150		8				Temp. is max. recommended	DuPont Tefzel®
Polyvinyl Alcohol			150		8				Temp. is max. recommended	DuPont Tefzel®
Potassium Aluminum Chloride			150		8				Temp. is max. recommended	DuPont Tefzel®
Potassium Aluminum Sulfate			150		8				Temp. is max. recommended	DuPont Tefzel®
Potassium Bicarbonate			150		8				Temp. is max. recommended	DuPont Tefzel®
Potassium Borate			150		8				Temp. is max. recommended	DuPont Tefzel®
Potassium Bromate			150		8				Temp. is max. recommended	DuPont Tefzel®
Potassium Bromide			150		8				Temp. is max. recommended	DuPont Tefzel®
Potassium Carbonate			150		8				Temp. is max. recommended	DuPont Tefzel®
Potassium Chlorate			150		8				Temp. is max. recommended	DuPont Tefzel®
Potassium Chloride			150		8				Temp. is max. recommended	DuPont Tefzel®
Potassium Chromate			150		8				Temp. is max. recommended	DuPont Tefzel®
Potassium Cyanide			150		8				Temp. is max. recommended	DuPont Tefzel®
Potassium Dichromate			150		8				Temp. is max. recommended	DuPont Tefzel®
Potassium Ferrocyanide			150		8				Temp. is max. recommended	DuPont Tefzel®
Potassium Fluoride			150		8				Temp. is max. recommended	DuPont Tefzel®
Potassium Hydroxide		20	100	7	9	0	100	100	Temp. is max. recommended	DuPont Tefzel®
		50	100		8				Temp. is max. recommended	DuPont Tefzel®
Potassium Hypochlorite			135		8				Temp. is max. recommended	DuPont Tefzel®
Potassium Nitrate			150		8				Temp. is max. recommended	DuPont Tefzel®
Potassium Perborate			135		8				Temp. is max. recommended	DuPont Tefzel®
Potassium Perchlorate			100		8				Temp. is max. recommended	DuPont Tefzel®
Potassium Permanganate			150		8				Temp. is max. recommended	DuPont Tefzel®
Potassium Persulfate			65		8				Temp. is max. recommended	DuPont Tefzel®
Potassium Sulfate			150		8				Temp. is max. recommended	DuPont Tefzel®
Potassium Sulfide			150		8				Temp. is max. recommended	DuPont Tefzel®
Propane			135		8				Temp. is max. recommended	DuPont Tefzel®
Propionic Acid			100		8				Temp. is max. recommended	DuPont Tefzel®
Propyl Alcohol	1-propanol		100		8				Temp. is max. recommended	DuPont Tefzel®
Propylene Dibromide			100		8				Temp. is max. recommended	DuPont Tefzel®
Propylene Dichloride			100		8				Temp. is max. recommended	DuPont Tefzel®
Propylene Glycol Methyl Ether			100		8				Temp. is max. recommended	DuPont Tefzel®
Propylene Oxide			65		8				Temp. is max. recommended	DuPont Tefzel®
Pyridine			65	7	8	1.5	100	100	Temp. is max. recommended	DuPont Tefzel®
			116		8					DuPont Tefzel®
Pyrogallic Acid	pyrogallol		65		8				Temp. is max. recommended	DuPont Tefzel®
Salicylaldehyde			100		8				Temp. is max. recommended	DuPont Tefzel®
Salicylic Acid			120		8				Temp. is max. recommended	DuPont Tefzel®
Salt Brine			150		8				Temp. is max. recommended	DuPont Tefzel®
Sea Water			150		8				Temp. is max. recommended	DuPont Tefzel®
Sewage	sewage water		135		8				Temp. is max. recommended	DuPont Tefzel®

(Cont'd.)

Table I.4 Chemical Resistance of Ethylene Tetrafluoroethylene Copolymer (ETFE) *(Cont'd.)*

Reagent	Reagent Note	Conc.	Temp (°C)	Time (days)	PDL Rating	% Change Weight	% Retained Tensile Strength	Elongation	Resistance Note	Material Note
Silicon Tetrachloride			60	7	9		100	100		DuPont Tefzel®
			120		8				Temp. is max. recommended	DuPont Tefzel®
Silver Chloride			150		8				Temp. is max. recommended	DuPont Tefzel®
Silver Cyanide			150		8				Temp. is max. recommended	DuPont Tefzel®
Silver Nitrate			150		8				Temp. is max. recommended	DuPont Tefzel®
Skydrol			149	7	7	3	100	95		DuPont Tefzel®
Sodium Acetate			150		8				Temp. is max. recommended	DuPont Tefzel®
Sodium Benzenesulfonate			150		8				Temp. is max. recommended	DuPont Tefzel®
Sodium Benzoate			150		8				Temp. is max. recommended	DuPont Tefzel®
Sodium Bicarbonate			150		8				Temp. is max. recommended	DuPont Tefzel®
Sodium Bisulfate			150		8				Temp. is max. recommended	DuPont Tefzel®
Sodium Bisulfite			150		8				Temp. is max. recommended	DuPont Tefzel®
Sodium Borate			100		8				Temp. is max. recommended	DuPont Tefzel®
Sodium Bromide			150		8				Temp. is max. recommended	DuPont Tefzel®
Sodium Carbonate			150		8				Temp. is max. recommended	DuPont Tefzel®
Sodium Chlorate			150		8				Temp. is max. recommended	DuPont Tefzel®
Sodium Chloride			150		8				Temp. is max. recommended	DuPont Tefzel®
Sodium Chromate			150		8				Temp. is max. recommended	DuPont Tefzel®
Sodium Cyanide	alkaline		150		8				Temp. is max. recommended	DuPont Tefzel®
Sodium Dichromate			100		8				Temp. is max. recommended	DuPont Tefzel®
Sodium Ferricyanide			160		8				Temp. is max. recommended	DuPont Tefzel®
Sodium Ferrocyanide			150		8				Temp. is max. recommended	DuPont Tefzel®
Sodium Fluoride			150		8				Temp. is max. recommended	DuPont Tefzel®
Sodium Glutamate			135		8				Temp. is max. recommended	DuPont Tefzel®
Sodium Hydroxide		10	110		8				Temp. is max. recommended	DuPont Tefzel®
		50	110	7	8	0.2	94	80		DuPont Tefzel®
		50	120		8				Temp. is max. recommended	DuPont Tefzel®
Sodium Hypochlorite			150		8				Temp. is max. recommended	DuPont Tefzel®
Sodium Hyposulfite			150		8				Temp. is max. recommended	DuPont Tefzel®
Sodium Iodide			150		8				Temp. is max. recommended	DuPont Tefzel®
Sodium Lignosulfonate			150		8				Temp. is max. recommended	DuPont Tefzel®
Sodium Metasilicate			150		8				Temp. is max. recommended	DuPont Tefzel®
Sodium Nitrate			150		8				Temp. is max. recommended	DuPont Tefzel®
Sodium Nitrite			150		8				Temp. is max. recommended	DuPont Tefzel®
Sodium Perborate			100		8				Temp. is max. recommended	DuPont Tefzel®
Sodium Perchlorate			65		8				Temp. is max. recommended	DuPont Tefzel®
Sodium Peroxide			150		8				Temp. is max. recommended	DuPont Tefzel®
Sodium Persulfate			80		8				Temp. is max. recommended	DuPont Tefzel®
Sodium Phosphate			150		8				Temp. is max. recommended	DuPont Tefzel®
Sodium Silicate			150		8				Temp. is max. recommended	DuPont Tefzel®
Sodium Silicofluoride			150		8				Temp. is max. recommended	DuPont Tefzel®
Sodium Sulfate			150		8				Temp. is max. recommended	DuPont Tefzel®

(Cont'd.)

Table I.4 Chemical Resistance of Ethylene Tetrafluoroethylene Copolymer (ETFE) *(Cont'd.)*

Reagent	Reagent Note	Conc.	Temp (°C)	Time (days)	PDL Rating	% Change Weight	% Retained Tensile Strength	% Retained Elongation	Resistance Note	Material Note
Sodium Sulfide			150		8				Temp. is max. recommended	DuPont Tefzel®
Sodium Sulfite			150		8				Temp. is max. recommended	DuPont Tefzel®
Sodium Thiosulfate			150		8				Temp. is max. recommended	DuPont Tefzel®
Sorbic Acid			135		8				Temp. is max. recommended	DuPont Tefzel®
Stannic Chloride			150		8				Temp. is max. recommended	DuPont Tefzel®
Stannous Chloride			150	7	8				Temp. is max. recommended	DuPont Tefzel®
Stannous Fluoride			120		8				Temp. is max. recommended	DuPont Tefzel®
Stearic Acid			150		8				Temp. is max. recommended	DuPont Tefzel®
Stoddard Solvents	white spirits		135		8				Temp. is max. recommended	DuPont Tefzel®
Stripper Solution	A-20		140		8		90	90		DuPont Tefzel®
Styrene			100		8				Temp. is max. recommended	DuPont Tefzel®
Succinic Acid			135		8				Temp. is max. recommended	DuPont Tefzel®
Sulfamic Acid			100		8				Temp. is max. recommended	DuPont Tefzel®
Sulfur	molten		120		8				Temp. is max. recommended	DuPont Tefzel®
Sulfur Dioxide			110		8				Temp. is max. recommended	DuPont Tefzel®
Sulfur Trioxide	liquid		25		8				Temp. is max. recommended	DuPont Tefzel®
Sulfuric Acid		60	150		8				Temp. is max. recommended	DuPont Tefzel®
	Fuming		50		8				Temp. is max. recommended	DuPont Tefzel®
	1/1 w/ nitric acid		100		8				Temp. is max. recommended	DuPont Tefzel®
	Concentrated		100		9	0	100	100		DuPont Tefzel®
	Concentrated		120		9	0	98	95		DuPont Tefzel®
	Concentrated		150	0.25	8				Temp. is max. recommended	DuPont Tefzel®
	Concentrated		150		9	0	98	90		DuPont Tefzel®
Sulfurous Acid			110	7	8				Temp. is max. recommended	DuPont Tefzel®
Sulfuryl Chloride			68		6	8	86	100		DuPont Tefzel®
Tall Oil			150		8				Temp. is max. recommended	DuPont Tefzel®
Tannic Acid			135		8				Temp. is max. recommended	DuPont Tefzel®
Tartaric Acid			135		8				Temp. is max. recommended	DuPont Tefzel®
Tetrachlorophenol (2,3,4,6-)			100		8				Temp. is max. recommended	DuPont Tefzel®
Tetraethyl lead			150		8				Temp. is max. recommended	DuPont Tefzel®
Tetrahydrofuran			66	7	6	3.5	86	93		DuPont Tefzel®
			100		8				Temp. is max. recommended	DuPont Tefzel®
Tetramethylammonium Hydroxide		50	100		8				Temp. is max. recommended	DuPont Tefzel®
Thionyl Chloride			100		8				Temp. is max. recommended	DuPont Tefzel®
Tin Tetrachloride			110		8				Temp. is max. recommended	DuPont Tefzel®
Titanium Dioxide			150		8				Temp. is max. recommended	DuPont Tefzel®
Titanium Tetrachloride			100		8				Temp. is max. recommended	DuPont Tefzel®
Toluene			120		8				Temp. is max. recommended	DuPont Tefzel®
Tributyl Phosphate			65		8				Temp. is max. recommended	DuPont Tefzel®
Tributylamine	tri-n-butylamine		110	7	8				Temp. is max. recommended	DuPont Tefzel®
	tri-n-butylamine		120	7	6		81	80		DuPont Tefzel®
	tri-n-butylamine		120	30	9		100	100		DuPont Tefzel®

(Cont'd.)

Table I.4 Chemical Resistance of Ethylene Tetrafluoroethylene Copolymer (ETFE) (*Cont'd.*)

Reagent	Reagent Note	Conc.	Temp (°C)	Time (days)	PDL Rating	% Change Weight	% Retained Tensile Strength	Elongation	Resistance Note	Material Note
Trichloroacetic Acid			100	7	8				Temp. is max. recommended	DuPont Tefzel®
			100	30	7	0	90	70		DuPont Tefzel®
			120		9		100	100		DuPont Tefzel®
Trichloroethane	methyl chloroform		65		8				Temp. is max. recommended	DuPont Tefzel®
Trichloroethylene			135		8				Temp. is max. recommended	DuPont Tefzel®
Trichloromethane			100		8				Temp. is max. recommended	DuPont Tefzel®
Trichlorophenol (2,4,5-)			100		8				Temp. is max. recommended	DuPont Tefzel®
Triethanolamine		50	65		8				Temp. is max. recommended	DuPont Tefzel®
Triethylamine			110		8				Temp. is max. recommended	DuPont Tefzel®
Trisodium Phosphate			135		8				Temp. is max. recommended	DuPont Tefzel®
Turpentine			135		8				Temp. is max. recommended	DuPont Tefzel®
UDMH			50		8				Temp. is max. recommended	DuPont Tefzel®
Urea	1/1 w/ Hydrazine		135		8				Temp. is max. recommended	DuPont Tefzel®
Varsol			135		8				Temp. is max. recommended	DuPont Tefzel®
Vinyl Acetate			135		8				Temp. is max. recommended	DuPont Tefzel®
Vinyl Chloride	monomer		65		8				Temp. is max. recommended	DuPont Tefzel®
Water			100	7	9	0	100	100	Temp. is max. recommended	DuPont Tefzel®
	sewage water		135		8				Temp. is max. recommended	DuPont Tefzel®
Waxes			150		8				Temp. is max. recommended	DuPont Tefzel®
Xylene			120		8				Temp. is max. recommended	DuPont Tefzel®
Zinc Acetate			120		8				Temp. is max. recommended	DuPont Tefzel®
Zinc Chloride		25	100	7	9	0	100	100	Temp. is max. recommended	DuPont Tefzel®
			150		8					DuPont Tefzel®
Zinc Hydrosulfite		10	120		8				Temp. is max. recommended	DuPont Tefzel®
Zinc Nitrate			150		8				Temp. is max. recommended	DuPont Tefzel®
Zinc Sulfate			150		8				Temp. is max. recommended	DuPont Tefzel®
Zinc Sulfide			159		8				Temp. is max. recommended	DuPont Tefzel®

Table I.5 Chemical Resistance of Fluorinated Ethylene Propylene Copolymer (FEP)

Reagent	Reagent Note	Conc.	Temp (°C)	Time (days)	PDL Rating	% Change Weight	% Retained Tensile Strength	Resistance Note	Material Note
Abietic Acid	up to boiling point				8			compatble	DuPont Teflon® FEP
Acetic Acid	up to boiling point	10	149		8			compatble	DuPont Teflon® FEP
Acetic Anhydride	up to boiling point				8			compatble	DuPont Teflon® FEP
Acetone	up to boiling point				8			compatble	DuPont Teflon® FEP
			25	365	9	0.3		no significant change	DuPont Teflon® FEP
			50	365	9	0.4		no significant change	DuPont Teflon® FEP
			70	14	9	0		no significant change	DuPont Teflon® FEP
Acetophenone	up to boiling point		201	7	8			compatble	DuPont Teflon® FEP
					6	0.6-0.8		no significant change	DuPont Teflon® FEP
Acrylic Anhydride	up to boiling point				8			compatible	DuPont Teflon® FEP
Acrylonitrile	up to boiling point				8			compatible	DuPont Teflon® FEP
Allyl Acetate	up to boiling point				8			compatible	DuPont Teflon® FEP
Allyl Methacrylate	up to boiling point				8			compatible	DuPont Teflon® FEP
Aluminum Chloride	up to boiling point				8			compatible	DuPont Teflon® FEP
Ammonia	up to boiling point				8			compatible	DuPont Teflon® FEP
Ammonium Chloride	up to boiling point				8			compatible	DuPont Teflon® FEP
Ammonium Hydroxide		10	25	365	9	0		no significant change	DuPont Teflon® FEP
		10	70	365	9	0.1		no significant change	DuPont Teflon® FEP
Aniline	up to boiling point		185	7	8			compatible	DuPont Teflon® FEP
					9	0.3-0.4		no significant change	DuPont Teflon® FEP 160
Animal Oils	up to boiling point				8			compatible	DuPont Teflon® FEP
Benzaldehyde			185	7	9	0.4-0.5		no significant change	DuPont Teflon® FEP 160
Benzene			78	4	9	0.5		no significant change	DuPont Teflon® FEP
			100	0.33	8	0.6		no significant change	DuPont Teflon® FEP
			200	0.33	7	1		no significant change	DuPont Teflon® FEP
Benzonitrile	up to boiling point				8			Compatible	DuPont Teflon® FEP
Benzoyl Chloride	up to boiling point				8			compatible	DuPont Teflon® FEP
Benzyl Alcohol	up to boiling point		204	7	8			Compatible	DuPont Teflon® FEP
					9	0.3-0.4		no significant change	DuPont Teflon® FEP 160
Borax	up to boiling point				8			Compatible	DuPont Teflon® FEP
Boric Acid					8			compatible	DuPont Teflon® FEP
Bromine	anhydrous		22	7	8			Compatible	DuPont Teflon® FEP DuPont Teflon® FEP 160
					9	0.5		no significant change	
Butyl Acetate	up to boiling point				8			Compatible	DuPont Teflon® FEP
Butyl Methacrylate	up to boiling point				8			Compatible	DuPont Teflon® FEP
Butylamine	up to boiling point		78	7	6			Compatible	DuPont Teflon® FEP 160
					9	0.3-0.4		no significant change	DuPont Teflon® FEP
Calcium Chloride	up to boiling point				8			compatible	DuPont Teflon® FEP
Carbon Tetrachloride			25	365	8	0.6		no significant change	DuPont Teflon® FEP DuPont Teflon® FEP DuPont
			50	365	5	1.6		no significant change	Teflon® FEP DuPont Teflon®
			70	14	5	1.99		no significant change	FEP DuPont Teflon® FEP 160
			78	7	4	2.3-2.4		no significant change	DuPont Teflon® FEP
			100	0.33	4	2.5		no significant change	DuPont Teflon® FEP
			200	0.33	3	3.7		no significant change	

(Cont'd.)

274

Table I.5 Chemical Resistance of Fluorinated Ethylene Propylene Copolymer (FEP) *(Cont'd.)*

Reagent	Reagent Note	Conc.	Temp (°C)	Time (days)	PDL Rating	% Change Weight	% Retained Tensile Strength	Resistance Note	Material Note
Carbon Disulfide	up to boiling point				8			Compatible	DuPont Teflon® FEP
Cetane	up to boiling point				8			compatible	DuPont Teflon® FEP
Chlorine	Anhydrous				8			Compatible	DuPont Teflon® FEP
			120	7	8	0.5-0.6		no significant change	DuPont Teflon® FEP 160
Chloroform	up to boiling point				8			Compatible	DuPont Teflon® FEP
Chlorosulfonic Acid	up to boiling point							Compatible	DuPont Teflon® FEP
			150	7	8	0.7-0.8		no significant change	DuPont Teflon® FEP 160
Chromic Acid		50						Compatible	DuPont Teflon® FEP
	up to boiling point		120	7	9	0.00-0.01		compatible	DuPont Teflon® FEP
Cyclohexane	up to boiling point				8			Compatible	DuPont Teflon® FEP DuPont
Detergents	up to boiling point				8			Compatible	Teflon® FEP DuPont Teflon®
Dibutyl Phthalate	up to boiling point				8			Compatible	FEP DuPont Teflon® FEP
Dibutyl Sebacate	up to boiling point				8			Compatible	DuPont Teflon® FEP DuPont
Diethyl Carbonate	up to boiling point				8			Compatible	Teflon® FEP DuPont Teflon®
Diisobutyl Adipate	up to boiling point				8			Compatible	FEP DuPont Teflon® FEP 160
Dimethyl Ether	up to boiling point				8			Compatible	DuPont Teflon® FEP DuPont
Dimethyl Sulfoxide			190	7	9	0.1-0.2		no significant change	Teflon® FEP DuPont Teflon®
Dimethyl formamide	up to boiling point				8			Compatible	FEP
Dimethylhydrazine	up to boiling point				8			Compatible	
Dioxane	up to boiling point				8			Compatible	
Ethyl Acetate	up to boiling point				8			Compatible	DuPont Teflon® FEP
			25	365	9	0.5		no significant change	DuPont Teflon® FEP DuPont
			50	365	8	0.7		no significant change	Teflon® FEP DuPont Teflon®
			70	14	8	0.7		no significant change	FEP
Ethyl Alcohol		95	25	365	9	0		Compatible	DuPont Teflon® FEP DuPont
		95	50	365	9	0		Compatible	Teflon® FEP DuPont Teflon®
		96	70	14	9	0		Compatible	FEP DuPont Teflon® FEP
		95	100	0.33	9	0.1		Compatible	DuPont Teflon® FEP DuPont
		95	200	0.33	9	0.3		Compatible	Teflon® FEP FEP
	up to boiling point				8			Compatible	
Ethyl Ether	up to boiling point							compatible	DuPont Teflon® FEP
Ethyl Hexoate	Up to boiling point				8			compatible	DuPont Teflon® FEP
Ethylene Bromide	Up to boiling point				8			compatible	DuPont Teflon® FEP
Ethylene Glycol	Up to boiling point				8		100	compatible	DuPont Teflon® FEP
Ferric Chloride	Up to boiling point	25	100	7	9	0.00-0.01		no significant change	DuPont Teflon® FEP 160
					8			compatible	DuPont Teflon® FEP

(Cont'd.)

Table I.5 Chemical Resistance of Fluorinated Ethylene Propylene Copolymer (FEP) (Cont'd.)

Reagent	Reagent Note	Conc.	Temp (°C)	Time (days)	PDL Rating	% Change Weight	% Retained Tensile Strength	Resistance Note	Material Note
Ferric Phosphate	Up to boiling point				8			Compatible	DuPont Teflon® FEP
Fluoronaphthalene	Up to boiling point				8			Compatible	DuPont Teflon® FEP
Fluoronitrobenzene	Up to boiling point				8			Compatible	DuPont Teflon® FEP
Formaldehyde	Up to boiling point				8			Compatible	DuPont Teflon® FEP
Formic Acid	Up to boiling point				8			Compatible	DuPont Teflon® FEP
Furan	Up to boiling point				8			Compatible	DuPont Teflon® FEP
Gasoline	Up to boiling point				8			Compatible	DuPont Teflon® FEP
Hexachloroethane	Up to boiling point				8			Compatible	DuPont Teflon® FEP
Hexane	Up to boiling point				8			Compatible	DuPont Teflon® FEP
Hydrazine	Up to boiling point				8			Compatible	DuPont Teflon® FEP
Hydrochloric Acid		10	25	365	9	0		no significant change	DuPont Teflon® FEP
		10	50	365	9	0		no significant change	DuPont Teflon® FEP
		10	70	365	9	0		no significant change	DuPont Teflon® FEP
Hydrochloric Acid		20	100	0.33	9	0		no significant change	DuPont Teflon® FEP
		20	200	0.33	9	0		no significant change	DuPont Teflon® FEP
		37	120	7	9	0.00-0.03		no significant change	DuPont Teflon® FEP 160
	Up to boiling point				8			compatible	DuPont Teflon® FEP
Hydrofluoric Acid	Up to boiling point				8			Compatible	DuPont Teflon® FEP
Hydrogen Peroxide	Up to boiling point				8			Compatible	DuPont Teflon® FEP DuPont
Isooctane	Up to boiling point		99	7	8	0.7-0.8		no significant change	Teflon® FEP 180
Lead	Up to boiling point				8			compatible	DuPont Teflon® FEP
Magnesium Chloride	up to boiling point				8			Compatible	DuPont Teflon® FEP
Mercury	up to boiling point				8			Compatible	DuPont Teflon® FEP
Methacrylic Acid	up to boiling point				8			Compatible	DuPont Teflon® FEP
Methyl Alcohol	up to boiling point				8			Compatible	DuPont Teflon® FEP
Methyl Ethyl Ketone	up to boiling point				8			Compatible	DuPont Teflon® FEP
Methyl Methacrylate	up to boiling point				8			Compatible	DuPont Teflon® FEP
Naphthalene	up to boiling point				8			Compatible	DuPont Teflon® FEP
Naphthols	up to boiling point				8			Compatible	DuPont Teflon® FEP
Nitric Acid		10	25	365	9	0		no significant change	DuPont Teflon® FEP
		10	70	365	9	0.1		no significant change	DuPont Teflon® FEP
	up to boiling point	10			8			Compatible	DuPont Teflon® FEP
Nitro-2-Methylpropanol (2)	up to boiling point				8			compatible	DuPont Teflon® FEP
Nitrobenzene	up to boiling point		210	7	7	0.7-0.9		Compatible	DuPont Teflon® FEP 160
Nitrobutanol (2-)	up to boiling point				8			no significant change	DuPont Teflon® FEP
Nitrogen Tetraoxide	up to boiling point				8			Compatible	DuPont Teflon® FEP
Nitromethane	up to boiling point				8			Compatible	DuPont Teflon® FEP
Octadecyl Alcohol	up to boiling point				8			Compatible	DuPont Teflon® FEP
Ozone	up to boiling point				8			Compatible	DuPont Teflon® FEP
Pentachlorobenzamide	up to boiling point				8			Compatible	DuPont Teflon® FEP
Perchloroethylene	up to boiling point		121	7	4	2.0-2.3		Compatible	DuPont Teflon® FEP 160
Perfluoroxylene	up to boiling point				8			no significant change	DuPont Teflon® FEP
Phenol	up to boiling point				8			Compatible	DuPont Teflon® FEP
								Compatible	DuPont Teflon® FEP

(Cont'd.)

Table I.5 Chemical Resistance of Fluorinated Ethylene Propylene Copolymer (FEP) (Cont'd.)

Reagent	Reagent Note	Conc.	Temp (°C)	Time (days)	PDL Rating	% Change Weight	% Retained Tensile Strength	Resistance Note	Material Note
Phosphoric Acid	up to boiling point				8			Compatible	DuPont Teflon® FEP
	Concentrated		100	7	9	0.00-0.01		no significant change	DuPont Teflon® FEP 160
Phosphorous Pentachloride	up to boiling point				8			Compatible	DuPont Teflon® FEP
Phthalic Acid	up to boiling point				8			Compatible	DuPont Teflon® FEP
Pinene	up to boiling point				8			Compatible	DuPont Teflon® FEP
Piperidine	up to boiling point				8			Compatible	DuPont Teflon® FEP
Potassium Acetate	up to boiling point				8			Compatible	DuPont Teflon® FEP
Potassium Hydroxide	up to boiling point				8			Compatible	DuPont Teflon® FEP
Potassium Permanganate	up to boiling point				8			Compatible	DuPont Teflon® FEP
Pyridine	up to boiling point				8			Compatible	DuPont Teflon® FEP
Soap	up to boiling point				8			Compatible	DuPont Teflon® FEP
Sodium Hydroxide		10	25	365	9	0		no significant change	DuPont Teflon® FEP
		10	70	365	9	0.1		no significant change	DuPont Teflon® FEP
		50	100	0.33	9	0		no significant change	DuPont Teflon® FEP
	up to boiling point				8			Compatible	DuPont Teflon® FEP
Sodium Hypochlorite					8			Compatible	DuPont Teflon® FEP
Sodium Peroxide					8			Compatible	DuPont Teflon® FEP
Solvents	aliphatic				8			Compatible	DuPont Teflon® FEP
	up to boiling pt.				8			Compatible	DuPont Teflon® FEP
Stannous Chloride	up to boiling point				8			Compatible	DuPont Teflon® FEP
Sulfur					8			Compatible	DuPont Teflon® FEP
Sulfuric Acid		30		365	9	0		no significant change	DuPont Teflon® FEP
		30	25	366	9	0		no significant change	DuPont Teflon® FEP
		30	70	0.33	9	0		no significant change	DuPont Teflon® FEP
		30	100	0.33	9	0.1		no significant change	DuPont Teflon® FEP
	up to boiling point				8			Compatible	DuPont Teflon® FEP
Sulfuryl Chloride	up to boiling point		68	7	4	1.7-2.7		no significant change	DuPont Teflon® FEP 160
Tetrabromoethane	up to boiling point				8			Compatible	DuPont Teflon® FEP
Tetrachloroethylene	up to boiling point				8			Compatible	DuPont Teflon® FEP
Toluene			25	365	9	0.3		no significant change	DuPont Teflon® FEP
			50	365	8	0.6		no significant change	DuPont Teflon® FEP
			70	14	8	0.6		no significant change	DuPont Teflon® FEP
			110	7	8	0.7-0.8		no significant change	DuPont Teflon® FEP 160
Tributyl Phosphate	up to boiling point		200	7	5	1.8-2.0		no significant change	DuPont Teflon® FEP 160
Trichloroacetic Acid	up to boiling point				8			Compatible	DuPont Teflon® FEP
Trichloroethylene	up to boiling point				8			Compatible	DuPont Teflon® FEP
Tricresyl Phosphate	up to boiling point				8			Compatible	DuPont Teflon® FEP
Triethanolamine	up to boiling point				8			Compatible	DuPont Teflon® FEP
Vegetable Oils	up to boiling point				8			Compatible	DuPont Teflon® FEP
Vinyl Methacrylate	up to boiling point				8			Compatible	DuPont Teflon® FEP
Water	up to boiling point				8			Compatible	DuPont Teflon® FEP
Xylene	up to boiling point				8			Compatible	DuPont Teflon® FEP
Zinc Chloride		25	100	7	9	0.00-0.03		no significant change	DuPont Teflon® FEP 160
	up to boiling point				8			Compatible	DuPont Teflon® FEP

Table I.6 Chemical Resistance of Polychlorotrifluoroethylene (PCTFE)

Reagent	Reagent Note	Conc.	Temp (°C)	Time (days)	PDL Rating	% Change Weight	Resistance Note	Material Note
Acetic Acid		3	23	14	9	0	no visible change	Allied Signal Aclar® 22; film
		3	23	14	9	0	no visible change	Allied Signal Aclar® 28; film
		3	23	14	9	0	no visible change	Allied Signal Aclar® 33; film
		5	25	7	9	0.1		3M Kel-F® 81; amorphous form
		50	175	7	9	0.1		3M Kel-F® 81; amorphous form
	Glacial		23	14	9	0.09	no visible change	Allied Signal Aclar® 22; film
	Glacial		23	14	9	0.09	no visible change	Allied Signal Aclar® 28; film
	Glacial		23	14	9	0.03	no visible change	Allied Signal Aclar® 33; film
			25	7	9	0		3M Kel-F® 81; amorphous form
	Glacial		70	7	9	0.2		3M Kel-F® 81; amorphous form
	Glacial		80	7	6	1.5		3M Kel-F® 81; amorphous form
			118	11	4	2.7		3M Kel-F® 81; amorphous form
	Glacial		175	7	4	2.5		3M Kel-F® 81; amorphous form
Acetic Anhydride			25	7	9	0		3M Kel-F® 81; amorphous form
			70	7	9	0.1		3M Kel-F® 81; amorphous form
			139	1	3	3.6		3M Kel-F® 81; amorphous form
Acetone			23	14	3	5.17	clouded, extremely flexible	Allied Signal Aclar® 22; film
			23	14	3	5.17	clouded, extremely flexible	Allied Signal Aclar® 20; film
			23	14	9	0.5	no visible change	Allied Signal Aclar® 33; film
			25	7	9	0.1		3M Kel-F® 81; amorphous form
			56	1	7	1		3M Kel-F® 81; amorphous form
Acetophenone			23	14	9	0	no visible change	Allied Signal Aclar® 22; film
			23	14	9	0	no visible change	Allied Signal Aclar® 28; film
			23	14	9	0	no visible change	Allied Signal Aclar® 33; film
			25	7	9	0		3M Kel-F® 81; amorphous form
Acetyl Chloride			25	7	9	0.1		3M Kel-F® 81; amorphous form
Allyl Chloride			26	7	9	0.2		3M Kel-F® 81; amorphous form
Aluminum Chloride	Saturated		175	7	9	0		3M Kel-F® 81; amorphous form
Ammonia	Anhydrous		25	7	9	0		3M Kel-F® 81; amorphous form
Ammonium Chloride	Saturated		175	7	9	0.1		3M Kel-F® 81; amorphous form
Ammonium Hydroxide		10	25	7	9	0		3M Kel-F® 81; amorphous form
		28	90	7	9	0.3		3M Kel-F® 81; amorphous form
		28	175	7	8	0.6		3M Kel-F® 81; amorphous form
			23	14	9	0	no visible change	Allied Signal Aclar® 22; film
			23	14	9	0	no visible change	Allied Signal Aclar® 28; film
			23	14	9	0	no visible change	Allied Signal Aclar® 33; film
Ammonium Persulfate	Saturated		175	7	9	0		3M Kel-F® 81; amorphous form
Ammonium Sulfate	saturated		175	7	9	0.1		3M Kel-F® 81; amorphous form
Amyl Acetate			25	7	9	0		3M Kel-F® 81; amorphous form
			70	7	7	0.9		3M Kel-F® 81; amorphous form
Amyl Acid Phosphate			25	7	9	0		3M Kel-F® 81; amorphous form

(Cont'd.)

Table I.6 Chemical Resistance of Polychlorotrifluoroethylene (PCTFE) (Cont'd.)

Reagent	Reagent Note	Conc.	Temp (°C)	Time (days)	PDL Rating	% Change Weight	Resistance Note	Material Note
Aniline			23	14	9	0.01	no visible change	Allied Signal Aclar® 22; film
			23	14	9	0.01	no visible change	Allied Signal Aclar® 28; film
			23	14	9	0	no visible change	Allied Signal Aclar® 33; film
			25	7	9	0		3M Kel-F® 81; amorphous form
			70	7	9	0		3M Kel-F® 81; amorphous form
Antimony Pentachloride			25	7	9	0		3M Kel-F® 81; amorphous form
Aqua Regia	boiling point			7	9	0.3	Clear yellow discoloration	3M Kel-F® 81; amorphous form
			23	14	8	0.1	Clear yellow discoloration	Allied Signal Aclar® 22; film
			23	14	8	0.1		Allied Signal Aclar® 28; film
Aqua Regia			23	14	9	0.04	no visible change	Allied Signal Aclar® 33; film
			25	7	9	0		3M Kel-F® 81; amorphous form
Aroclor 1242			25	7	9	0		3M Kel-F® 81; amorphous form
Aroclor 1248	Monsanto		25	7	9	0		3M Kel-F® 81; amorphous form
Aroclor 1254	Monsanto		25	7	9	0		3M Kel-F® 81; amorphous form
Arsenic Acid			175	7	9	-0.1		3M Kel-F® 81; amorphous form
Benzaldehyde			23	14	9	0.02	no visible change	Allied Signal Aclar® 22; film
			23	14	9	0.02	no visible change	Allied Signal Aclar® 28; film
			23	14	9	0	no visible change	Allied Signal Aclar® 33; film
			25	7	9	0		3M Kel-F® 81; amorphous form
Benzene			23	14	4	2.4	clouded, flexible	Allied Signal Aclar® 22; film
			23	14	4	2.4		Allied Signal Aclar® 28; film
			2a	14	9	0.6	no visible change	Allied Signal Aclar® 33; film
Benzene			25	7	9	0.2		3M Kel-F® 81; amorphous form
			81	1	1	6.6		3M Kel-F® 81; amorphous form
			90	7	1	7		3M Kel-F® 81; amorphous form
			135	7	1	107		3M Kel-F® 81; amorphous form
Benzoic Acid	saturated		90	7	9	0.1		3M Kel-F® 81; amorphous form
Benzonitrile			25	7	9	0		3M Kel-F® 81; amorphous form
Benzoyl Chloride			23	14	9	0.14	no visible change	Allied Signal Aclar® 22; film
			23	14	9	0.14	no visible change	Allied Signal Aclar® 28; film
			23	14	9	0	no visible change	Allied Signal Aclar® 33; film
			25	7	9	0		3M Kel-F® 81; amorphous form
Benzyl Alcohol			25	7	9	0		3M Kel-F® 81; amorphous form
Bleach Lye			25	30	9	0		3M Kel-F® 81; amorphous form
Bromine			23	14	8	0.15		Allied Signal Aclar® 22; film
			23	14	8	0.15		Allied Signal Aclar® 28; film
			23	14	8	0.1		Allied Signal Aclar® 33; film
			25	7	9	0	Clear, amber discoloration	3M Kel-F® 81; amorphous form
Bromobenzene			25	7	5	0		3M Kel-F® 81; amorphous form
			70	7		1.9		3M Kel-F® 81; amorphous form

(Cont'd.)

Table I.6 Chemical Resistance of Polychlorotrifluoroethylene (PCTFE) *(Cont'd.)*

Reagent	Reagent Note	Conc.	Temp (°C)	Time (days)	PDL Rating	% Change Weight	Resistance Note	Material Note
Butyl Acetate			25	7	9	0.3		3M Kel-F® 81; amorphous form
			80	7	2	5.1		3M Kel-F® 81; amorphous form
			90	7	2	5.8		3M Kel-F® 81; amorphous form
			125	1	1	6.7		3M Kel-F® 81; amorphous form
			135	7	1	6.5		3M Kel-F® 81; amorphous form
Butyl Alcohol	Butanol		23	14	9	0	no visible change	Allied Signal Aclar® 33; film
	Butanol		25	7	9	0		3M Kel-F® 81; amorphous form
	Butanol		70	7	9	0		3M Kel-F® 81; amorphous form
	Butanol		117	1	8	0.6		3M Kel-F® 81; amorphous form
Butyl Ether	n-butyl ether		25	7	9	0		3M Kel-F® 81; amorphous form
Butyl Sebacate	n-butyl sebacate		25	7	9	0		3M Kel-F® 81; amorphous form
Calcium Chloride	saturated solution		25	7	9	0		3M Kel-F® 81; amorphous form
	saturated		80	7	9	0		3M Kel-F® 81; amorphous form
	saturated		175	7	3	3.9		3M Kel-F® 81; amorphous form
Carbitol Acetate			25	7	9	0		3M Kel-F® 81; amorphous form
Carbon Disulfide	ACS		23	14	7	0.4	clouded	Allied Signal Aclar® 22; film
	ACS		23	14	7	0.4	clouded	Allied Signal Aclar® 28; film
	ACS		23	14	9	0.2	no visible change	Allied Signal Aclar® 33; film
Carbon Disulfide			25	7	9	0.1		3M Kel-F® 81; amorphous form
			25	30	9	0.5		3M Kel-F® 81; amorphous form
Carbon Tetrachloride			23	14	3	4.1	flexible	Allied Signal Aclar® 22; film
			23	14	3	4.1	flexible	Allied Signal Aclar® 28; film
			23	14	5	1.6	slightly flexible	Allied Signal Aclar® 33; film
			25	7	9	0.4		3M Kel-F® 81; amorphous form
			25	60	7	0.9		3M Kel-F® 81; amorphous form
			70	7	1	9.7		3M Kel-F® 81; amorphous form
			90	7	1	18		3M Kel-F® 81; amorphous form
			135	7	1	600		3M Kel-F® 81; amorphous form
Cellosolve Acetate			25	7	9	0		3M Kel-F® 81; amorphous form
Chlorine	gas		-40	0.083	1		tends to plasticize the film	Allied Signal Aclar® 22; film
	liquid		25		9	0		3M Kel-F® 81; amorphous form
	liquid		25	6	1	12.3		3M Kel-F® 81; amorphous form
Chlorine	gas		25	60	9	0	tends to plasticize the film	3M Kel-F® 81; amorphous form
	liquid		50	6	1	9		3M Kel-F® 81; amorphous form
Chloroacetic Acid	boiling point		140	7	5	1.7		3M Kel-F® 81; amorphous form
Chlorobenzene			132	1	1	21.8		3M Kel-F® 81; amorphous form
Chloroform			90	7	1	8.5		3M Kel-F® 81; amorphous form
Chloronitropropane			25	7	9	0		3M Kel-F® 81; amorphous form
Chloropropane (2-)			25	7	9	0.3		3M Kel-F® 81; amorphous form
Chlorosulfonic Acid			25	30	9	0		3M Kel-F® 81; amorphous form
			140	7	9	0.2		3M Kel-F® 81; amorphous form

(Cont'd.)

Table I.6 Chemical Resistance of Polychlorotrifluoroethylene (PCTFE) (*Cont'd.*)

Reagent	Reagent Note	Conc.	Temp (°C)	Time (days)	PDL Rating	% Change Weight	Resistance Note	Material Note
Chlorotoluene (p.)			25	7	9	0		3M Kel-F® 81; amorphous form
Chlorotrifluoroethylene			25	7	1	9.1		3M Kel-F® 81; amorphous form
	oil		25	7	9	0		3M Kel-F® 81; amorphous form
Chromic Acid	boiling point	50	25	7	9	0		3M Kel-F® 81; amorphous form
	cleaning solution		80	7	9	0		3M Kel-F® 81; amorphous form
			175	7	9	-0.1		3M Kel-F® 81; amorphous form
			175	7	9	0		3M Kel-F® 81; amorphous form
Chromosulfuric Acid	saturated		140	7	4		Slight swelling	3M Kel-F® 81; amorphous form
Citric Acid		3	23	14	9	0	no visible change	Allied Signal Aclar® 22; film
		3	23	14	9	0	no visible change	Allied Signal Aclar® 28; film
		3	23	14	9	0	no visible change	Allied Signal Aclar® 33; film
Cresol			25	7	9	0		3M Kel-F® 81; amorphous form
			140	7	5	2		3M Kel-F® 81; amorphous form
Cupric Chloride	Saturated		175	7	9	0		3M Kel-F® 81; amorphous form
Cupric Sulfate	Saturated		175	7	9	0		3M Kel-F® 81; amorphous form
Cyclohexanone			23	14	7	0.35	clouded	Allied Signal Aclar® 22; film
			23	14	7	0.35	clouded	Allied Signal Aclar® 28; film
			23	14	9	0	no visible change	Allied Signal Aclar® 33; film
			25	7	9	0		3M Kel-F® 81; amorphous form
			155	1	1	10.5		3M Kel-F® 81; amorphous form
Dibutyl Phthalate			25	7	9	0		3M Kel-F® 81; amorphous form
Dibutyl Sebacate			25	7	9	0		3M Kel-F® 81; amorphous form
Dichlorobutane (1,2-)			25	7	9	0		3M Kel-F® 81; amorphous form
Dichloroethane (1,2-)			23	14	8	0.11	clouded	Allied Signal Aclar® 22; film
			23	14	8	0.11	clouded	Allied Signal Aclar® 28; film
			23	14	9	0.03	no visible change	Allied Signal Aclar® 33; film
Dichloroethyl Ether			26	7	9	0		3M Kel-F® 81; amorphous form
Dichloroethylene			25	7	9	0		3M Kel-F® 81; amorphous form
			70	7	6	1.2		3M Kel-F® 81; amorphous form
Dichlorohexafluorocyclobutane (1,2-)			25	7	9	0.1		3M Kel-F® 81; amorphous form
Dichloropropylene			25	7	9	0		3M Kel-F® 81; amorphous form
Dichlorotoluene (2,4-)			23	14	8	0.15	clouded	Allied Signal Aclar® 22; film
			23	14	8	0.15	clouded	Allied Signal Aclar® 28; film
			23	14	9	0.06	No visible change	Allied Signal Aclar® 33; film
			25	7	9	0		3M Kel-F® 81; amorphous form
Dichlorotoluene (3,4-)			25	7	9	0		3M Kel-F® 81; amorphous form
Dicyclopentadiene			25	7	9	0		3M Kel-F® 81; amorphous form
Diethyl Carbitol			26	7	9	0.1		3M Kel-F® 81; amorphous form
Diethyl Cellosolve			25	7	7	0.8		3M Kel-F® 81; amorphous form
Diethyl Phthalate			23	14	18	0	clouded	Allied Signal Acler® 22; film
Diethyl Phthalate			23	14	8	0	clouded	Allied Signal Aclar® 28; film
			23	14	9	0	No visible change	Allied Signal Aclar® 33; film

(*Cont'd.*)

Table I.6 Chemical Resistance of Polychlorotrifluoroethylene (PCTFE) *(Cont'd.)*

Reagent	Reagent Note	Conc.	Temp (°C)	Time (days)	PDL Rating	% Change Weight	Resistance Note	Material Note
Diethylamine			25	7	5	1.9		3M Kel-F® 81; amorphous form
Diethylenetriamine	DETA		25	7	9	0		3M Kel-F® 81; amorphous form
Diisobutyl Ketone			25	7	6	1.2		3M Kel-F® 81; amorphous form
Dimethylhydrazine	anhydrous		23	14	3	3.9	Blistered	Allied Signal Aclar® 92; film
	anhydrous		23	14	3	3.9	Blistered	Allied Signal Aclar® 28; film
	anhydrous		23	14	5	1.8	No visible change	Allied Signal Aclar® 33; film
	unsymmetrical		25	7	9	0.1		3M Kel-F® 81; amorphous form
Dioxane			23	14	5	1.9	Flexible	Allied Signal Aclar® 22; film
			23	14	5	1.9	Flexible	Allied Signal Abler 28; like
			23	14	9	0.15	no visible Change	Allied Signal Aclar® 33; film
			25	7	9	0		3M Kel-F® 81; amorphous form
			90	7	9	0		3M Kel-F® 81; amorphous form
Ethyl Acetate			23	14	2	7.65	extremely flexible	Allied Signal Aclar® 22; film
			23	14	2	7.65	extremely flexible	Allied Signal Aclar® 28; film
			23	14	3	6	very flexible	Allied Signal Aclar® 33; film
			25	7	6	1.2		3M Kel-F® 81; amorphous form
			25	30	2	5.5		3M Kel-F® 81; amorphous form
			70	7	1	6.5		3M Kel-F® 81; amorphous form
			77	1	2	5.9		3M Kel-F® 81; amorphous form
Ethyl Alcohol		50	25	7	2	5.7		3M Kel-F® 81; amorphous form
		95	25	7	9	0		3M Kel-F® 81; amorphous form
		95	135	7	9	0.4		3M Kel-F® 81; amorphous form
	anhydrous, denatured		23	14	9	0	No visible change	Allied Signal Aclar® 22; film
	anhydrous, denatured		23	14	9	0	No visible change	Allied Signal Aclar® 28; film
	anhydrous, denatured		23	14	9	0	No visible change	Allied Signal Aclar® 33; film
	absolute		78	1	9	0.1		3M Kel-F® 61; amorphous form
	absolute		80	7	9	0.2		3M Kel-F® 61; amorphous form
Ethyl Butyrate			25	7	9	0.6		3M Kel-F® 61; amorphous form
Ethyl Ether			23	14	3	5.6	Clouded, very flexible	Allied Signal Aclar® 22; film
			23	14	3	5.6	Clouded, very flexible	Allied Signal Aclar® 28; film
			23	14	3	5.2	very flexible	Allied Signal Aclar® 33; film
			25	7	3	3.8		3M Kel-F® 81; amorphous form
			35	1	2	5.2		3M Kel-F® 81; amorphous form
Ethyl Formate			25	7	9	0.2		3M Kel-F® 81; amorphous form
Ethyl Propionate			25	7	7	1		3M Kel-F® 81; amorphous form
Ethylene Bromide			131	1	1	6.6		3M Kel-F® 81; amorphous form
Ethylene Chloride			25	7	9	0		3M Kel-F® 81; amorphous form
			70	7	6	1.2		3M Kel-F® 81; amorphous form
Ethylene Glycol			175	7	9	0		3M Kel-F® 81; amorphous form
Ethylene Oxide			23	14	3	5.8	Clouded, extremely flexible	Allied Signal Aclar® 22; film
			23	14	3	5.8	Clouded, extremely flexible	Allied Signal Aclar® 28; film
			23	14	3	4	Very flexible	Allied Signal Aclar® 33; film
			25		3		Swelling	3M Kel-F® 81; amorphous form

(Cont'd.)

Table I.6 Chemical Resistance of Polychlorotrifluoroethylene (PCTFE) *(Cont'd.)*

Reagent	Reagent Note	Conc.	Temp (°C)	Time (days)	PDL Rating	% Change Weight	Resistance Note	Material Note
Ferric Chloride	saturated		175	7	9	0		3M Kel-F® 81; amorphous form
Ferrous Chloride	saturated		175	7	9	0		3M Kel-F® 81; amorphous form
Ferrous Sulfate	saturated		175	7	9	0		3M Kel-F® 81; amorphous form
Fluorine	gas		85	14	9	0		3M Kel-F® 81; amorphous form
Formaldehyde			135	7	8	0.7		3M Kel-F® 81; amorphous form
Formic Acid		87	90	7	9	0		3M Kel-F® 81; amorphous form
		67	135	0.21	4	2.9		3M Kel-F® 81; amorphous form
			23	14	9	0	no visible change	Allied Signal Aclar® 22; film
			23	14	9	0	no visible change	Allied Signal Aclar® 28; film
			23	14	9	0	no visible change	Allied Signal Aclar® 33; film
			25	7	9	0		3M Kel-F® 81; amorphous form
			25	12	9	0		3M Kel-F® 81; amorphous form
	up to the boiling point		101	12	6	0.7		3M Kel-F® 81; amorphous form
Freon® 11			25	7	1	6.4		3M Kel-F® 81; amorphous form
Freon® 113			25	7	6	1.2		3M Kel-F® 81; amorphous form
			90	7	1	22.4		3M Kel-F® 81; amorphous form
Freon® 12			25	7	4	3		3M Kel-F® 81; amorphous form
Freon® 22			25	7	4	2.1		3M Kel-F® 81; amorphous form
Furan	boiling point 31-32 °C		23	14	3	5.4	discolored, extremely flexible	Allied Signal Aclar® 22; film
	boiling point 31-32 °C		23	14	3	5.4	discolored, extremely flexible	Allied Signal Aclar® 28; film
	boiling point 31-32 °C		23	14	3	3.7	very flexible	Allied Signal Aclar® 33; film
			25	7	4	2.4		3M Kel-F® 81; amorphous form
Gallic Acid			175	7	9	0.2		3M Kel-F® 81; amorphous form
Gasoline	premium grade		23	14	6	0.83	clear, amber discoloration	Allied Signal Aclar® 22; film
	premium grade		23	14	6	0.83	clear, amber discoloration	Allied Signal Aclar® 28; film
	premium grade		23	14	9	0.2	no visible change	Allied Signal Aclar® 33; film
			60-95	1	9	0.5		3M Kel-F® 81; amorphous form
Glycerin			175	7	9	-0.1		3M Kel-F® 81; amorphous form
Halowax 1000			25	7	9	0		3M Kel-F® 81; amorphous form
Heptane			23	14	8	0	Slightly clouded	Allied Signal Aclar® 22; film
			23	14	8	0	Slightly clouded	Allied Signal Aclar® 28; film
			23	14	9	0	no visible change	Allied Signal Aclar® 33; film
			25	7	9	0		3M Kel-F® 81; amorphous form
			80	7	4	2.8		3M Kel-F® 81; amorphous form
			90	7	5	1.8		3M Kel-F® 81; amorphous form
Hexachloroacetone	20% heavy cycle oil		23	14	9	0	no visible change	Allied Signal Aclar® 22; film
	20% kerosene		23	14	9	0	no visible change	Allied Signal Aclar® 22; film
	20% heavy cycle oil		23	14	9	0	no visible change	Allied Signal Aclar® 28; film
	20% kerosene		23	14	9	0	no visible change	Allied Signal Aclar® 28; film
Hexachloroacetone	20% heavy cycle oil		23	14	9	0	no visible change	Allied Signal Aclar® 33; film
	20% kerosene		23	14	9	0	no visible change	Allied Signal Aclar® 33; film

(Cont'd.)

Table I.6 Chemical Resistance of Polychlorotrifluoroethylene (PCTFE) *(Cont'd.)*

Reagent	Reagent Note	Conc.	Temp (°C)	Time (days)	PDL Rating	% Change Weight	Resistance Note	Material Note
Hexane			80	7	2	4.3		3M Kel-F® 81; amorphous form
			90	7	2	4.6		3M Kel-F® 81; amorphous form
Hydraulic Fluids	Monsanto fluid 08-45		23	14	9	0	no visible change	Allied Signal Aclar® 22; film
	Pydraul F9, Monsanto		23	14	9	0	no visible change	Allied Signal Aclar® 22; film
	Monsanto fluid OS-45		23	14	9	0	no visible change	Allied Signal Aclar® 28; film
	Pydraul F9. Monsanto		23	14	9	0	no visible change	Allied Signal Aclar® 28; film
	Monsanto fluid OS-45		23	14	9	0	no visible change	Allied Signal Aclar® 33; film
	Pydraul F9, Monsanto		23	14	9	0	no visible change	Allied Signal Aclar® 33; film
Hydrobromic Acid	boiling point	48		7	9	0.2		3M Kel-F® 81; amorphous form
Hydrochloric Acid		10	23	14	9	0	no visible change	Allied Signal Aclar® 22; film
		10	23	14	9	0	no visible change	Allied Signal Aclar® 28; film
		10	23	14	9	0	no visible change	Allied Signal Aclar® 33; film
	boiling point	10	25	7	9	0		3M Kel-F® 81; amorphous form
		20		7	9	0		3M Kel-F® 81; amorphous form
Hydrochloric Acid	concentrated	36	23	14	9	0	no visible change	Allied Signal Aclar® 22; film
		36	23	14	9	0	no visible change	Allied Signal Aclar® 28; film
		36	23	14	9	0	no visible change	Allied Signal Aclar® 33; film
		37	175	7	9	0.3		3M Kel-F® 81; amorphous form
Hydrofluoric Acid	concentrated	50	25	7	9	0		3M Kel-F® 81; amorphous form
	concentrated	so	23	14	9	0	no visible change	Allied Signal Aclar® 22; film
		60	23	14	9	0	no visible change	Allied Signal Aclar® 28; ffilm
		60	23	14	9	0	no visible change	Allied Signal Aclar® 33; film
	anhydrous		25	7	9	0		3M Kel-F® 81; amorphous form
	anhydrous		50	60	9	0		3M Kel-F® 81; amorphous form
Hydrogen Peroxide		3	25	7	9	0		3M Kel-F® 81; amorphous form
		30	23	14	8	0.23	Clouded	Allied Signal Aclar® 22; film
		so	23	14	9	0.23	Clouded	Allied Signal Aclar® 28; film
		30	23	14	9	0	no visible change	Allied Signal Aclar® 33; film
		30	25	7	9	0		3M Kel-F® 81; amorphous form
		30	25	30	9	0		3M Kel-F® 81; amorphous form
Hydrogen Sulfide	saturated		175	7	9	0.1		3M Kel-F® 81; amorphous form
Hydrolube	Hollingshead H-2		25	8	9	0		3M Kel-F® 81; amorphous form
	Hollingshead H-2		80	8	9	0		3M Kel-F® 81; amorphous form
Isoamyl Alcohol			135	7	6	1.4		3M Kel-F® 81; amorphous form
Isopropyl Ether			25	7	9	0.2		3M Kel-F® 81; amorphous form
Jet Aircraft Fuels	JP 4 flight grade		23	14	9	0.02	no visible change	Allied Signal Aclar® 22; film
	JP 4 referee grade		23	14	9	0.09	no visible change	Allied Signal Aclar® 22; film
	JP 4 flight grade		23	14	9	0.02	no visible change	Allied Signal Aclar® 28; film
	JP 4 referee grade		23	14	9	0.09	no visible change	Allied Signal Aclar® 28; film
	JP 4 flight grade		23	14	9	0.01	no visible change	Allied Signal Aclar® 33; film
	JP 4 referee grade		23	14	9	0.03	no visible change	Allied Signal Aclar® 33; film

(Cont'd.)

Table I.6 Chemical Resistance of Polychlorotrifluoroethylene (PCTFE) *(Cont'd.)*

Reagent	Reagent Note	Conc.	Temp (°C)	Time (days)	PDL Rating	% Change Weight	Resistance Note	Material Note
Lactic Acid		3	23	14	9	0		Allied Signal Aclar® 22; film
		3	23	14	9	0		Allied Signal Aclar® 22; film
		3	23	14	9	0		Allied Signal Aclar® 28; film
Machining Oils			100	1	9	0		3M Kel-F® 81; amorphous form
Malathion	EM-J		23	14	9	0.05	no visible change	Allied Signal Aclar® 22; film
	EM-J		23	14	9	0.05	no visible change	Allied Signal Aclar® 28; film
	EM-J		23	14	9	0	no visible change	Allied Signal Aclar® 33; film
Mercuric Chloride	saturated		175	7	2	-5.6		3M Kel-F® 81; amorphous form
Methallyl Chloride			25	7	9	0.1		3M Kel-F® 81; amorphous form
Methyl Acetate			25	7	7	1		3M Kel-F® 81; amorphous form
Methyl Alcohol			23	14	9	0.1	no visible change	Allied Signal Aclar® 22; film
			23	14	9	0.1	no visible change	Allied Signal Aclar® 28; film
			23	14	9	0	no visible change	Allied Signal Aclar® 33; film
			25	7	9	0		3M Kel-F® 81; amorphous form
Methyl Butyrate			25	7	7	0.8		3M Kel-F® 81; amorphous form
Methyl Ether			25	7	9	0.2		3M Kel-F® 81; amorphous form
Methyl Ethyl Ketone			23	14	3	5.9	extremely flexible	Allied Signal Aclar® 22; film
			23	14	3	5.9	extremely flexible	Allied Signal Aclar® 28; film
			23	14	6	1.2	slightly flexible	Allied Signal Aclar® 33; film
			25	7	9	0.2		3M Kel-F® 81; amorphous form
			90	7	2	4.6		3M Kel-F® 81; amorphous form
Methyl Formate			25	7	9	0.1		3M Kel-F® 81; amorphous form
Methyl Propionate			25	7	6	1.4		3M Kel-F® 81; amorphous form
Methylal			25	7	6	1.3		3M Kel-F® 81; amorphous form
Mineral Oils			25	7	9	0		3M Kel-F® 81; amorphous form
Motor Oils	premium grade		23	14	9	0.01	no visible change	Allied Signal Aclar® 22; film
	premium grade		23	14	9	0.01	no visible change	Allied Signal Aclar® 28; film
	premium grade		23	14	9	0.01	no visible change	Allied Signal Aclar® 33; film
Naphtha	solvent		25	7	13	0		3M Kel-F® 81; amorphous form
Nickel Ammonium Sulfate	saturated		175	7	9	0.3		3M Kel-F® 81; amorphous form
Nitric Acid		10	23	14	9	0	no visible change	Allied Signal Aclar® 22; film
		10	23	14	9	0	no visible change	Allied Signal Acker 28; film
		10	23	14	9	0	no visible change	Allied Signal Aclar® 33; film
		10	25	7	9	0		3M Kel-F® 81; amorphous form
	boiling point	30	175	7	9	0.1		3M Kel-F® 81; amorphous form
	concentrated	60		7	9	0		3M Kel-F® 81; amorphous form
	concentrated	70	23	14	9	0	no visible change	Allied Signal Aclar® 22; film
	concentrated	70	23	14	9	0	no visible change	Allied Signal Aclar® 28; film
	concentrated	70	23	14	9	0	no visible change	Allied Signal Aclar® 33; film
		70	25	7	9	0		3M Kel-F® 81; amorphous form
		70	70	7	9	0		3M Kel-F® 81; amorphous form

(Cont'd.)

Table I.6 Chemical Resistance of Polychlorotrifluoroethylene (PCTFE) *(Cont'd.)*

Reagent	Reagent Note	Conc.	Temp (°C)	Time (days)	PDL Rating	% Change Weight	Resistance Note	Material Note
Nitric Acid	fuming	95	25	7	9	0		3M Kel-F® 81; amorphous form
	with hydrofluoric acid		23	14	9	0	no visible change	Allied Signal Aclar® 22; film
	red fuming		23	14	9	0.07	no visible change	Allied Signal Aclar® 22; film
	with hydrofluoric acid		23	14	9	0	no visible change	Allied Signal Aclar® 28; film
	red fuming		23	14	9	0.07	no visible change	Allied Signal Aclar® 28; film
	with hydrofluoric acid		23	14	9	0	no visible change	Allied Signal Aclar® 33; film
	red fuming		23	14	9	0.04	no visible change	Allied Signal Aclar® 33; film
	white fuming		90	7	9	0.3		3M Kel-F® 81; amorphous form
Nitrobenzene			25	7	9	0		3M Kel-F® 81; amorphous form
			140	7	6	1.5		3M Kel-F® 81; amorphous form
Nitrogen Tetraoxide			5	7	1		tends to plasticize this film	Allied Signal Aclar® 22; film
			23	14	1	9.9		3M Kel-F® 81; amorphous form
			23	14	5		Flexible, yellows	Allied Signal Aclar® 22; film
			23	14	5		Flexible, yellows	Allied Signal Aclar® 28; film
			23	14	5		Flexible, yellows	Allied Signal Aclar® 33; film
Nitromethane			25	7	9	0		3M Kel-F® 81; amorphous form
Oleic Acid			25	7	9	0		3M Kel-F® 81; amorphous form
Oleum			25	2	9	0.1		3M Kel-F® 81; amorphous form
Orthochlorotoluene			25	7	9	0		3M Kel-F® 81; amorphous form
Oxalic Acid			175	7	9	-0.2		3M Kel-F® 81; amorphous form
Oxygen	Liquid		23	14	8		passes lox impact test	Allied Signal Aclar® 22; film
	Liquid		23	14	8		passes lox impact test	Allied Signal Aclar® 28; film
	Liquid		23	14	8		passes lox impact test	Allied Signal Aclar® 33; film
Ozone	5% in oxygen		150	2	9	0	No molecular degradation	3M Kel-F® 81; amorphous form
Pentachloroethane			25	7	9	0		3M Kel-F® 81; amorphous form
Pentanedione (2,4-)			23	14	8	0.17	Clouded	Allied Signal Aclar® 22; film
			23	14	8	0.17	Clouded	Allied Signal Aclar® 28; film
			23	14	9	0.2	no visible change	Allied Signal Aclar® 33; film
Perchloric Acid		70	25	30	9	0		3M Kel-F® 81; amorphous film
			25	14	9	-0.2		3M Kel-F® 81; amorphous film
Perfluorotriethylamine			25	7	9	0		3M Kel-F® 81; amorphous film
Phenol		5	70	7	9	0		3M Kel-F® 81; amorphous film
Phosphoric Acid		30	175	7	9	0.1		3M Kel-F® 81; amorphous form
		85	175	7	9	0		3M Kel-F® 81; amorphous film
			140	7	9	0		3M Kel-F® 81; amorphous film
Piperidine			25	7	9	0		3M Kel-F® 81; amorphous form
Potassium Dichromate	saturated solution		175	7	9	0		3M Kel-F® 81; amorphous form
Potassium Hydroxide		10	25	7	9	-0.2		3M Kel-F® 81; amorphous form
	boiling point	10	80	7	9	0.1		3M Kel-F® 81; amorphous film
		50		7	9	0.1		3M Kel-F® 81; amorphous film

(Cont'd.)

Table I.6 Chemical Resistance of Polychlorotrifluoroethylene (PCTFE) (Cont'd.)

Reagent	Reagent Note	Conc.	Temp (°C)	Time (days)	PDL Rating	% Change Weight	Resistance Note	Material Note
Potassium Permanganate	Saturated		25	30	9	0		3M Kel-F® 81; amorphous form
Potassium Persulfate	Saturated		25	30	9	0		3M Kel-F® 81; amorphous film
Propyl Acetate	n-propyl acetate		25	7	6	0.6		3M Kel-F® 81; amorphous film
Propyl Ether	n-propyl ether		25	7	9	0.3		3M Kel-F® 81; amorphous form
Propyl Formate	n-propyl Formate		25	7	9	0.1		3M Kel-F® 81; amorphous film
Propyl Propionate	n-propyl propionate		25	7	9	0.4		3M Kel-F® 81; amorphous film
Propylene Chloride			25	7	9	0		3M Kel-F® 81; amorphous film
Pyridine			23	14	7	0.55	Clouded	Allied Signal Aclar® 22; film
			23	14	7	0.55	Clouded	Allied Signal Aclar® 28; film
			23	14	9	0.1	no visible change	Allied Signal Aclar® 33; film
			25	7	9	0		3M Kel-F® 81; amorphous form
			115	1	1	7.4		3M Kel-F® 81; amorphous form
Pyrogallic Acid	Saturated		175	7	9	0.1		3M Kel-F® 81; amorphous form
Salicylic Acid	Saturated		175	7	9	0.2		3M Kel-F® 81; amorphous film
Santicizer 8			25	7	9	0		3M Kel-F® 81; amorphous form
Santicizer B16			25	7	9	0		3M Kel-F® 81; amorphous film
Santicizer E15			25	7	9	0		3M Kel-F® 81; amorphous film
Santicizer M17			25	7	9	0		3M Kel-F® 81; amorphous film
Santolube 31			25	7	9	0		3M Kel-F® 81; amorphous film
Silicone Oils	DC-200		70	7	9	0.1		3M Kel-F® 81; amorphous form
	DC-200		75	7	9	0.1		3M Kel-F® 81; amorphous film
	DC-200		190	7	9	-0.3		3M Kel-F® 81; amorphous film
Sodium Bisulfite	Saturated		175	7	8	-0.7		3M Kel-F® 81; amorphous form
Sodium Borate	Saturated		175	7	9	0.2		3M Kel-F® 81; amorphous film
Sodium Carbonate		2	25	7	9	0		3M Kel-F® 81; amorphous film
Sodium Chloride	saturated	10	25	7	9	0		3M Kel-F® 81; amorphous form
			175	7	9	0		3M Kel-F® 81; amorphous form
Sodium Hydroxide		1	25	7	9	0		3M Kel-F® 81; amorphous form
		10	25	7	9	0		3M Kel-F® 81; amorphous form
		30	175	7	6	-1.2		3M Kel-F® 81; amorphous form
	boiling point	50		7	9	0.1		3M Kel-F® 81; amorphous form
		50	23	14	9	0	no visible change	Allied Signal Aclar® 22; film
		50	23	14	9	0	no visible change	Allied Signal Aclar® 28; film
		50	23	14	9	0	no visible change	Allied Signal Aclar® 33; film
Sodium Hypochlorite			23	14	9	0	no visible change	Allied Signal Aclar® 28; film
			23	14	9	0	no visible change	Allied Signal Aclar® 33; film
			23	14	9	0	no visible change	Allied Signal Aclar® 22; film
Sodium Phosphate	saturated		175	7	9	0		3M Kel-F® 81; amorphous form
Stannic Chloride			25	7	9	0		3M Kel-F® 81; amorphous form
			175	7	9	0.1		3M Kel-F® 81; amorphous form
Sulfur Dioxide	anhydrous		25	7	9	0.1		3M Kel-F® 81; amorphous form

(Cont'd.)

Table I.6 Chemical Resistance of Polychlorotrifluoroethylene (PCTFE) *(Cont'd.)*

Reagent	Reagent Note	Conc.	Temp (°C)	Time (days)	PDL Rating	% Change Weight	Resistance Note	Material Note
Sulfuric Acid		3	25	7	9	0		3M Kel-F® 81; amorphous form
	fuming, 20% oleum	20	23	14	9	0.03	no visible change	Allied Signal Aclar® 22; film
	fuming, 20% oleum	20	23	14	9	0.03	no visible change	Allied Signal Aclar® 28; film
	fuming, 20% oleum	20	23	14	9	0.02	no visible change	Allied Signal Aclar® 33; film
		20	25	7	9	0	no visible change	3M Kel-F® 81; amorphous form
		30	23	14	9	0	no visible change	Allied Signal Aclar® 22; film
		30	23	14	9	0	no visible change	Allied Signal Aclar® 28; film
		30	23	14	9	0		Mlied Signal Aclar® 33; film
		30	25	7	9	0		3M Kel-F® 81; amorphous form
		30	175	7	9	0		3M Kel-F® 81; amorphous form
		50	40	30	9	0		3M Kel-F® 81; amorphous form
		92	40	30	9	0		3M Kel-F® 81; amorphous form
		95	175	7	9	0		3M Kel-F® 81; amorphous form
		96	70	7	9	0		3M Kel-F® 81; amorphous form
Tetrachloroethane	symmetrical		25	7	9	0		3M Kel-F® 81; amorphous form
Tetrachloroethylene			25	7	7	0.8		3M Kel-F® 81; amorphous form
Tetrahydrofuran			25	7	1	8.5		3M Kel-F® 81; amorphous form
			64	1	1	8.2		3M Kel-F® 81; amorphous form
Thionyl Chloride			90	7	1	8.5		3M Kel-F® 81; amorphous form
Titanium Tetrachloride			90	7	4	2.6		3M Kel-F® 81; amorphous form
Toluene			23	14	4	2.8	flexible	Allied Signal War 22; film
			23	14	4	2.8	flexible	Allied Signal War 28; film
			23	14	6	1.1	slightly flexible	Allied Signal Aclar® 33; film
			25	7	9	0.4		3M Kel-F® 81; amorphous form
			110	7	2	5		3M Kel-F® 81; amorphous form
Toluene Diisocyanate			23	14	9	0.44	no visible change	Allied Signal Aclar® 22; film
			23	14	9	0.44	no visible change	Allied Signal Aclar® 28; film
Trichloroacetic Acid			70	7	9	0		3M Kel-F® 81; amorphous form
Trichloroethane			25	7	9	0.1		3M Kel-F® 81; amorphous form
Trichloroethane (1,1,2-)	Technical		23	14	8	0.04	Clouded	Allied Signal Aclar® 22; film
	Technical		23	14	8	0.04	Clouded	Allied Signal Aclar® 20; film
	Technical		23	14	9	0.02	no visible change	Allied Signal Aclar® 33; film
			25	7	9	0		3M Kel-F® 81 amorphous form
Trichloroethylene			23	14	2	10.9	Clouded, extremely flexible	Allied Signal Aclar® 22; film
			23	14	2	10.9	Clouded, extremely flexible	Allied Signal Aclar® 28; film
			23	14	2	7.8	clear, very flexible	Allied Signal Aclar® 33; film
			25	7	4	2.3		3M Kel-F® 81; amorphous form
			80	7	1	9.2		3M Kel-F® 81; amorphous form
Trichloropropane (1,2,3-)			25	7	9	0		3M K6F 81; amorphous form

(Cont'd.)

Table I.6 Chemical Resistance of Polychlorotrifluoroethylene (PCTFE) *(Cont'd.)*

Reagent	Reagent Note	Conc.	Temp (°C)	Time (days)	PDL Rating	% Change Weight	Resistance Note	Material Note
Trichlorotrifluoroethane	Genesolv D		23	14	4		Clouded, extremely flexible	Allied Signal Aclar® 22; film
	Genesolv D		23	14	4		Clouded, extremely flexible	Allied Signal Aclar® 28; film
	Genesolv D		23	14	4		Clouded, extremely flexible	Allied Signal Aclar® 33; film
Tricresyl Phosphate			25	7	9	0		3M Kel-F® 81; amorphous form
			140	7	9	0		3M Kel-F® 81; amorphous form
Triethylaluminum			23	14	7	0.13	slightly crazed	Allied Signal Aclar® 22; film
			23	14	7	0.13	slightly crazed	Allied Signal Aclar® 28; film
			23	14	7	0.01	slightly crazed	Allied Signal Aclar® 33; film
Triethylamine			25	7	9	0.2		3M Kel-F® 81; amorphous form
Water			77	21	9	0		3M Kel-F® 81; amorphous form
Xylene			25	7	9	0.4		3M Kel-F® 81; amorphous form
			90	7	1	6.5		3M Kel-F® 81; amorphous form
			138	7	1	27		3M Kel-F® 81; amorphous form
Zinc Sulfate	saturated		174	7	9	0.4		3M Kel-F® 81; amorphous form

Table I.7 Chemical Resistance of Perfluoroalkoxy Copolymer (PFA)

Reagent	Reagent Note	Conc.	Temp (°C)	Time (days)	PDL Rating	% Change Weight	% Retained		Resistance Note	Material Note
							Tensile Strength	Elongation		
Abietic Acid	up to boiling point				8				Compatible	DuPont Teflon® PFA
Acetic Acid	up to boiling point				8				Compatible	DuPont Teflon® PFA
	glacial		118	7	9	0.4	95	100		Teflon® PFA; 1.27 mm thick
Acetic Anhydride	up to boiling point				8				Compatible	DuPont Teflon® PFA
			139	7	9	0.3	91	99		DuPont Teflon® PFA
Acetone	up to boiling point				8				Compatible	DuPont Teflon® PFA
			25	365	9	0.3			no significant change	DuPont Teflon® PFA
			50	365	9	0.4			no significant change	DuPont Teflon® PFA
			70	14	9	0			no significant change	DuPont Teflon® PFA
Acetophenone	up to boiling point				8				Compatible	DuPont Teflon® PFA
			201	7	8	0.7				DuPont Teflon® PFA
			201	7	8	0.6-0.8	90	100	no significant change	DuPont Teflon® PFA 350
			202		8	0.6			no significant change	DuPont Teflon® PFA
Acrylic Anhydride	up to boiling point				8				Compatible	DuPont Teflon® PFA
Acrylonitrile	up to boiling point				8				Compatible	DuPont Teflon® PFA
Allyl Acetate	up to boiling point				8				Compatible	DuPont Teflon® PFA
Allyl Methacrylate	up to boiling point				8				Compatible	DuPont Teflon® PFA
Aluminum Chloride	up to boiling point				8				Compatible	DuPont Teflon® PFA
Ammonia	liquid to boiling pt				8				Compatible	DuPont Teflon® PFA
Ammonium Chloride	up to boiling point				8				Compatible	DuPont Teflon® PFA
Ammonium Hydroxide		10	25 7	365	9	0			no significant change	DuPont Teflon® PFA
		10	70	365	9	0.1			no significant change	DuPont Teflon® PFA.
Ammonium Hydroxide	concentrated		66	7	9	0	98	100		DuPont Teflon® PFA
Aniline	up to boiling point				8				Compatible	DuPont Teflon® PFA
			185	7	9	0.5				DuPont Teflon® PFA
			185	7	9	0.3	94	100		DuPont Teflon® PFA
			185		9	0.3-0.4			no significant change	DuPont Teflon® PFA 350
Animal Oils	up to boiling point				8				Compatible	DuPont Teflon® PFA
Aqua Regia			120	7	9	0	99	100	Compatible	DuPont Teflon® PFA
Benzaldehyde			179	7	9	0.5				DuPont Teflon® PFA
			179	7	9	0.5	90	99		DuPont Teflon® PFA
			179	7	9	0.4-0.5			no significant change	DuPont Teflon® PFA 360
Benzene			78	4	9	0.5			no significant change	DuPont Teflon® PFA
			100	0.33	8	0.6			no significant change	DuPont Teflon® PFA
			200	0.33	7	1			no significant change	DuPont Teflon® PFA
Benzonitrile	up to boiling point				8				Compatible	DuPont Teflon® PFA
Benzoyl Chloride	up to boiling point				8				Compatible	DuPont Teflon® PFA
Benzyl Alcohol	up to boiling point				8					DuPont Teflon® PFA
			204	7	9	0.4				DuPont Teflon® PFA
			204	7	9	0.3-0.4	93	99		DuPont Teflon® PFA 350
			205		9	0.3			no significant change	DuPont Teflon® PFA
Borax	up to boiling point				8				Compatible	DuPont Teflon® PFA
Boric Acid	up to boiling point				8				Compatible	DuPont Teflon® PFA

(Cont'd.)

Table I.7 Chemical Resistance of Perfluoroalkoxy Copolymer (PFA) (Cont'd.)

Reagent	Reagent Note	Conc.	Temp (°C)	Time (days)	PDL Rating	% Change Weight	% Retained		Resistance Note	Material Note
							Tensile Strength	Elongation		
Bromine	up to boiling point		22		8				Compatible	DuPont Teflon® PFA
	anhydrous		22		9	0.5			Compatible	DuPont Teflon® PFA DuPont Teflon® PFA 350
	anhydrous		23	7	9	0.5	99	100	no significant change	DuPont Teflon® PFA
			59	7	9	0.5	95	95		DuPont Teflon® PFA
Butyl Acetate	up to boiling point				8				Compatible	DuPont Teflon® PFA
			125	7	9	0.5	93	100		DuPont Teflon® PFA
Butyl Methacrylate	up to boiling point				8				Compatible	DuPont Teflon® PFA
Butylamine	up to boiling point				8				Compatible	DuPont Teflon® PFA
	n-butylamine		78		9	0.4				DuPont Teflon® PFA
	n-butylamine		78	7	9	0.4	86	97		DuPont Teflon® PFA
	n-butylamine		78	7	9	0.3-0.4			no significant change	DuPont Teflon® PFA 350
Calcium Chloride	up to boiling point				8				compatible	DuPont Teflon® PFA
Carbon Tetrachloride			25	365	8	0.6			no significant change	DuPont Teflon® PFA
			50	365	5	1.6			no significant change	DuPont Teflon® PFA
			70	14	5	1.99			no significant change	DuPont Teflon® PFA
			77	7	7	2.3				DuPont Teflon® PFA
			78		3	3.4	87	100		DuPont Teflon® PFA
			78	7	4	2.3-2.4			no significant change	DuPont Teflon® PFA
			100	0.33	4	2.5			no significant change	DuPont Teflon® PFA 350
			200	0.33	3	3.7			no significant change	DuPont Teflon® PFA
Carbon Disulfide	up to boiling point				8				Compatible	DuPont Teflon® PFA
Cetane	up to boiling point				8				Compatible	DuPont Teflon® PFA
Chlorine	up to boiling point		120		8	0.6			Compatible	DuPont Teflon® PFA
	anhydrous		120	7	9	0.5	92	100		DuPont Teflon® PFA
	anhydrous		120	7	8	0.5-0.6			no significant change	DuPont Teflon® PFA DuPont Teflon® PFA 350
Chloroform	up to boiling point				8				Compatible	DuPont Teflon® PFA
Chlorosulfonic Acid	up to boiling point				8				Compatible	DuPont Teflon® PFA
			150		7	0.8				DuPont Teflon® PFA
			150	7	8	0.7-0.8			no significant change	DuPont Teflon® PFA 350
			151	7	a	0.7	91	100		DuPont Teflon® PFA
Chromic Acid		50	120		9	0.01	93	97		DuPont Teflon® PFA
		50	120	7	9	0			no significant change	DuPont Teflon® PFA
		50	120	7	9	0.00-0.01				DuPont Teflon® PFA 350
	up to boiling point				8				compatible	DuPont Teflon® PFA

(Cont'd.)

Table I.7 Chemical Resistance of Perfluoroalkoxy Copolymer (PFA) *(Cont'd.)*

Reagent	Reagent Note	Conc.	Temp (°C)	Time (days)	PDL Rating	% Change Weight	% Retained Tensile Strength	Elongation	Resistance Note	Material Note
Cresol (o-)	up to boiling point		191	7	9	0.2	92	96		DuPont Teflon® PFA
Cyclohexane					8				Compatible	DuPont Teflon® PFA
Cyclohexanone	up to boiling point		156	7	9	0.4	92	100	Compatible	DuPont Teflon® PFA DuPont Teflon® PFA
Detergents	up to boiling point				8				Compatible	DuPont Teflon® PFA
Dibutyl Phthalate	up to boiling point				8				Compatible	DuPont Teflon® PFA
Dibutyl Sebacate	up to boiling point				8				Compatible	DuPont Teflon® PFA DuPont Teflon® PFA
Diethyl Carbonate	up to boiling point				8				Compatible	DuPont Teflon® PFA
Diisobutyl Adipate	up to boiling point				8				Compatible	DuPont Teflon® PFA
Dimethyl Ether	up to boiling point				8				Compatible	DuPont Teflon® PFA
Dimethyl Phthalate	up to boiling point		220	7	9	0.3	98	100		DuPont Teflon® PFA
Dimethyl Sulfoxide			189	7	9	0.1	95	100		DuPont Teflon® PFA
			190	7	9	0.1–0.2		100	no significant change	DuPont Teflon® PFA 350
Dimethyl Formamide	up to boiling point		154	7	8	0.2	96	100	Compatible	DuPont Teflon® PFA
Dimethylhydrazine	up to boiling point				9				Compatible	DuPont Teflon® PFA
Dioxane	up to boiling point		101	7	8	0.6	92	100	Compatible	DuPont Teflon® PFA
Ethyl Acetate	up to boiling point				8				Compatible	DuPont Teflon® PFA
			25	365	9	0.5			no significant change	DuPont Teflon® PFA
			50	365	8	0.7			no significant change	DuPont Teflon® PFA DuPont Teflon® PFA DuPont
			70	14	8	0.7			no significant change	Teflon® PFA
Ethyl Alcohol		95	25	365	9	0			Compatible	DuPont Teflon® PFA
		96	50	365	9	0			Compatible	DuPont Teflon® PFA
		95	70	14	9	0			Compatible	DuPont Teflon® PFA DuPont Teflon® PFA DuPont
		95	100	0.33	9	0.1			Compatible	Teflon® PFA
		95	200	0.33	9	03			Compatible	DuPont Teflon® PFA
	up to boiling point				8				Compatible	DuPont Teflon® PFA
Ethyl Ether	up to boiling point				8				Compatible	DuPont Teflon® PFA
Ethyl Hexoate	up to boiling point				8				Compatible	DuPont Teflon® PFA
Ethylene Bromide	up to boiling point				8				Compatible	DuPont Teflon® PFA
Ethylene Glycol	up to boiling point				8				Compatible	DuPont Teflon® PFA
Ethylenediamine			117	7	9	0.1	96	100	Compatible	DuPont Teflon® PFA
Ferric Chloride		25	100		9	0.01			no significant change	DuPont Teflon® PFA DuPont Teflon® PFA 350
		25	100	7	9	<0.01				DuPont Teflon® PFA
	up to boiling point		100	7	8	0	93	98	Compatible	DuPont Teflon® PFA
Ferric Phosphate	up to boiling point				8				Compatible	DuPont Teflon® PFA
Fluoronaphthalene	up to boiling point				8				Compatible	DuPont Teflon® PFA
Fluoronitrobenzene	up to boiling point				8				Compatible	DuPont Teflon® PFA
Formaldehyde	up to boiling point				8				Compatible	DuPont Teflon® PFA
Formic Acid	up to boiling point				8				Compatible	DuPont Teflon® PFA

(Cont'd.)

Table I.7 Chemical Resistance of Perfluoroalkoxy Copolymer (PFA) (Cont'd.)

Reagent	Reagent Note	Conc.	Temp (°C)	Time (days)	PDL Rating	% Change Weight	% Retained		Resistance Note	Material Note
							Tensile Strength	Elongation		
Freon 113			47		6	1.2				DuPont Teflon® PFA 350
			47	7	6	1.2				DuPont Teflon® PFA 350
Furan	up to boiling point				8				Compatible	DuPont Teflon® PFA
Gasoline	up to boiling point				8				Compatible	DuPont Teflon® PFA
Hexachloroethane	up to boiling point				8				Compatible	DuPont Teflon® PFA
Hexane	up to boiling point				8				Compatible	DuPont Teflon® PFA
Hydrazine	up to boiling point				8				Compatible	DuPont Teflon® PFA
Hydrochloric Acid		10	25	365	9	0			no significant change	DuPont Teflon® PFA
		10	50	366	9	0			no significant change	DuPont Teflon® PFA
		10	70	365	9	0			no significant change	DuPont Teflon® PFA
Hydrochloric Acid		20	100	0.33	9	0			no significant change	DuPont Teflon® PFA
		20	200	0.33	9	0			no significant change	DuPont Teflon® PFA
		37	120		9	0.03				DuPont Teflon® PFA 350
		37	120	7	9	0.00-0.03			no significant change	DuPont Teflon® PFA
	up to boiling point concentrated		120	7	8	0	98	100	Compatible	DuPont Teflon® PFA
Hydrofluoric Acid	up to boiling point	60	23	7	9	0	99	99	Compatible	DuPont Teflon® PFA
Hydrogen Peroxide	up to boiling point	30	23	7	9	0	93	95	Compatible	DuPont Teflon® PFA
Isooctane			99		7	0.8				DuPont Teflon® PFA
			99	7	9	0.7				DuPont Teflon® PFA
			99	7	8	0.7-0.8	94	100	no significant change	DuPont Teflon® PFA 350
Lead	up to boiling point				8				Compatible	DuPont Teflon® PFA
Magnesium Chloride	up to boiling point				8				Compatible	DuPont Teflon® PFA
Mercury	up to boiling point				8				Compatible	DuPont Teflon® PFA
Methacrylic Acid	up to boiling point				8				Compatible	DuPont Teflon® PFA
Methyl Alcohol	up to boiling point				8					DuPont Teflon® PFA
Methyl Ethyl Ketone	up to boiling point		80	7	9	0.4	90	100	Compatible	DuPont Teflon® PFA
Methyl Methacrylate	up to boiling point		40	7	8	0.8	94	100	Compatible	DuPont Teflon® PFA
Methylene Chloride			180	7	8	0	87	95		DuPont Teflon® PFA
Mineral Oils					8					DuPont Teflon® PFA
Naphtha	up to boiling point		100	7	9	0.5	91	100		DuPont Teflon® PFA
Naphthalene	up to boiling point				8				Compatible	DuPont Teflon® PFA
Naphthols					8				Compatible	DuPont Teflon® PFA
Nitric Acid		10	25	365	9	0			no significant change	DuPont Teflon® PFA
		10	70	365	9	0.1			no significant change	DuPont Teflon® PFA
Nitric Acid	up to boiling point				8					DuPont Teflon® PFA
	fuming		23	7	9	0	99	99		DuPont Teflon® PFA
	concentrated		120	7	9	0	95	98	compatible	DuPont Teflon® PFA

(Cont'd.)

Table I.7 Chemical Resistance of Perfluoroalkoxy Copolymer (PFA) (Cont'd.)

Reagent	Reagent Note	Conc.	Temp (°C)	Time (days)	PDL Rating	% Change Weight	% Retained Tensile Strength	% Retained Elongation	Resistance Note	Material Note
Nitro-2-Methylpropanol (2-)	up to boiling point				8				Compatible	DuPont Teflon® PFA
Nitrobenzene	up to boiling point				8				Compatible	DuPont Teflon® PFA
			210		7	0.9				DuPont Teflon® PFA
			210	7	8	0.7	90			DuPont Teflon® PFA
			210	7	7	0.7-0.9		100	no significant change	DuPont Teflon® PFA 350
Nitrobutanol (2-)	up to boiling point				8				Compatible	DuPont Teflon® PFA
Nitrogen Tetraoxide	up to boiling point				8				Compatible	DuPont Teflon® PFA
Nitromethane	up to boiling point				8				Compatible	DuPont Teflon® PFA DuPont Teflon® PFA
Octadecyl Alcohol	up to boiling point				8				Compatible	DuPont Teflon® PFA
Ozone	up to boiling point				8				Compatible	DuPont Teflon® PFA
Pentachlorobenzamide	up to boiling point				8				Compatible	DuPont Teflon® PFA
Perchloroethylene	up to boiling point				8				Compatible	DuPont Teflon® PFA
			121		4	2.2				DuPont Teflon® PFA
			121	7	7	2	86			DuPont Teflon® PFA
			121	7	4	2.0-2.3		100	no significant change	DuPont Teflon® PFA 350
Perfluoroxylene	up to boiling point				8				Compatible	DuPont Teflon® PFA
Phenol	up to boiling point				8				Compatible	DuPont Teflon® PFA
Phosphoric Acid	up to boiling point				8				Compatible	DuPont Teflon® PFA
	concentrated		100		9	0.01				DuPont Teflon® PFA
	concentrated		100	7	9	0	93			DuPont Teflon® PFA
	concentrated		100	7	9	0.00-0.01		100	no significant change	DuPont Teflon® PFA 350
Phosphorous Pentachloride	up to boiling point				8				Compatible	DuPont Teflon® PFA
Phthalic Acid	up to boiling point				8				Compatible	DuPont Teflon® PFA
Pinene	up to boiling point				8				Compatible	DuPont Teflon® PFA DuPont Teflon® PFA
Piperidine	up to boiling point				8				Compatible	DuPont Teflon® PFA
Potassium Acetate	up to boiling point				8				Compatible	DuPont Teflon® PFA
Potassium Hydroxide	up to boiling point				8				Compatible	DuPont Teflon® PFA
Potassium Permanganate	up to boiling point				8				Compatible	DuPont Teflon® PFA
Pyridine	up to boiling point				8				Compatible	DuPont Teflon® PFA
Soap	up to boiling point				8				Compatible	DuPont Teflon® PFA
Sodium Hydroxide		10	25	365	9	0			no significant change	DuPont Teflon® PFA
		10	70	365	9	0.1			no significant change	DuPont Teflon® PFA
		50	100	0.33	9	0			no significant change	DuPont Teflon® PFA
		50	120	7	9	0.4	93	99		DuPont Teflon® PFA
	up to boiling point				8				Compatible	DuPont Teflon® PFA
Sodium Hypochlorite	up to boiling point				8				Compatible	DuPont Teflon® PFA
Sodium Peroxide	up to boiling point				8				Compatible	DuPont Teflon® PFA
	aliphatic				8				Compatible	DuPont Teflon® PFA
	up to boiling point				8				Compatible	DuPont Teflon® PFA
Stannous Chloride	up to boiling point				8				Compatible	DuPont Teflon® PFA
Sulfur	up to boiling point				8				Compatible	DuPont Teflon® PFA

(Cont'd.)

Table I.7 Chemical Resistance of Perfluoroalkoxy Copolymer (PFA) (Cont'd.)

Reagent	Reagent Note	Conc.	Temp (°C)	Time (days)	PDL Rating	% Change Weight	% Retained Tensile Strength	% Retained Elongation	Resistance Note	Material Note
Sulfuric Acid		30	25	365	9	0			no significant change	DuPont Teflon® PFA
		30	70	365	9	0			no significant change	DuPont Teflon® PFA
		30	100	0.33	9	0			no significant change	DuPont Teflon® PFA
	up to boiling point	30	200	0.33	8	0.1			no significant change	DuPont Teflon® PFA
	fuming, 20% oleum		23	7	9	0	95	96	Compatible	DuPont Teflon® PFA
	concentrated		120	7	9	0	95	96		DuPont Teflon® PFA
Sulfuryl Chloride			68	7	4	2.2				DuPont Teflon® PFA
			68	7	4	1.7-2.7	83	100		DuPont Teflon® PFA 350
			69		6	2.7				DuPont Teflon® PFA
Tetrabromomethane	up to boiling point				8				Compatible	DuPont Teflon® PFA
Tetrachloroethylene	up to boiling point				8				Compatible	DuPont Teflon® PFA
Tetrahydrofuran			66	7	8	0.7	88	100	Compatible	DuPont Teflon® PFA
Toluene			25	365	9	0.3				DuPont Teflon® PFA
			50	365	8	0.6				DuPont Teflon® PFA
			70	14	8	0.6				DuPont Teflon® PFA
			110	7	7	0.8				DuPont Teflon® PFA
			110	7	8	0.7	88	100		DuPont Teflon® PFA
			110		8	0.7-0.8				DuPont Teflon® PFA 360
Tributyl Phosphate	tri-n-butyl phosphate		200	7	5	1.9				DuPont Teflon® PFA
			200	7	7	2	91	100		DuPont Teflon® PFA DuPont
	phosphate		200		5	1.8-2.0			no significant change	Teflon® PFA 350
Trichloroacetic Acid	up to boiling point		196	7	8	2.2	90	100	Compatible	DuPont Teflon® PFA
					7					DuPont Teflon® PFA
Trichloroethylene	up to boiling point				8				Compatible	DuPont Teflon® PFA
Tricresyl Phosphate	up to boiling point				8				Compatible	DuPont Teflon® PFA
Triethanolamine	up to boiling point				8				Compatible	DuPont Teflon® PFA
Vegetable Oils	up to boiling point				8				Compatible	DuPont Teflon® PFA
Vinyl Methacrylate	up to boiling point				8				Compatible	DuPont Teflon® PFA
Water	up to boiling point				8				Compatible	DuPont Teflon® PFA
Xylene	up to boiling point				8					DuPont Teflon® PFA
Zinc Chloride		25	100	7	9	0.03	96	100	no significant change	DuPont Teflon® PFA
		25	100	7	9	0				DuPont Teflon® PFA
		25	100		9	0.00-0.03				DuPont Teflon® PFA 350
	up to boiling point				8				Compatible	DuPont Teflon® PFA

Table I.8 Chemical Resistance of Polyvinylidene Fluoride (PVDF)

Reagent	Reagent Note	Conc.	Temp (°C)	Time (days)	PDL Rating	% Change Weight	% Retained Tensile Strength	Resistance Note	Material Note
Acetaldehyde			23		2			not recommended for use	Atochem Kynar
Acetamide			24		8			temp. is max. recommended	Atochem Kynar
Acetic Acid	in water	10	107		8			temp. is max. recommended	Atochem Kynar
	in water	50	93		8			temp. is max. recommended	Atochem Kynar
		50	125	7	7	<1	>90	no visual change	Atochem Foraflon
		50	130	365	8			satisfactory resistance	Atochem Foraflon
	in water	80	79		8			temp. is max. recommended	Atochem Kynar
			25	7	7	<1	>90	no visual change	Atochem Foraflon
			49		8			temp. is max. recommended	Atochem Kynar
Acetic Anhydride			23	7	2			not recommended for use	Atochem Foraflon
			23	90	2			not recommended for use	Atochem Kynar
					5			questionable	Atochem Foraflon
Acetone	in water	10	52		8			temp. is max. recommended	Atochem Kynar
			23	7	2			not recommended for use	Atochem Foraflon
					2			not recommended for use	Atochem Kynar
Acetonitrile			25	7	7	<1	>90	no visual change	Atochem Foraflon
			52		8			temp. is max. recommended	Atochem Kynar
Acetophenone			23	7	2			not recommended for use	Atochem Foraflon
					2			not recommended for use	Atochem Kynar
Acetyl Bromide			52		8			temp. is max. recommended	Atochem Kynar
Acetyl Chloride			50	7	7	<1	>90	no visual change	Atochem Foraflon
			52		8			temp. is max. recommended	Atochem Kynar
Acetylacetone			23	7	2			not recommended for use	Atochem Foraflon
					2			not recommended for use	Atochem Kynar
Acetylene			121		8			temp. is max. recommended	Atochem Kynar
Acrolein					2			not recommended for use	Atochem Foraflon
Acrylonitrile			24		8			temp. is max. recommended	Atochem Kynar
			25	7	7	<1	>90	no visual change	Atochem Foraflon
			25	7	7	<1	>90	no visual change	Atochem Foraflon
Adipic Acid			66		8			temp. is max. recommended	Atochem Kynar
Air			141		8			temp. is max. recommended	Atochem Kynar
Alcoholic Spirits	with ethyl alcohol	40	93		8			temp. is max. recommended	Atochem Kynar
Allyl Alcohol			52		8			temp. is max. recommended	Atochem Kynar
Allyl Chloride			100		8			temp. is max. recommended	Atochem Kynar
			100	7	7	<1	>90	no visual change	Atochem Foraflon
Aluminum Acetate	aqueous solution/solid		141		8			temp. is max. recommended	Atochem Kynar
Aluminum Bromide			141		8			temp. is max. recommended	Atochem Kynar
Aluminum Chloride	In water	=40	141		8			temp. is max. recommended	Atochem Kynar
Aluminum Fluoride	aqueous solution/solid		135		8			temp. is max. recommended	Atochem Kynar
Aluminum Hydroxide			135		8			temp. is max. recommended	Atochem Kynar
Aluminum Nitrate	aqueous solution/solid		135		8			temp. is max. recommended	Atochem Kynar
Aluminum Oxychloride	aqueous solution/solid		135		8			temp. is max. recommended	Atochem Kynar

(Cont'd.)

Table I.8 Chemical Resistance of Polyvinylidene Fluoride (PVDF) (Cont'd.)

Reagent	Reagent Note	Conc.	Temp (°C)	Time (days)	PDL Rating	% Change Weight	% Retained Tensile Strength	Resistance Note	Material Note
Aluminum Sulfate	aqueous solution/solid		135		8			Temp. is max. recommended	Atochem Kynar
Ammonia	Gas		23		2			not recommended for use	Atochem Kynar
	liquefied		23		2			not recommended for use	Atochem Kynar
	gas		150	7	7	<1	>90	no visual change	Atochem Foraflon
Ammonium Acetate	aqueous solution/solid		79		8			Temp. is max. recommended	Atochem Kynar
Ammonium Alum	aqueous solution/solid		135		8			Temp. is max. recommended	Atochem Kynar
Ammonium Bifluoride	aqueous solution/solid		66		8			Temp. is max. recommended	Atochem Kynar
Ammonium Bromide	aqueous solution/solid		121		8			Temp. is max. recommended	Atochem Kynar
Ammonium Carbonate	aqueous solution/solid		136		8			Temp. is max. recommended	Atochem Kynar
Ammonium Chloride	aqueous solution/solid		135		8			Temp. is max. recommended	Atochem Kynar
Ammonium Dichromate	aqueous solution/solid		121		8			Temp. is max. recommended	Atochem Kynar
Ammonium Fluoride	aqueous solution/solid		135		8			Temp. is max. recommended	Atochem Kynar
Ammonium Hydroxide		20	23	180	8			Satisfactory resistance	Atochem Foraflon
		20	50	90	8			Satisfactory resistance	Atochem Foraflon
		20	90	30	2			not recommended for use	Atochem Foraflon
		29	75	14	8			Satisfactory resistance	Atochem Foraflon
		30	150	7	7	<1	>90	no visual change	Atochem Foraflon
Ammonium Hydroxide	aqueous solution/solid	29	110		8			Temp. is max. recommended	Atochem Kynar
Ammonium Metaphosphate	aqueous solution/solid		135		8			Temp. is max. recommended	Atochem Kynar
Ammonium Nitrate	aqueous solution/solid		135		8			Temp. is max. recommended	Atochem Kynar
Ammonium Persulfate	aqueous solution/solid		25		8			Temp. is max. recommended	Atochem Kynar
Ammonium Phosphate	aqueous solution/solid		135		8			Temp. is max. recommended	Atochem Kynar
Ammonium Sulfate	aqueous solution/solid		135		8			Temp. is max. recommended	Atochem Kynar
Ammonium Sulfide	aqueous solution/solid		52		8			Temp. is max. recommended	Atochem Kynar
Ammonium Thiocyanate	aqueous solution/solid		135		8			Temp. is max. recommended	Atochem Kynar
Amyl Acetate			50	7	7	<1	>90	:no visual change	Atochem Foraflon
			52		8			Temp. is max. recommended	Atochem Kynar
Amyl Alcohol			135		8			Temp. is max. recommended	Atochem Kynar
	Pentanol		150	7	7	<1	>90	no visual change	Atochem Foraflon
	Pentanol		160	7	7	<1	>90	no visual change	Atochem Foraflon
Amyl Alcohol (sec-)			52		8			Temp. is max. recommended	Atochem Kynar
Amyl Chloride			100	7	7	<1	>90	Temp. is max. recommended	Atochem Kynar
			141		8			Temp. is max. recommended	Atochem Kynar
Aniline			38		8			Temp. is max. recommended	Atochem Kynar
			38	7	7	<1	>90	no visual change	Atochem Foraflon
Aniline Hydrochloride	aqueous solution/solid		24		8			Temp. is max. recommended	Atochem Kynar
Animal Oils	lard oil		141		8			Temp. is max. recommended	Atochem Kynar
Antimony Trichloride			75	7	7	<1	>90	no visual change	Atochem Foraflon
Aqua Regia			24		8			Temp. is max. recommended	Atochem Kynar
			100	7	7	<1	>90	no visual change	Atochem Foraflon

(Cont'd.)

Table I.8 Chemical Resistance of Polyvinylidene Fluoride (PVDF) (Cont'd.)

Reagent	Reagent Note	Conc.	Temp (°C)	Time (days)	PDL Rating	% Change Weight	% Retained Tensile Strength	Resistance Note	Material Note
Arsenic Acid	aqueous solution		135		8			temp. is max. recommended	Atochem Kynar
Asphalt	aqueous solution/solid		121		8			temp. is max. recommended	Atochem Kynar
Barium Carbonate			141		8			temp. is max. recommended	Atochem Kynar
Barium Chloride			141		8			temp. is max. recommended	Atochem Kynar
Barium Hydroxide			135		8			temp. is max. recommended	Atochem Kynar
Barium Sulfate	aqueous solution/solid		135		8			temp. is max. recommended	Atochem Kynar
Barium Sulfide			141		8			temp. is max. recommended	Atochem Kynar
Beer			135		8			temp. is max. recommended	Atochem Kynar
Beet Sugar Liquors			100		8			temp. is max. recommended	Atochem Kynar
			107		8			temp. is max. recommended	Atochem Kynar
Benzaldehyde			21		8			temp. is max. recommended	Atochem Kynar
			25	7	2			not recommended for use	Atochem Foraflon
Benzene	1/1 with Chlorobenzene	50	130	180	5			questionable	Atochem Foraflon
			38	7	7	<1	>90	no visual charge	Atochem Foraflon
			77		8			temp. is max. recommended	Atochem Kynar
Benzenesulfonic Acid	concentrated		25	7	7	<1	>90	no visual change	Atochem Foraflon
	aqueous solution/solid		52		8			temp. is max. recommended	Atochem Kynar
Benzoic Acid			107		8			temp. is max. recommended	Atochem Kynar
	saturated		125	7	7	<1	>90	no visual change	Atochem Foraflon
Benzoyl Chloride			75	7	7	<1	>90	no visual change	Atochem Foraflon
			77		8			temp. is max. recommended	Atochem Kynar
Benzoyl Peroxide			77		8			temp. is max. recommended	Atochem Kynar
Benzyl Alcohol			121		8			temp. is max. recommended	Atochem Kynar
Benzyl Chloride			100	7	7	<1	>90	no visual change	Atochem Foraflon
			141		8			temp. is max. recommended	Atochem Kynar
Benzyl Ether			38		8			temp. is max. recommended	Atochem Kynar
Benzylamine	aqueous solution/liquid		24		8			temp. is max. recommended	Atochem Kynar
Benzylic Alcohol			125	7	7	<1	>90	no visual change	Atochem Foraflon
Black Liquor			79		8			temp. is max. recommended	Atochem Kynar
Bleach	liquid		90	15	8			Satisfactory resistance	Atochem Foraflon
	liquid		90	90	8			Satisfactory resistance	Atochem Foraflon
	liquid		130	90	5			questionab e	Atochem Foraflon
	bleaching agents		135		8			temp. is max. recommended	Atochem Foraflon
Borax			135		8			temp. is max. recommended	Atochem Kynar
Boric Acid			135		8			temp. is max. recommended	Atochem Kynar
			150	7	7	<1	>90	no visual change	Atochem Foraflon
Boron Trifluoride			24		8			temp. is max. recommended	Atochem Kynar
			25	7	7	<1	>90	no visual change	Atochem Foraflon
Brines	chlorinated acid		93		8			temp. is max. recommended	Atochem Kynar
	Acid		141		8			temp. is max. recommended	Atochem Kynar
			141		8			temp. is max. recommended	Atochem Kynar
	basic		141		8			temp. is max. recommended	Atochem Kynar

(Cont'd.)

Table I.8 Chemical Resistance of Polyvinylidene Fluoride (PVDF) (*Cont'd.*)

Reagent	Reagent Note	Conc.	Temp (°C)	Time (days)	PDL Rating	% Change Weight	% Retained Tensile Strength	Resistance Note	Material Note
Bromine	bromine liquid		60	365	8			Satisfactory	Atochem Foraflon
Bromine	dry gas		66		8			temp. is max recommended	Atochem Kynar
	bromine water		66		8			temp. is max recommended	Atochem Kynar
			100		8			temp. is max recommended	Atochem Kynar
	dry		100	7	7	<1	>90	no visual change	Atochem Foraflon
	moist		100	7	7	<1	>90	no visual change	Atochem Foraflon
Bromobenzene			66		8			temp. is max recommended	Atochem Kynar
Bromoform			66		8			temp. is max recommended	Atochem Kynar
Butadiene			100	7	7	<1	>90	no visual change	Atochem Foraflon
			121		8			temp. is max recommended	Atochem Kynar
Butane			121		8			temp. is max recommended	Atochem Kynar
Butanediol	aqueous solution/liquid		135		8			temp. is max recommended	Atochem Kynar
Butanone					2			not recommended for use	Atochem Foraflon
Butane			150	7	7	<1	>90	no visual change	Atochem Foraflon
Butyl Acetate			25	7	7	<1	>90	no visual change	Atochem Foraflon
			27		8			temp. is max recommended	Atochem Kynar
Butyl Acrylate			25	7	7	<1	>90	no visual change	Atochem Foraflon
			52		8			temp. is max recommended	Atochem Kynar
Butyl Alcohol	n-butanol		75	7	7	<1	>90	no visual change	Atochem Foraflon
			107		8			temp. is max recommended	Atochem Kynar
Butyl Alcohol (sec-)	aqueous solution/liquid		75	7	7	<1	>90	no visual change	Atochem Foraflon
			93		8			temp. is max recommended	Atochem Kynar
Butyl Alcohol (tert-)	aqueous solution/liquid		75	7	7	<1	>90	no visual change	Atochem Foraflon
			93		8			temp. is max recommended	Atochem Kynar
Butyl Bromide			100	7	7	<1	>90	no visual change	Atochem Foraflon
			141		8			temp. is max recommended	Atochem Kynar
Butyl Chloride			141		8			temp. is max recommended	Atochem Kynar
Butyl Ether			38		8			temp. is max recommended	Atochem Kynar
Butyl-2-Hydroxybenzone-butylphenol (l-)			100	7	7	<1	>90	no visual change	Atochem Foraflon
Butyl Mercaptan			141		8			temp. is max recommended	Atochem Kynar
Butyl Methyl Ether (tert-)			50	120	8			Satisfactory resistance	Atochem Foraflon
Butyl Phenol			107		8			Satisfactory resistance	Atochem Foraflon
Butyl Stearate			38		8			Satisfactory resistance	Atochem Foraflon
Butylamine	n-butylamine		23	7	2			not recommended for use	Atochem, Kynar
	aqueous solution/liquid		23		2			not recommended for use	Atochem Kynar
Butylamine (see-)	aqueous solution/liquid		21		8			not recommended for use	Atochem Kynar
	aqueous solution/liquid		25	7	7	<1	>90	no visual change	Atochem Foraflon
Butylamine (tert-)	aqueous solution/solid		21	7	7	1	>90	temp. is max recommended	Atochem Kynar
			25	7	7		>90	no visual change	Atochem Foraflon

(Cont'd.)

Table I.8 Chemical Resistance of Polyvinylidene Fluoride (PVDF) (*Cont'd.*)

Reagent	Reagent Note	Conc.	Temp (°C)	Time (days)	PDL Rating	% Change Weight	% Retained Tensile Strength	Resistance Note	Material Note
Butylene			141		8			temp. is max recommended	Atochem Foraflon
Butyraldehyde			66		8			temp. is max recommended	Atochem Foraflon
Butyric Acid			107		8			temp. is max recommended	Atochem Foraflon
Calcium Acetate	aqueous solution/solid		141		8			temp. is max recommended	Atochem Foraflon
Calcium Bisulfate	aqueous solution/solid		141		8			temp. is max recommended	Atochem Foraflon
Calcium Bisulfite	aqueous solution/solid		93		8			temp. is max recommended	Atochem Foraflon
Calcium Bromide	aqueous solution/solid		141		8			temp. is max recommended	Atochem Foraflon
Calcium Carbonate			141		8			temp. is max recommended	Atochem Foraflon
Calcium Chlorate	aqueous solution/solid		141		8			temp. is max recommended	Atochem Foraflon
Calcium Chloride	aqueous solution/solid		141		8			temp. is max recommended	Atochem Foraflon
Calcium Hydroxide			135		8			temp. is max. recommended	Atochem Foraflon
Calcium Hypochlorite	aqueous solution/solid		93		8			temp. is max. recommended	Atochem Foraflon
Calcium Nitrate	aqueous solution/solid		135		8			temp. is max. recommended	Atochem Foraflon
Calcium Oxide			121		8			temp. is max. recommended	Atochem Foraflon
Calcium Phosphate			141		8			temp. is max. recommended	Atochem Foraflon
Calcium Sulfate			141		8			temp. is max. recommended	Atochem Foraflon
Cane Sugar	can sugar liquors		141		8			temp. is max. recommended	Atochem Foraflon
Caprylic Acid			79		8			temp. is max. recommended	Atochem Foraflon
Carbon Dioxide			141		8			temp. is max. recommended	Atochem Foraflon
Carbon Disulfide			24		8			temp. is max. recommended	Atochem Foraflon
			25	7	7	<1	>90	no visual change	Atochem Foraflon
Carbon Monoxide			141		8			temp. is max. recommended	Atochem Foraflon
Carbon Tetrachloride			90	180	5			questionable	Atochem Kynar
			135		8			temp. is max. recommended	Atochem Foraflon
			150	7	7	<1	>90	no visual change	Atochem Foraflon
Carbonic Acid			135		8		>90	temp. is max. recommended	Atochem Kynar
Casein			121		8		>90	temp. is max. recommended	Atochem Kynar
Castor Oil			41		8			temp. is max. recommended	Atochem Kynar
Chloral			25	7	7			no visual change	Atochem Foraflon
Chloral Hydrate			24		8			temp. is max. recommended	Atochem Kynar
Chlorine	in carbon tetrachloride	5	93		8			temp. is max. recommended	Atochem Kynar
	saturated with H_2SO_4	65-98	23	240	8			Satisfactory resistance	Atochem Foraflon
	chlorine gas w/UV light		30	11	8			Satisfactory resistance	Atochem Foraflon
	gas		93		8			temp. is max. recommended	Atochem Kynar
	liquid		93		8			temp. is max. recommended	Atochem Kynar
	dry		100	7	7	<1	>90	no visual change	Atochem Foraflon
	moist		100	7	7	<1	>90	no visual change	Atochem Foraflon
	chlorine gas w/o light		100	11	8			Satisfactory resistance	Atochem Foraflon
	chlorine water		107		8			temp. is max. recommended	Atochem Kynar
Chlorine Dioxide			66		8			temp. is max. recommended	Atochem Kynar

(*Cont'd.*)

Table I.8 Chemical Resistance of Polyvinylidene Fluoride (PVDF) *(Cont'd.)*

Reagent	Reagent Note	Conc.	Temp (°C)	Time (days)	PDL Rating	% Change Weight	% Retained Tensile Strength	Resistance Note	Material Note
Chloroacetic Acid	Aqueous solution/pure	75	75	7	7	<1	>90	no visual change	Atochem Foraflon
			23		2			not recommended for use	Atochem Kynar
			100	7	2			not recommended for use	Atochem Foraflon
Chloroacetyl Chloride			25	7	7	<1	>90	no visual change	Atochem Foraflon
			52		8			not recommended for use	Atochem Kynar
Chlorobenzene	50/50 w/benzene	50	130	180	5			questionable	Atochem Foraflon
			50	7	7	<1	>90	no visual change	Atochem Foraflon
			77		8			temp. is max. recommended	Atochem Kynar
			130	120	5			questionable	Atochem Kynar
Chlorobenzenesulfonic Acid	Aqueous solution/pure		93		8			temp. is max. recommended	Atochem Kynar
Chlorobenzyl Chloride			52		8			temp. is max. recommended	Atochem Kynar
Chloroform	with sulfuric acid (98%)	10	50	180	8			Satisfactory resistance	Atochem Foraflon
	with methanol HCl		50	180	8			Satisfactory resistance	Atochem Foraflon
Chloroform			52	7	8			temp. is max. recommended	Atochem Kynar
Chloroform	Trichloromethane		100	7	7	<1	>90	no visual change	Atochem Foraflon
Chlorohexanol (6-)			77		8			no visual change	Atochem Foraflon
Chlorohydrin			52	7	8			temp. is max. recommended	Atochem Kynar
Chloromethyl Methyl Ether					5			questionable	Atochem Foraflon
Chloropicrin			66		8			temp. is max. recommended	Atochem Foraflon
Chlorosulfonic Acid		98	25	7	7	<1	>90	no visual change	Atochem Foraflon
			23		2			not recommended for use	Atochem Kynar
Chlorotrimethylsilane	aqueous solution/solid		52		8			temp. is max. recommended	Atochem Foraflon
Chrome Alum			93		8			temp. is max. recommended	Atochem Kynar
Chromic Acid	in water	=40	79		8			temp. is max. recommended	Atochem Kynar
	in water	50	52		8			temp. is max. recommended	Atochem Kynar
	930 g/l + surfactant		90	120	8			Satisfactory resistance	Atochem Foraflon
Chromic Anhydride	saturated		100	7	7	<1	>90	no visual change	Atochem Foraflon
Chromyl Chloride			52		8			temp. is max. recommended	Atochem Kynar
Cider			60		8			no visual change	Atochem Foraflon
Citric Acid	aqueous solution/solid	50	150	7	7	<1	>90	no visual change	Atochem Foraflon
			135		8			temp. is max. recommended	Atochem Kynar
Coal Gas			107		8		i	temp. is max. recommended	Atochem Kynar
Coconut Oil			141		8			temp. is max. recommended	Atochem Kynar
Copper Acetate	aqueous solution/solid		121		8			temp. is max. recommended	Atochem Kynar
Copper Carbonate	basic		141		8			temp. is max. recommended	Atochem Kynar
Copper Chloride	aqueous solution/solid		141		8			temp. is max. recommended	Atochem Kynar
Copper Cyanide	aqueous solution/solid		135		8			temp. is max. recommended	Atochem Kynar
Copper Fluoride			135		8			temp. is max. recommended	Atochem Kynar
Copper Nitrate	aqueous solution/solid		135		8			temp. is max. recommended	Atochem Kynar
Copper Sulfate	aqueous solution/solid		141		8			temp. is max. recommended	Atochem Kynar
Corn Oil			141		8			temp. is max. recommended	Atochem Kynar
Corn Syrup			121		8			temp. is max. recommended	Atochem Kynar
Cottonseed Oil			141		8			temp. is max. recommended	Atochem Kynar

(Cont'd.)

Table I.8 Chemical Resistance of Polyvinylidene Fluoride (PVDF) (Cont'd.)

Reagent	Reagent Note	Conc.	Temp (°C)	Time (days)	PDL Rating	% Change Weight	% Retained Tensile Strength	Resistance Note	Material Note
Cresol			66		8			temp. is max. recommended	Atochem Kynar
			75	7	7	<1	>90	no visual change	Atochem Foraflon
Cresylic Acid			66		8			temp. is max. recommended	Atochem Kynar
Crotonaldehyde			50	7	7	<1	>90	no visual change	Atochem Foraflon
			52		8			temp. is max. recommended	Atochem Kynar
Crude Oils	sour crude oil		90	365	8			Satisfactory resistance	Atochem Foraflon
			130	365	8			Satisfactory resistance	Atochem Foraflon
			141		8			temp. is max. recommended	Atochem Kynar
			141		8			temp. is max. recommended	Atochem Kynar
			150	7	7	<1	>90	no visual Change	Atochem Foraflon
			150	365	8			Satisfactory resistance	Atochem Foraflon
Cryolite			121		8			temp. is max. recommended	Atochem Kynar
Cuprous Chloride			121		8			temp. is max. recommended	Atochem Kynar
Cyclohexane			141		8			temp. is max. recommended	Atochem Kynar
			150	7	7	<1	>90	no visual change	Atochem Foraflon
Cyclohexanone			24	7	2			not recommended for use	Atochem Foraflon
					8			temp. is max. recommended	Atochem Kynar
Cyclohexyl Acetate			50	7	7	<1	>90	no visual change	Atochem Foraflon
			93		8			temp. is max. recommended	Atochem Kynar
Cyclohexyl Alcohol			75	7	7	<1	>90	no visual change	Atochem Foraflon
			121		8			temp. is max. recommended	Atochem Kynar
Decalin			100	7	7	<1	>90	no visual change	Atochem Foraflon
Decane			121		8			temp. is max. recommended	Atochem Kynar
Dextrin	aqueous solution/solid		121		8			temp. is max. recommended	Atochem Kynar
Diacetone Alcohol			24	7	5			questionable	Atochem Foraflon
					8			temp. is max. recommended	Atochem Kynar
Dibromobenzene (p-)			93		8			temp. is max. recommended	Atochem Kynar
Dibromoethane (1,2-)			50	7	7	<1	>90	no visual Change	Atochem Foraflon
Dibromopropane (1,2-)			93		8			temp. is max. recommended	Atochem Kynar
Dibutyl Phthalate			23		2			not recommended for use	Atochem Kynar
Dibutyl Sebacate			23		2			not recommended for use	Atochem Kynar
Dibutylamine	aqueous solution/liquid		21		8			temp. is max. recommended	Atochem Kynar
Dichloroacetic Acid	aqueous solution/liquid		52		8			temp. is max. recommended	Atochem Kynar
Dichlorobenzene (o-)			66		8			temp. is max. recommended	Atochem Kynar
			130	120	8			satisfactory resistance	Atochem Foraflon
Dichlorodifluoromethane	R12		100	7	7	<1	>90	no visual change	Atochem Foraflon
Dichlorodimethylsilane			52		8			temp. is max. recommended	Atochem Kynar
Dichloroethane	with hydrochloric acid	10	90	180	8			satisfactory resistance	Atochem Foraflon
		10	130	180	8			satisfactory resistance	Atochem Foraflon
Dichloroethane (1,2-)			90	365	8			satisfactory resistance	Atochem Foraflon

(Cont'd.)

Table I.8 Chemical Resistance of Polyvinylidene Fluoride (PVDF) (Cont'd.)

Reagent	Reagent Note	Conc.	Temp (°C)	Time (days)	PDL Rating	% Change Weight	% Retained Tensile Strength	Resistance Note	Material Note
Dichloroethylene			107		8			temp. is max. recommended	Atochem Kynar
Dichloropropionic Acid (2,2-)			52		8			temp. is max. recommended	Atochem Kynar
Dichlorotetrafluoroethane	R114; Freon® 114		50	7	7	<1	>90	no visual change	Atochem Foraflon
Dichlorotoluene (a)			66		8			temp. is max. recommended	Atochem Kynar
Diesel Fuels			141		8			temp. is max. recommended	Atochem Kynar
Diethanolamine	aqueous solution/liquid		23		2			not recommended for use	Atochem Kynar
Diethyl Ether	with sulfuric acid (98%)	10	50	180	8			satisfactory resistance	Atochem Foraflon
			25	7	7	<1	>90	no visual change	Atochem Foraflon
Diethyl Malonate			23		2			not recommended for use	Atochem Kynar
Diethylamine	aqueous solution/liquid		24		8		>90	temp. is max. recommended	Atochem Kynar
			25	7	7	<1		no visual change	Atochem Foraflon
Diethylenetriamine	DETA		50	7	7	<1	>90	no visual change	Atochem Foraflon
	DETA		52	7	8			temp. is max. recommended	Atochem Kynar
Diglycolic Acid			24		8			temp. is max. recommended	Atochem Kynar
Diisobutyl Ketone			.50	7	7	41	>90	no visual change	Atochem Foraflon
			52		8			temp. is max. recommended	Atochem Kynar
Diisobutylene			141		8			temp. is max. recommended	Atochem Kynar
Diisopropyl Ketone			21		8			temp. is max. recommended	Atochem Kynar
Dimethyl Phthalate			24		8			temp. is max. recommended	Atochem Kynar
			25	7	7	<1	>90	no visual change	Atochem Foraflon
Dimethyl Sulfate			24		8			temp. is max. recommended	Atochem Kynar
Dimethyl Sulfoxide			23		2			not recommended for use	Atochem Kynar
Dimethyl-1,5-Hexediene (2,5-)			121		8			not recommended for use	Atochem Kynar
Dimethyl-4-Heptanof (2,6-)			93		8			temp. is max. recommended	Atochem Kynar
Dimethylacetamide			23		2			not recommended for use	Atochem Kynar
Dimethylamine	aqueous solution or gas		24	7	2			not recommended for use	Atochem Foraflon
			24		8			temp. is max. recommended	Atochem Kynar
			25	7	7	<1	>90	no visual change	Atochem Foraflon
Dimethyl Formamide			23		2			not recommended for use	Atochem Kynar
Dioctyl Phthalate			24	7	8			temp. is max. recommended	Atochem Kynar
Dioxane (1,4-)			23		2			not recommended for use	Atochem Kynar
Dioxolane			23		2			not recommended for use	Atochem Kynar
Dipropylene Glycol Methyl Ether			24		8			temp. is max. recommended	Atochem Kynar
Dishwashing Detergents			90	42	8			satisfactory resistance	Atochem Foraflon
			90	42	8			satisfactory resistance	Atochem Foraflon
Disodium Phosphate	aqueous solution/solid		93		8			temp. is max. recommended	Atochem Kynar
Divinylbenzene			52		8			temp. is max. recommended	Atochem Kynar
Epichlorohydrin			23	7	2			not recommended for use	Atochem Foraflon
			23		2			not recommended for use	Atochem Kynar

(Cont'd.)

Table I.8 Chemical Resistance of Polyvinylidene Fluoride (PVDF) (Cont'd.)

Reagent	Reagent Note	Conc.	Temp (°C)	Time (days)	PDL Rating	% Change Weight	% Retained Tensile Strength	Resistance Note	Material Note
EPSOM Salts	aqueous solution/solid		93		8			temp. is max. recommended	Atochem Kynar
Ethanethiol			24		8			temp. is max. recommended	Atochem Kynar
Ethoxyethyl Acetate (2-)	aqueous solution/liquid		50	120	5			questionable	Atochem Foraflon
			93		8			temp. is max. recommended.	Atochem Kynar
Ethyl Acetate			23		2			not recommended for use	Atochem Kynar
			25	7	7	<1	>90	no visual change	Atochem Foraflon
Ethyl Acetoacetate			24		8			temp. is max. recommended.	Atochem Kynar
Ethyl Acrylate			24		8			temp. is max. recommended	Atochem Kynar
			25	7	7	<1	>90	no visual change	Atochem Foraflon
Ethyl Alcohol	in alcoholic spirits	40	93		8			temp. is max. recommended	Atochem Kynar
				7	2			not recommended for use	Atochem Foraflon
	aqueous solution/liquid		23		2			not recommended for use	Atochem Kynar
			100	7	7	<1	>90	no visual change	Atochem Foraflon
	aqueous solution/liquid		141		8			temp. is max. recommended.	Atochem Kynar
Ethyl Benzene			52		8			temp. is max. recommended.	Atochem Kynar
Ethyl Chloride			100	7	7	<1	>90	no visual change	Atochem Foraflon
			141		8			temp. is max. recommended.	Atochem Kynar
Ethyl Chloroacetate			24		8			temp. is max. recommended	Atochem Kynar
Ethyl Chloroformate			52		8			temp. is max. recommended	Atochem Kynar
Ethyl Cyanoacetate			24		8			temp. is max. recommended	Atochem Kynar
Ethyl Ether			52		8			temp. is max. recommended	Atochem Kynar
Ethyl Formate			24		8			temp. is max. recommended	Atochem Kynar
Ethylene Chloride			100	7	7	<1	>90	no visual change	Atochem Foraflon
Ethylene Chlorohydrin	aqueous solution/solid		24		8			temp. is max. recommended	Atochem Kynar
Ethylene Dichloride			135		8			temp. is max. recommended	Atochem Kynar
Ethylene Glycol	aqueous solution/solid		130	365	8			satisfactory resistance	Atochem Foraflon
			141		8			temp. is max. recommended	Atochem Kynar
			150	7	7	<1	>90	no visual change	Atochem Foraflon
Ethylene Oxide			50	7	7	<1	>90	no visual change	Atochem Foraflon
			93		8			temp. is max. recommended	Atochem Kynar
Ethylenediamine	aqua us solution/solid		107	7	2			not recommended for use	Atochem Foraflon
					8			temp. is max. recommended	Atochem Kynar
Ethylhexyl Alcohol (2-)	2-ethylhexanol		121		8			temp. is max. recommended	Atochem Kynar
Fatty Acids	sultanates		79		8			temp. is max. recommended	Atochem Kynar
			141		8			temp. is max. recommended	Atochem Kynar

(Cont'd.)

304

Table I.8 Chemical Resistance of Polyvinylidene Fluoride (PVDF) (*Cont'd.*)

Reagent	Reagent Note	Conc.	Temp (°C)	Time (days)	PDL Rating	% Change Weight	% Retained Tensile Strength	Resistance Note	Material Note
Ferric Chloride	aqueous solution/solid		141		8			temp. is max. recommended	Atochem Kynar
Ferric Hydroxide			121		8			temp. is max. recommended	Atochem Kynar
Ferric Nitrate	aqueous solution/solid		135		8			temp. is max. recommended	Atochem Kynar
Ferric Sulfate			141		8			temp. is max. recommended	Atochem Kynar
Ferrous Sulfide			121		8			temp. is max. recommended	Atochem Kynar
Ferrous Chloride			141		8			temp. is max. recommended	Atochem Kynar
Ferrous Hydroxide			121		8			temp. is max. recommended	Atochem Kynar
Ferrous Nitrate			135		8			temp. is max. recommended	Atochem Kynar
Ferrous Sulfate			141		8			temp. is max. recommended	Atochem Kynar
Fluoboric Acid	aqueous solution		135		8			temp. is max. recommended	Atochem Kynar
Fluorine			24		8			temp. is max. recommended	Atochem Kynar
			25	7	7	<1	>90	no visual change	Atochem Foraflon
Fluosilicic Acid		97	135		8			temp. is max. recommended	Atochem Kynar
Fluosulfonic Acid			25	7	7	<1	>90	no visual change	Atochem Foraflon
Formaldehyde	in water	30	50	7	7	<1	>90	no visual change	Atochem Foraflon
		37	52		8			temp. is max. recommended	Atochem Kynar
Formic Acid		98	75	7	7	<1	>90	no visual change	Atochem Foraflon
	aqueous solution/solid		121		8			temp. is max. recommended	Atochem Kynar
Freon® 11	chlorofluorocarbon 11		93		8			temp. is max. recommended	Atochem Kynar
			100	7	7	<1	>90	no visual change	Atochem Foraflon
Freon® 113	chlorofluorocarbon 113		93		8			temp. is max. recommended	Atochem Kynar
Freon® 114	chlorofluorocarbon 114		93		8			temp. is max. recommended	Atochem Kynar
Freon® 12	chlorofluorocarbon 12		93		8			temp. is max. recommended	Atochem Kynar
Freon® 13	chlorofluorocarbon 13		93		8			temp. is max. recommended	Atochem Kynar
Freon® 14	chlorofluorocarbon 14		93		8			temp. is max. recommended	Atochem Kynar
Freon® 21	chlorofluorocarbon 21		93		8			temp. is max. recommended	Atochem Kynar
Freon® 22	chlorofluorocarbon 22		93		8			temp. is max. recommended	Atochem Kynar
	chlorodifluoromethane		100	7	7	<1	>90	no visual change	Atochem Foraflon
Fructose	aqueous solution/solid and pulp		141		8			temp. is max. recommended	Atochem Kynar
Fruit Juices			100		8			temp. is max. recommended	Atochem Kynar
Fuel Oils			141		8			temp. is max. recommended	Atochem Kynar
Fuels	light fuel		125	7	7	<1	>90	no visual change	Atochem Foraflon
	diesel		141		8			temp. is max. recommended	Atochem Kynar
Fumaric Acid			77		8			temp. is max. recommended	Atochem Kynar
Furan			23	7	2			not recommended for use	Atochem Foraflon
					2			not recommended for use	Atochem Kynar
Furfural			24	7	5			questionable	Atochem Foraflon
					8			temp. is max. recommended	Atochem Kynar
Furfural Alcohol	aqueous solution/liquid		38		8			temp. is max. recommended	Atochem Kynar

(Cont'd.)

Table I.8 Chemical Resistance of Polyvinylidene Fluoride (PVDF) (Cont'd.)

Reagent	Reagent Note	Conc.	Temp (°C)	Time (days)	PDL Rating	% Change Weight	% Retained Tensile Strength	Resistance Note	Material Note
Gallic Acid	saturated		24		8			temp. is max. recommended	Atochem Kynar
			25	7	7	<1	>90	no visual change	Atochem Foraflon
Gases	manufactured		141		8			temp. is max. recommended	Atochem Kynar
Gasoline	E		125	7	7	<1	>90	no visual change	Atochem Foraflon
	leaded		141		8			temp. is max. recommended	Atochem Kynar
	sour		141		8			temp. is max. recommended	Atochem Kynar
	unleaded		141		8			temp. is max. recommended	Atochem Kynar
Gelatins			121		8			temp. is max. recommended	Atochem Kynar
Gin			93		8			temp. is max. recommended	Atochem Kynar
Glucose	aqueous solution/solid		141		8			temp. is max. recommended	Atochem Kynar
			150	7	7	<1	>90	no visual change	Atochem Foraflon
Glues			121		8			temp. is max. recommended	Atochem Kynar
Glutamic Acid			93		8			temp. is max. recommended	Atochem Kynar
Glycerin	aqueous solution/liquid		125	7	7	<1	>90	no visual change	Atochem Foraflon
			130	365	8			satisfactory resistance	Atochem Foraflon
			141		8			temp. is max. recommended	Atochem Kynar
Glycine	aqueous solution/solid		24		8			temp. is max. recommended	Atochem Kynar
Glycolic Acid	saturated		24		8			temp. is max. recommended	Atochem Kynar
			25	7	7	<1	>90	no visual change	Atochem Foraflon
Heptane			141		8			temp. is max. recommended	Atochem Kynar
			150	7	7	<1	>90	no visual change	Atochem Foraflon
Hexachloro-1,3-Butadione			52		8			temp. is max. recommended	Atochem Kynar
Hexamethylenediamine			23		2			not recommended for use	Atochem Kynar
Hexamethylphosphoric Triamide			23		2			not recommended for use	Atochem Kynar
Hexane			141		8			temp. is max. recommended	Atochem Kynar
			150	7	7	<1	>90	no visual change	Atochem Foraflon
Hexyl Alcohol			79		8			temp. is max. recommended	Atochem Kynar
Hydrazine	aqueous solution/liquid		93		8			temp. is max. recommended	Atochem Kynar
Hydrazine Dihydrochloride	aqueous Solution/solid		24		8			temp. is max. recommended	Atochem Kynar
Hydrazine Hydrate	aqueous solution/liquid		52		8			temp. is max. recommended	Atochem Kynar
Hydriodic Acid	aqueous solution		135		8			temp. is max. recommended	Atochem Kynar
Hydrobromic Acid	in water	=50	135		8			temp. is max. recommended	Atochem Foraflon
		50	150	7	7	<1	>90	no visual change	Atochem Foraflon
		66	90	365	8			satisfactory resistance	Atochem Foraflon
Hydrochloric Acid		35	130	365	8			satisfactory resistance	Atochem Foraflon
	with 10% methanol	35	150	7	7	<1	>90	no visual change	Atochem Foraflon
	w/ MeOH + chlorofom		50	180	8			satisfactory resistance	Atochem Foraflon
	w/10% dichloroethane		50	180	8			satisfactory resistance	Atochem Foraflon
	w/10% dichloroethane		90	180	8			satisfactory resistance	Atochem Foraflon
	up to concentrated		130	180	8			satisfactory resistance	Atochem Foraflon
			141		8			temp. is max. recommended	Atochem Kynar
	gas		150	7	7			no visual change	Atochem Foraflon

(Cont'd.)

Table I.8 Chemical Resistance of Polyvinylidene Fluoride (PVDF) (Cont'd.)

Reagent	Reagent Note	Conc.	Temp (°C)	Time (days)	PDL Rating	% Change Weight	% Retained Tensile Strength	Resistance Note	Material Note
Hydrocyanic Acid	aqueous solution		25	7	7	<1	>90	no visual change	Atochem Foraflon
			135		8			temp. is max. recommended	Atochem Kynar
Hydrofluoric Acid	in water	=40	121		8			temp. is max. recommended	Atochem Kynar
		40	100	7	7	<1	>90	no visual change	Atochem Foraflon
		70	75	7	7	<1	>90	no visual change	Atochem Foraflon
	in water	41-100	93		8			temp. is max. recommended	Atochem Foraflon
			50	7	7	<1	>90	no visual change	Atochem Foraflon
Hydrofluosilicic Acid			100	7	7	<1	>90	no visual change	Atochem Foraflon
Hydrogen			141		8			temp. is max. recommended	Atochem Kynar
			150	7	7	<1	>90	no visual change	Atochem Foraflon
Hydrogen Chloride			141		8			temp. is max. recommended	Atochem Kynar
Hydrogen Cyanide			135		8			temp. is max. recommended	Atochem Kynar
Hydrogen Fluoride			93		8			temp. is max. recommended	Atochem Kynar
Hydrogen Peroxide	in water	=30	93		8			temp. is max. recommended	Atochem Kynar
		50	100	7	7	<1	>90	no Visual change	Atochem Kynar
	in water	90	21		6			temp. is max. recommended	Atochem Foraflon
Hydrogen Sulfide			100	7	7	<1	>90	no visual change	Atochem Foraflon
	aqueous solution		107		8			temp. is max. recommended	Atochem Kynar
			135		8			temp. is max. recommended	Atochem Kynar
Hydroquinone			121		8			temp. is max. recommended	Atochem Kynar
Hypochlorous Acid	aqueous solution		21		8			temp. is max. recommended	Atochem Kynar
Iodine	in non-aqueous solvent	10	66		8			temp. is max. recommended	Atochem Kynar
	gas		66		8			temp. is max. recommended	Atochem Kynar
	dry		75	7	7	<1	>90	no visual change	Atochem Foraflon
	moist		75	7	7	<1	>90	no visual change	Atochem Foraflon
Iodoform			75	7	7	<1	>90	no visual change	Atochem Foraflon
			93		8			temp. is max. recommended	Atochem Kynar
Isoamyl Ether			121		8			temp. is max. recommended	Atochem Kynar
Isobutyl Alcohol			121		8			temp. is max. recommended	Atochem Kynar
Isooctane			121		8			temp. is max. recommended	Atochem Kynar
Isophorone			79		8			temp. is max. recommended	Atochem Kynar
Isopropyl Alcohol			60		8			temp. is max. recommended	Atochem Kynar
Isopropyl Benzene			38		8			temp. is max. recommended	Atochem Kynar
Isopropyl Chloride			38		8			temp. is max. recommended	Atochem Kynar
Isopropyl Ether			52		8			temp. is max. recommended	Atochem Kynar
Jet Aircraft Fuels	JP 4 and JP 5		93		8			temp. is max. recommended	Atochem Kynar

(Cont'd.)

Table I.8 Chemical Resistance of Polyvinylidene Fluoride (PVDF) (Cont'd.)

Reagent	Reagent Note	Conc.	Temp (°C)	Time (days)	PDL Rating	% Change Weight	% Retained Tensile Strength	Resistance Note	Material Note
Kerosene			141		8			temp. is max. recommended	Atochem Kynar
			150	7	7	<1	>90	no visual change	Atochem Foraflon
Lactic Acid	aqueous solution/pure	50	25	7	7	<1	>90	no visual change	Atochem Foraflon
			52		8			temp. is max. recommended	Atochem Kynar
Lanolin			121		8			temp. is max. recommended	Atochem Kynar
Lauric Acid			100	7	7	<1	>90	no visual change	Atochem Foraflon
			107		8			temp. is max. recommended	Atochem Kynar
Lauryl Chloride			121		8			temp. is max. recommended	Atochem Kynar
			150	7	7	<1	>90	no visual change	Atochem Foraflon
Lauryl Mercaptan			93		8			temp. is max. recommended	Atochem Kynar
Lauryl Sulfate	aqueous solution/solid		121		8			temp. is max. recommended	Atochem Kynar
Lead Acetate			135		8			temp. is max. recommended	Atochem Kynar
Lead Chloride	aqueous solution/solid		121		8			temp. is max. recommended	Atochem Kynar
Lead Nitrate			121		8			temp. is max. recommended	Atochem Kynar
Lead Sulfate			121		8			temp. is max. recommended	Atochem Kynar
Lemon Oil			121		8			temp. is max. recommended	Atochem Kynar
Linoleic Acid			121		8			temp. is max. recommended	Atochem Kynar
			125	7	7	<1	>90	no visual change	Atochem Foraflon
Linseed Oil			141		8			temp. is max. recommended	Atochem Kynar
Liquors	cane sugar		141		8			temp. is max. recommended	Atochem Kynar
Lithium Bromide	aqueous solution/solid		107		8			temp. is max. recommended	Atochem Kynar
Lithium Chloride	aqueous solution/solid		121		8			temp. is max. recommended	Atochem Kynar
Lubricating Oils			141		8			temp. is max. recommended	Atochem Kynar
Magnesium Carbonate			141		8			temp. is max. recommended	Atochem Kynar
Magnesium Chloride	aqueous solution/solid		141		8			temp. is max. recommended	Atochem Kynar
Magnesium Citrate			121		8			temp. is max. recommended	Atochem Kynar
Magnesium Hydroxide			135		8			temp. is max. recommended	Atochem Kynar
Magnesium Nitrate	aqueous solution/solid		135		8			temp. is max. recommended	Atochem Kynar
Magnesium Sulfate	aqueous solution/solid		135		8			temp. is max. recommended	Atochem Kynar
Maleic Acid	aqueous solution/solid		121		8			temp. is max. recommended	Atochem Kynar
	saturated		125	7	7	<1	>90	no visual change	Atochem Foraflon
Maleic Anhydride			24		8			temp. is max. recommended	Atochem Kynar
Malic Acid	aqueous solution/solid		121		8			temp. is max. recommended	Atochem Kynar
	saturated		125	7	7	<1	>90	no visual change	Atochem Foraflon
Manganese Sulfate	aqueous solution/solid		121		8			temp. is max. recommended	Atochem Kynar
Mercuric Chloride			121		8			temp. is max. recommended	Atochem Kynar
Mercuric Cyanide			121		8			temp. is max. recommended	Atochem Kynar
Mercuric Nitrate	aqueous solution/solid		135		8			temp. is max. recommended	Atochem Kynar
Mercury			141		8			temp. is max. recommended	Atochem Kynar
			150	7	7		>90	no visual change	Atochem Foraflon

(Cont'd.)

Table I.8 Chemical Resistance of Polyvinylidene Fluoride (PVDF) (Cont'd.)

Reagent	Reagent Note	Conc.	Temp (°C)	Time (days)	PDL Rating	% Change Weight	% Retained Tensile Strength	Resistance Note	Material Note
Metabromotoluene			79		8			temp. is max. recommended	Atochem Kynar
Methacrylic Acid			52		8			temp. is max. recommended	Atochem Kynar
Methane			100	7	7	<1	>90	no visual change	Atochem Foraflon
			141		8			temp. is max. recommended	Atochem Kynar
Methanesulfonic Acid	aqueous solution/liquid	50	75	7	7	<1	>90	no visual change	Atochem Foraflon
			93		8			temp. is max. recommended	Atochem Kynar
Methyl Acetate			38		8			temp. is max. recommended	Atochem Kynar
Methyl Acrylate			38		8			temp. is max. recommended	Atochem Kynar
Methyl Alcohol	with hydrochloric acid	10	50	180	8			satisfactory resistance	Atochem Foraflon
	w/ HCl and chloroform		50	180	8			satisfactory resistance	Atochem Foraflon
	methanol		75	7	7	<1	>90	no visual change	Atochem Foraflon
	aqueous solution/liquid		141		8			temp. is max. recommended	Atochem Kynar
Methyl Bromide			100	7	7	<1	>90	no visual change	Atochem Foraflon
			141		8			temp. is max. recommended	Atochem Kynar
Methyl Chloride			100	7	7	<1	>90	no visual change	Atochem Foraflon
			141		8			temp. is max. recommended	Atochem Kynar
Methyl Chloroacetate			24		8			temp. is max. recommended	Atochem Kynar
Methyl Chloroform			50	7	7	<1	>90	no visual change	Atochem Foraflon
			52		8			temp. is max. recommended	Atochem Kynar
Methyl Chloromethyl Ether			24		8			temp. is max. recommended	Atochem Kynar
Methyl Ethyl Ketone			23		2			not recommended for use	Atochem Foraflon
			23		2			not recommended for use	Atochem Kynar
Methyl Isobutyl Ketone			23	7	2			not recommended for use	Atochem Foraflon
Methyl Isopropyl Ketone					2			not recommended for use	Atochem Foraflon
Methyl Methacrylate			52		8			temp. is max. recommended	Atochem Kynar
Methyl Salicylate			66		8			temp. is max. recommended	Atochem Kynar
Methyl Sulfuric Acid	aqueous solution/liquid		52		8			temp. is max. recommended	Atochem Kynar
Methyl Trichlorosilane			66		8			temp. is max. recommended	Atochem Kynar
Methylamine			23		2			not recommended for use	Atochem Kynar
Methylene Bromide			79		8			temp. is max. recommended	Atochem Kynar
Methylene Chloride			52	7	5			questionable	Atochem Foraflon
					8			temp. is max. recommended	Atochem Kynar
Methylene Iodine			93		8			temp. is max. recommended	Atochem Kynar
Milk			100	7	7	<1	>90	no visual change	Atochem Foraflon
			121		8			temp. is max. recommended	Atochem Kynar
Mineral Oils			141	7	8	<1	>90	no visual change	Atochem Foraflon
			150		7			temp. is max. recommended	Atochem Kynar
Molasses			79		8			temp. is max. recommended	Atochem Kynar
Morpholine	aqueous solution/liquid		24	7	2			not recommended for use	Atochem Foraflon
					8			temp. is max. recommended	Atochem Kynar

(Cont'd.)

Table I.8 Chemical Resistance of Polyvinylidene Fluoride (PVDF) *(Cont'd.)*

Reagent	Reagent Note	Conc.	Temp (°C)	Time (days)	PDL Rating	% Change Weight	% Retained Tensile Strength	Resistance Note	Material Note
Naphtha			135		8			temp. is max. recommended	Atochem Kynar
			150	7	7	<1	>90	no visual change	Atochem Foraflon
Naphthalene			50	7	7	<1	>90	temp. is max. recommended	Atochem Foraflon
			93		8			no visual change	Atochem Kynar
Natural Gas			141		8			temp. is max. recommended	Atochem Kynar
Nickel Acetate	aqueous solution/solid		121		8			temp. is max. recommended	Atochem Kynar
Nickel Chloride	aqueous solution/solid		141		8			temp. is max. recommended	Atochem Kynar
Nickel Nitrate	aqueous solution/solid		141		8			temp. is max. recommended	Atochem Kynar
Nickel Sulfate	aqueous solution/solid		141		8			temp. is max. recommended	Atochem Kynar
Nicotine			21		8			temp. is max. recommended	Atochem Kynar
Nicotinic Acid			121		8			temp. is max. recommended	Atochem Kynar
Nitric Acid	in water	≤10	79		8			temp. is max. recommended	Atochem Kynar
		30	125	7	7	<1	>90	no visual change	Atochem Foraflon
		32	90	365	8			satisfactory resistance	Atochem Foraflon
		32	130	180	8			satisfactory resistance	Atochem Foraflon
	in water	11-50	52		8			temp. is max. recommended	Atochem Kynar
Nitric Acid		65	63	7	7	<1	>90	no visual change	Atochem Foraflon
		65	90	180	8	1-3	≥90	satisfactory resistance	Atochem Foraflon
		98	23	7	5			limited use possible	Atochem Foraflon
		98	75	120	8			satisfactory resistance	Atochem Foraflon
	concentrated	98	90	60	2			not recommended for use	Atochem Kynar
	fuming		23		2			not recommended for use	Atochem Kynar
			23		2			not recommended for use	Atochem Kynar
Nitrobenzene			24		8			temp. is max. recommended	Atochem Kynar
			25	7	7	<1	>90	no visual change	Atochem Foraflon
Nitroethane			21		8			temp. is max. recommended	Atochem Kynar
Nitrogen			141		8			temp. is max. recommended	Atochem Kynar
Nitrogen Dioxide			77		8			temp. is max. recommended	Atochem Kynar
Nitroglycerin			52		8			temp. is max. recommended	Atochem Kynar
Nitromethane			25	7	7	<1	>90	no visual change	Atochem Foraflon
			49		8			temp. is max. recommended	Atochem Kynar
Nitrotoluene			79		8			temp. is max. recommended	Atochem Kynar
Nitrous Oxide			23		2			not recommended for use	Atochem Kynar
Octane			141		8			temp. is max. recommended	Atochem Kynar
			150	7	7	<1	>90	no visual change	Atochem Foraflon
Octene			141		8			temp. is max. recommended	Atochem Kynar
			150	7	7	<1	>90	no visual change	Atochem Foraflon
Oils	PTFCE oil 3		125	7	7	<1	>90	no visual change	Atochem Foraflon
Oleic Acid			121		8	<1	>90	temp. is max. recommended	Atochem Kynar
	9 octadecenoic		125	7	7	<1	>90	no visual change	Atochem Foraflon

(Cont'd.)

Table I.8 Chemical Resistance of Polyvinylidene Fluoride (PVDF) (*Cont'd.*)

Reagent	Reagent Note	Conc.	Temp (°C)	Time (days)	PDL Rating	% Change Weight	% Retained Tensile Strength	Resistance Note	Material Note
Oleum			23		2			Not recommended for use	Atochem Kynar
Olive Oil			121		8			temp. is max. recommended	Atochem Kynar
Orthophenylphenol			79		8			temp. is max. recommended	Atochem Kynar
Oxalic Acid	saturated solution		50	7	7	<1	>90	no visual change	Atochem Foraflon
			52		8			temp. is max. recommended	Atochem Foraflon
Oxygen			141		8			temp. is max. recommended	Atochem Kynar
			150	7	7	<1	>90	no visual change	Atochem Foraflon
Ozone			107		8			temp. is max. recommended	Atochem Kynar
			150	7	7	<1	>90	no visual change	Atochem Foraflon
Palm Oil			93		8			temp. is max. recommended	Atochem Kynar
Palmitic Acid			121		8			temp. is max. recommended	Atochem Kynar
			125	7	7	<1	>90	no visual change	Atochem Foraflon
Paraffin	paraffin oil		121		8			temp. is max. recommended	Atochem Kynar
			121		8			temp. is max. recommended	Atochem Kynar
Peanut Oil			121		8			temp. is max. recommended	Atochem Kynar
Perchloric Acid	in water	10	93		8			temp. is max. recommended	Atochem Kynar
	in water	70	52		8			temp. is max. recommended	Atochem Kynar
		70	100	7	7	<1	>90	no visual change	Atochem Foraflon
Perchloroethylene	tetrachloroethylene		50	7	7	<1	>90	no visual change	Atochem Foraflon
			90	270	8			satisfactory resistance	Atochem Foraflon
			135		8			temp. is max. recommended	Atochem Kynar
Perchloromethylmercaptan			52		8			temp. is max. recommended	Atochem Kynar
Petrolatum			141		8			temp. is max. recommended	Atochem Kynar
Petroleum			135		8			temp. is max. recommended	Atochem Kynar
Phenol	in water	5	79		8			temp. is max. recommended	Atochem Kynar
	in water	10	75	7	7	<1	>90	no visual change	Atochem Foraflon
	in water	10	90	365	8			satisfactory resistance	Atochem Foraflon
	chlorinated		50	7	7	<1	>90	no visual change	Atochem Foraflon
	chlorinated phenol		52		8			temp. is max. recommended	Atochem Kynar
			66		8			temp. is max. recommended	Atochem Kynar
			66		8			temp. is max. recommended	Atochem Kynar
Phenyl Ether			52		8			temp. is max. recommended	Atochem Kynar
Phenol-2-Sulfonic Acid (1-)			52		8			temp. is max. recommended	Atochem Kynar
Phenylhydrazine			50	7	7	<1	>90	no visual change	Atochem Foraflon
			52		8			temp. is max. recommended	Atochem Kynar
Phenylhydrazine Hydrochloride	aqueous solution/solid		52		8			temp. is max. recommended	Atochem Kynar
Phosgene			79		6			temp. is max. recommended	Atochem Kynar

(*Cont'd.*)

Table I.8 Chemical Resistance of Polyvinylidene Fluoride (PVDF) *(Cont'd.)*

Reagent	Reagent Note	Conc.	Temp (°C)	Time (days)	PDL Rating	% Change Weight	% Retained Tensile Strength	Resistance Note	Material Note
Phosphoric Acid	aqueous solution	≤85	135		8			temp. is max. recommended	Atochem Kynar
		85	107		8			temp. is max. recommended	Atochem Kynar
		85	125	7	7	<1	>90	no visual change	Atochem Foraflon
		85	130	365	8			satisfactory resistance	Atochem Foraflon
		98	125	7	7	<1	>90	no visual change	Atochem Foraflon
Phosphoric Trichloride			50	365	8			satisfactory resistance	Atochem Foraflon
Phosphorous	red		24		8			temp. is max. recommended	Atochem Kynar
Phosphorous Oxychloride			23		2			not recommended for use	Atochem Kynar
			25	7	7	<1	>90	no visual change	Atochem Foraflon
Phosphorous Pentachloride			93		8			temp. is max. recommended	Atochem Kynar
Phosphorous Pentoxide			93		8			temp. is max. recommended	Atochem Kynar
Phosphorous Trichloride			75	7	7	<1	>90	no visual change	Atochem Foraflon
			93		8			temp. is max. recommended	Atochem Kynar
Phthalic Acid			93	7	8	<1	>90	temp. is max. recommended	Atochem Kynar
	saturated		100		7			no visual change	Atochem Foraflon
Picric Acid	2,4,6-trinitrophenol	10	75	7	7	1	>90	no visual change	Atochem Foraflon
			24		8			temp. is max. recommended	Atochem Kynar
Plating Solutions	brass		93		8			temp. is max. recommended	Atochem Kynar
	cadmium		93		3			temp. is max. recommended	Atochem Kynar
	chrome		93		8			temp. is max. recommended temp. is max. recommended	Atochem Kynar
	copper		93		8			temp. is max. recommended	Atochem Kynar
	iron		93		8			temp. is max. recommended	Atochem Kynar
	lead		93		8			temp. is max. recommended temp. is max. recommended	Atochem Kynar
	nickel		93		8			temp. is max. recommended	Atochem Kynar
	rhodium		93		8			temp. is max. recommended	Atochem Kynar
	silver		93		8			temp. is max. recommended temp. is max. recommended	Atochem Kynar
	speculum		93		8			temp. is max. recommended	Atochem Kynar
	tin		93		8			temp. is max. recommended	Atochem Kynar
	zinc		93		8			temp. is max. recommended	Atochem Kynar
Polyethylene Glycol			93		8			temp. is max. recommended	Atochem Kynar
Polyvinyl Acetate			135		8			temp. is max. recommended	Atochem Kynar
Polyvinyl Alcohol			135		8			temp. is max. recommended	Atochem Kynar
Potassium			23		2			not recommended for use	Atochem Kynar
Potassium Acetate	aqueous solution/solid		141		8			temp. is max. recommended	Atochem Kynar
Potassium Alum	aqueous solution/liquid		141		8			temp. is max. recommended	Atochem Kynar
Potassium Aluminum Chloride			141		8			temp. is max. recommended	Atochem Kynar
Potassium Bicarbonate	aqueous solution/solid		93		8			temp. is max. recommended	Atochem Kynar
Potassium Bisulfate	aqueous solution/solid		141		8			temp. is max. recommended	Atochem Kynar
Potassium Borate	aqueous solution/solid		141		8			temp. is max. recommended	Atochem Kynar

(Cont'd.)

Table I.8 Chemical Resistance of Polyvinylidene Fluoride (PVDF) (Cont'd.)

Reagent	Reagent Note	Conc.	Temp (°C)	Time (days)	PDL Rating	% Change Weight	% Retained Tensile Strength	Resistance Note	Material Note
Potassium Bromate	aqueous solution/solid		141		8			temp. is max. recommended	Atochem Kynar
Potassium Bromide	aqueous solution/solid		141		8			temp. is max. recommended	Atochem Kynar
Potassium Carbonate	aqueous solution/solid		141		8			temp. is max. recommended	Atochem Kynar
Potassium Chlorate			93		8			temp. is max. recommended	Atochem Kynar
Potassium Chloride	aqueous solution/solid		141		8			temp. is max. recommended	Atochem Kynar
Potassium Chromate	aqueous solution/solid		141		8			temp. is max. recommended	Atochem Kynar
Potassium Cyanide	aqueous solution/solid		141		8			temp. is max. recommended	Atochem Kynar
Potassium Dichromate			141		8			temp. is max. recommended	Atochem Kynar
Potassium Ferricyanide	aqueous solution/solid		141		8			temp. is max. recommended	Atochem Kynar
Potassium Ferrocyanide	aqueous solution/solid		141		8			temp. is max. recommended	Atochem Kynar
Potassium Fluoride	aqueous solution/solid		141		8			temp. is max. recommended	Atochem Kynar
Potassium Hydroxide	in water	<10	66		8			temp. is max. recommended	Atochem Kynar
		50	100	7	7	<1	>90	no visual change	Atochem Foraflon
	in water	>50	23		2			Not recommended for use	Atochem Kynar
Potassium Hypochlorite	aqueous solution		93		8			temp. is max. recommended	Atochem Kynar
Potassium Iodide	aqueous solution/solid		121		8			temp. is max. recommended	Atochem Kynar
Potassium Nitrate	aqueous solution/solid		141		8			temp. is max. recommended	Atochem Kynar
Potassium Perborate			141		8			temp. is max. recommended	Atochem Kynar
Potassium Perchlorate			93		8			temp. is max. recommended	Atochem Kynar
Potassium Permanganate	aqueous solution/solid		121		8			temp. is max. recommended	Atochem Kynar
Potassium Persulfate			52		8			temp. is max. recommended	Atochem Kynar
Potassium Sulfate	aqueous solution/solid		141		8			temp. is max. recommended	Atochem Kynar
Potassium Sulfide	aqueous solution/solid		141		8			temp. is max. recommended	Atochem Kynar
Propane			141		8			temp. is max. recommended	Atochem Kynar
			150	7	7	<1	>90	no visual change	Atochem Foraflon
Propyl Acetate			36		8			temp. is max. recommended	Atochem Kynar
Propyl Alcohol	1-propanol		50	7	7	<1	>90	no visual change	Atochem Foraflon
			66		8			temp. is max. recommended	Atochem Kynar
Propylamine			23		2			not recommended for use	Atochem Kynar
Propylene Dibromide			93		8			temp. is max. recommended	Atochem Kynar
Propylene Dichloride			93		8			temp. is max. recommended	Atochem Kynar
Propylene Glycol	aqueous solution/liquid		66		8			temp. is max. recommended	Atochem Kynar
Propylene Oxide				7	2			not recommended for use	Atochem Foraflon
			23		2			not recommended for use	Atochem Kynar
PTFCE Oil 3			125	7	7	<1	>90	no visual change	Atochem Foraflon
Pyridine				7	2			not recommended for use	Atochem Foraflon
			23		2			not recommended for use	Atochem Kynar
Pyrogallic Acid	Pyrogallol		49		8			temp. is max. recommended	Atochem Kynar
	Pyrogallol		50	7	7	<1	>90	no visual change	Atochem Foraflon

(Cont'd.)

Table I.8 Chemical Resistance of Polyvinylidene Fluoride (PVDF) *(Cont'd.)*

Reagent	Reagent Note	Conc.	Temp (°C)	Time (days)	PDL Rating	% Change Weight	% Retained Tensile Strength	Resistance Note	Material Note
Salicylaldehyde			52		8			temp. is max. recommended	Atochem Kynar
Salicylic Acid		50	50	7	7	<1	>90	no visual change	Atochem Foraflon
			93		8			temp. is max. recommended	Atochem Foraflon
Salicylic Aldehyde			50	7	7	<1	>90	no visual change	Atochem Foraflon
Sea Water		50	150	7	7	<1	>90	no visual change	Atochem Foraflon
			141		8			temp. is max. recommended	Atochem Kynar
			150	7	7	<1	>90	no visual change	Atochem Foraflon
Selenic Acid	aqueous solution/pure		66		8			no visual change	Atochem Foraflon
Sewage	Sewage water		121		8			temp. is max. recommended	Atochem Kynar
Silicon Tetrachloride			52		8			no visual change	Atochem Foraflon
Silicone Oils	S510		121		8			temp. is max. recommended	Atochem Kynar
			150	7	7	<1	>90	no visual change	Atochem Foraflon
Silver Cyanide			141		8			temp. is max. recommended	Atochem Kynar
Silver Nitrate	aqueous solution/solid		141		8			temp. is max. recommended	Atochem Kynar
Silver Sulfate			121		8			temp. is max. recommended	Atochem Kynar
Sodium			23		2			not recommended for use	Atochem Kynar
Sodium Acetate	aqueous solution/solid		141		8			temp. is max. recommended	Atochem Kynar
Sodium Amalgam			23		2			not recommended for use	Atochem Kynar
Sodium Benzoate	aqueous solution/solid		141		8			temp. is max. recommended	Atochem Kynar
Sodium Bicarbonate	aqueous solution/solid		141		8			temp. is max. recommended	Atochem Kynar
Sodium Bisulfate	aqueous solution/solid		141		8			temp. is max. recommended	Atochem Kynar
Sodium Bisulfite	aqueous solution/solid		141		8			temp. is max. recommended	Atochem Kynar
Sodium Bromate	aqueous solution/solid		93		8			temp. is max. recommended	Atochem Kynar
Sodium Bromide	aqueous solution/solid		141		8			temp. is max. recommended	Atochem Kynar
Sodium Carbonate	aqueous solution/solid	40	90	180	8			satisfactory resistance	Atochem Foraflon
			141		8			temp. is max. recommended	Atochem Kynar
Sodium Chlorate	500 g/l		90	365	8			satisfactory resistance	Atochem Foraflon
	aqueous solution/solid		121		8			temp. is max. recommended	Atochem Kynar
Sodium Chlorite	845 g/l		60	180	8			satisfactory resistance	Atochem Foraflon
	845 g/l		90	90	8			satisfactory resistance	Atochem Foraflon
	aqueous solution/solid		121		8			temp. is max. recommended	Atochem Kynar
Sodium Chromate	aqueous solution/solid		93		8			temp. is max. recommended	Atochem Kynar
Sodium Cyanide	aqueous solution/solid		135		8			temp. is max. recommended	Atochem Kynar
Sodium Dichromate	aqueous solution/solid		93		8			temp. is max. recommended	Atochem Kynar
Sodium Dithionite	aqueous solution/solid		38		8			temp. is max. recommended	Atochem Kynar
Sodium Ferricyanide	aqueous solution/solid		135		8			temp. is max. recommended	Atochem Kynar
Sodium Ferrocyanide	aqueous solution/solid		135		8			temp. is max. recommended	Atochem Kynar
Sodium Fluoride	aqueous solution/solid		141		8			temp. is max. recommended	Atochem Kynar
Sodium Fluorosilicate			93		8			temp. is max. recommended	Atochem Kynar
Sodium Hydrogen Phosphate	aqueous solution/solid		121		8			temp. is max. recommended	Atochem Kynar

(Cont'd.)

Table I.8 Chemical Resistance of Polyvinylidene Fluoride (PVDF) (*Cont'd.*)

Reagent	Reagent Note	Conc.	Temp (°C)	Time (days)	PDL Rating	% Change Weight	% Retained Tensile Strength	Resistance Note	Material Note
Sodium Hydroxide	pH 12	0.4	90	120	8			satisfactory resistance	Atochem Foraflon
	pH 14	3.8	50	120	8			satisfactory resistance	Atochem Foraflon
	pH 14	3.8	75	30	8			satisfactory resistance	Atochem Foraflon
	pH 14	3.8	90	120	8			satisfactory resistance	Atochem Foraflon
Sodium Hydroxide	aqueous solution with 1.7% of triton X 100	≤10	66		8			temp. is max. recommended	Atochem Kynar
		10	23	30	8			satisfactory resistance	Atochem Foraflon
		10	23	60	8			satisfactory resistance	Atochem Foraflon
		10	50	60	8			satisfactory resistance	Atochem Foraflon
		10	90	60	8			satisfactory resistance	Atochem Foraflon
		45	90	365	8			satisfactory resistance	Atochem Foraflon
		45	100	7	7	<1	>90	no visual change	Atochem Foraflon
		45	130	90	2			not recommended for use	Atochem Foraflon
Sodium Hydroxide	aqueous solution	>50	23	7	2			not recommended for use	Atochem Foraflon
		60	50		7	<1	>90	no visual change	Atochem Foraflon
	pH 11		130	30	8			satisfactory resistance	Atochem Foraflon
	pH 11		130	60	8			satisfactory resistance	Atochem Foraflon
Sodium Hypochlorite	in water	≤5	135		8			temp. is max. recommended	Atochem Kynar
	in water	6-15	93		8			temp. is max. recommended	Atochem Kynar
Sodium Iodide	aqueous solution/solid		141		8			temp. is max. recommended	Atochem Kynar
Sodium Nitrate	aqueous solution/solid		135		8			temp. is max. recommended	Atochem Kynar
Sodium Nitrite	aqueous solution/solid		135		8			temp. is max. recommended	Atochem Kynar
Sodium Palmitate			121		8			temp. is max. recommended	Atochem Kynar
Sodium Perchlorate	aqueous solution/solid		121		8			temp. is max. recommended	Atochem Kynar
Sodium Peroxide			93		8			temp. is max. recommended	Atochem Kynar
Sodium Phosphate	aqueous solution/solid		141		8			temp. is max. recommended	Atochem Kynar
Sodium Thiocyanate	aqueous solution/solid		121		8			temp. is max. recommended	Atochem Kynar
Sodium Thiosulfate	aqueous solution/solid		135		8			temp. is max. recommended	Atochem Kynar
Soybean Oil			121		8			temp. is max. recommended	Atochem Kynar
Stannic Chloride	aqueous solution/solid		141		8			temp. is max. recommended	Atochem Kynar
Stannous Chloride	aqueous solution/solid		141		8			temp. is max. recommended	Atochem Kynar
Starch			93		8			temp. is max. recommended	Atochem Kynar
Stearic Acid	saturated		125	7	7	1	>90	no visual change	Atochem Foraflon
			141		8			temp. is max. recommended	Atochem Kynar
Stilbene			79		8			temp. is max. recommended	Atochem Kynar
Styrene			82	7	2			not recommended for use	Atochem Foraflon
					8			temp. is max. recommended	Atochem Kynar
Succinic Acid			66		8			temp. is max. recommended	Atochem Kynar
Sugars	sugar syrups		141		8			temp. is max. recommended	Atochem Kynar
Sulfochromic Acid			90	120	8			satisfactory resistance	Atochem Foraflon
Sulfonates	fatty acids		79		8			temp. is max. recommended	Atochem Kynar

(*Cont'd.*)

Table I.8 Chemical Resistance of Polyvinylidene Fluoride (PVDF) *(Cont'd.)*

Reagent	Reagent Note	Conc.	Temp (°C)	Time (days)	PDL Rating	% Change Weight	% Retained Tensile Strength	Resistance Note	Material Note
Sulfonic p-Hydroxybenzene			90	120	8			satisfactory resistance	Atochem Foraflon
Sulfonic p-Toluene			90	120	8			satisfactory resistance	Atochem Foraflon
Sulfonitric Acid		40	125	7	7	<1	>90	no visual change	Atochem Foraflon
Sulfur			100	7	7	<1	>90	no visual change	Atochem Foraflon
			121		8			temp. is max. recommended	Atochem Kynar
Sulfur Chloride			24		8			temp. is max. recommended	Atochem Kynar
Sulfur Dichloride			24		8			temp. is max. recommended	Atochem Kynar
Sulfur Dioxide			79		8			temp. is max. recommended	Atochem Kynar
Sulfur Trioxide			23		2			not recommended for use	Atochem Kynar
Sulfuric Acid		50	130	7	8			satisfactory resistance	Atochem Foraflon
	aqueous solution	50	150	365	7			no visual change	Atochem Kynar
		≤60	121	7	8	<1	>90	temp. is max. recommended	Atochem Kynar
Sulfuric Acid		80	90	365	8			satisfactory resistance	Atochem Foraflon
		80	125	7	7	<1	>90	no visual change	Atochem Foraflon
		80	130	90	8			satisfactory resistance	Atochem Foraflon
		80	130	180	2			not recommended for use	Atochem Foraflon
	aqueous solution	80-93	93		8			temp. is max. recommended	Atochem Kynar
Sulfuric Acid		93	75	7	7	<1	>90	no visual change	Atochem Foraflon
		94	50	60	8			satisfactory resistance	Atochem Foraflon
		94	50	90	8			satisfactory resistance	Atochem Foraflon
		94	50	180	8			satisfactory resistance	Atochem Foraflon
		94	75	60	8			satisfactory resistance	Atochem Foraflon
Sulfuric Acid		94	75	90	8			satisfactory resistance	Atochem Foraflon
		94	75	180	8			satisfactory resistance	Atochem Foraflon
		94	90	60	8			satisfactory resistance	Atochem Foraflon
		94	90	90	8			satisfactory resistance	Atochem Foraflon
		96	23	180	8			satisfactory resistance	Atochem Foraflon
		96	50	180	8			satisfactory resistance	Atochem Foraflon
		96	50	365	8			satisfactory resistance	Atochem Foraflon
Sulfuric Acid		96	75	7	8			satisfactory resistance	Atochem Foraflon
		96	75	60	8			satisfactory resistance	Atochem Foraflon
		96	75	120	8			satisfactory resistance	Atochem Foraflon
		96	75	180	8			satisfactory resistance	Atochem Foraflon
		96	90	7	8			satisfactory resistance	Atochem Foraflon
		96	90	60	5			questionable	Atochem Foraflon
		96	90	120	2			not recommended for use	Atochem Foraflon
		98	23	180	8			satisfactory resistance	Atochem Foraflon

(Cont'd.)

Table I.8 Chemical Resistance of Polyvinylidene Fluoride (PVDF) (Cont'd.)

Reagent	Reagent Note	Conc.	Temp (°C)	Time (days)	PDL Rating	% Change Weight	% Retained Tensile Strength	Resistance Note	Material Note
Sulfuric Acid		98	50	7	7	<1	>90	no visual change	Atochem Foraflon
		98	50	60	8			satisfactory resistance	Atochem Foraflon
	with 10% chloroform	98	50	180	8			satisfactory resistance	Atochem Foraflon
	with diethylether (10%)	98	50	180	8			satisfactory resistance	Atochem Foraflon
	aqueous solution	98	50	180	8			satisfactory resistance	Atochem Foraflon
		98	66		8			satisfactory resistance	Atochem Kynar
Sulfuric Acid		98	75	7	8			satisfactory resistance	Atochem Foraflon
		98	75	49	8			satisfactory resistance	Atochem Foraflon
		98	75	90	5			questionable	Atochem Foraflon
		98	90	7	8			satisfactory resistance	Atochem Foraflon
		98	90	49	2			not recommended for use	Atochem Foraflon
Sulfuric Acid		99.2	23	180	8			satisfactory resistance	Atochem Foraflon
		99.2	23	365	8			satisfactory resistance	Atochem Foraflon
		99.2	50	7	8			satisfactory resistance	Atochem Foraflon
		99.2	50	49	8			satisfactory resistance	Atochem Foraflon
		99.2	50	90	8			satisfactory resistance	Atochem Foraflon
		99.2	50	180	8			satisfactory resistance	Atochem Foraflon
Sulfuric Acid		99.2	75	7	8			satisfactory resistance	Atochem Foraflon
		99.2	75	49	5			questionable	Atochem Foraflon
		99.2	75	90	5			questionable	Atochem Foraflon
		99.2	90	7	5			questionable	Atochem Foraflon
		99.2	90	49	2			not recommended for use	Atochem Foraflon
	Fuming, 20% oleum		23		2			not recommended for use	Atochem Kynar
	with chlorine		23	240	8			satisfactory resistance	Atochem Foraflon
Sulfuric Anhydride			23	7	2			not recommended for use	Atochem Foraflon
Sulfuryl Chloride			23		2			not recommended for use	Atochem Kynar
			25	7	7	<1	>90	no visual change	Atochem Foraflon
			50	120	8			satisfactory resistance	Atochem Foraflon
Sulfuryl Fluoride			24		8			temp. is max. recommended	Atochem Kynar
Surfactants			90	60	8			satisfactory resistance	Atochem Foraflon
	non ionic		90	60	8			satisfactory resistance	Atochem Foraflon
			130	60	8			satisfactory resistance	Atochem Foraflon
Tall Oil			141		8			temp. is max. recommended	Atochem Kynar
Tallow			141		8			temp. is max. recommended	Atochem Kynar
Tannic Acid			107		8			temp. is max. recommended	Atochem Kynar
Tar			121		8			temp. is max. recommended	Atochem Kynar
Tartaric Acid	Aq. solution of solid		121		3			temp. is max. recommended	Atochem Kynar
	saturated solution		125	7	7	<1	>90	no visual change	Atochem Foraflon

(Cont'd.)

Table I.8 Chemical Resistance of Polyvinylidene Fluoride (PVDF) *(Cont'd.)*

Reagent	Reagent Note	Conc.	Temp (°C)	Time (days)	PDL Rating	% Change Weight	% Retained Tensile Strength	Resistance Note	Material Note
Tetrabromoethane (1,1,2,2-)			121		8			temp. is max. recommended	Atochem Kynar
Tetrachloroethane (1,1,2,2-)			121		8			temp. is max. recommended	Atochem Kynar
Tetrachloroethylene	perchloroethylene		50	7	7	<1	>90	temp. is max. recommended	Atochem Kynar
Tetrachlorophenol (2,3,4,6-)			66		8			temp. is max. recommended	Atochem Kynar
Tetraethyllead			141		8			temp. is max. recommended	Atochem Kynar
Tetrahydrofuran	Aq. solution or solid		23	7	2			not recommended for use	Atochem Foraflon
					2			not recommended for use	Atochem Kynar
Tetramethylammonium Hydroxide	in water	≤10	93		8			temp. is max. recommended	Atochem Kynar
Tetramethylurea			23		2			not recommended for use	Atochem Kynar
Thioglycol			24		8			temp. is max. recommended	Atochem Kynar
Thioglycolic Acid			79		8			temp. is max. recommended	Atochem Kynar
Thionyl Chloride			23		2			not recommended for use	Atochem Kynar
			25	7	7	<1	>90	no visual change	Atochem Foraflon
			50	120	8			satisfactory resistance	Atochem Foraflon
Thlophosphoryl Chloride			23		2			not recommended for use	Atochem Kynar
Thread Cutting Oils			93		8			temp. is max. recommended	Atochem Kynar
Titanium Tetrachloride			66		8	<1	>90	temp. is max. recommended	Atochem Kynar
			100	7	7			no visual change	Atochem Foraflon
Toluene			75	7	7	<1	>90	no visual change	Atochem Foraflon
			79		8			temp. is max. recommended	Atochem Kynar
			90	365	8			satisfactory resistance	Atochem Foraflon
Toluene Diisocyanate			100	7	7	<1	>90	no visual change	Atochem Foraflon
Toluenesulfonyl Chloride			52		8			temp. is max. recommended	Atochem Kynar
Tomato Juice			100		8			temp. is max. recommended	Atochem Kynar
Tributyl Phosphate			24	7	8	<1	>90	temp. is max. recommended	Atochem Kynar
			50		7			no visual change	Atochem Foraflon
Trichloroacetic Acid	in water	≤10	93		8			temp. is max. recommended	Atochem Kynar
		50	50	7	7	<1	>90	no visual change	Atochem Foraflon
		50	75	120	8			satisfactory resistance	Atochem Foraflon
	50% in water to pure	50-100	52		8			temp. is max. recommended	Atochem Kynar
			75	7	7	<1	>90	no visual change	Atochem Foraflon
Trichloroacetyl Chloride			25	7	7	<1	>90	no visual change	Atochem Foraflon
Trichlorobenzene (1,2,4-)			93		8			satisfactory resistance	Atochem Kynar
Trichloroethane (1,1,2-)			66		8			temp. is max. recommended	Atochem Foraflon
Trichloroethylene			50	7	7	<1	>90	no visual change	Atochem Foraflon
			90	365	8			satisfactory resistance	Atochem Foraflon
			141		8			temp. is max. recommended	Atochem Kynar

(Cont'd.)

Table I.8 Chemical Resistance of Polyvinylidene Fluoride (PVDF) *(Cont'd.)*

Reagent	Reagent Note	Conc.	Temp (°C)	Time (days)	PDL Rating	% Change Weight	% Retained Tensile Strength	Resistance Note	Material Note
Trichlorofluoromethane	R113		50	7	7	<1	>90	no visual change	Atochem Foraflon
Trichloromethane	chloroform		100	7	7	<1	>90	no visual change	Atochem Foraflon
Trichlorophenol (2,4,5-)			66		8			temp. is max. recommended	Atochem Kynar
Tricresyl Phosphate			23		2			not recommended for use	Atochem Kynar
Triethanolamine	aqueous solution/liquid		52		8			temp. is max. recommended for use	Atochem Kynar
Triethyl Phosphate			3		2			not recommended for use	Atochem Kynar
Triethylamine			38	7	7	<1	>90	no visual change	Atochem Foraflon
			52		8			temp. is max. recommended	Atochem Kynar
Trifluoroacetic Acid	in water	50	93		8			temp. is max. recommended	Atochem Kynar
			52		8			temp. is max. recommended	Atochem Kynar
Trimethylamine	aqueous solution/gas		66		8			temp. is max. recommended	Atochem Kynar
Triton X 100	with 10% NaOH	1.7	23	30	8			satisfactory resistance	Atochem Foraflon
Turpentine			50	7	7	<1	>90	no visual change	Atochem Foraflon
			141		8			temp. is max. recommended	Atochem Kynar
Urea	aqueous solution/solid		121		8			temp. is max. recommended	Atochem Kynar
Varnish			121		8			temp. is max. recommended	Atochem Kynar
Varsol			121		8			temp. is max. recommended	Atochem Kynar
Vegetable Oils			141		8			temp. is max. recommended	Atochem Kynar
Vinegar			107		8			temp. is max. recommended	Atochem Kynar
Vinyl Acetate			121		8			temp. is max. recommended	Atochem Kynar
Vinyl Chloride			93		8			temp. is max. recommended	Atochem Kynar
Vinylidene Chloride			93		8			temp. is max. recommended	Atochem Kynar
Water	salt water	50	150	7	7	<1	>90	no visual change	Atochem Foraflon
	sewage water		121		8			temp. is max. recommended	Atochem Kynar
			141		8			temp. is max. recommended	Atochem Kynar
	salt water		141		8			temp. is max. recommended	Atochem Kynar
	sea water		150	7	7	<1	>90	no visual change	Atochem Foraflon
	tap water		150	7	7	<1	>90	no visual change	Atochem Foraflon
Whiskey			107		8			temp. is max. recommended	Atochem Kynar
Wines			93		8			temp. is max. recommended	Atochem Kynar
Xylene			93		8			temp. is max. recommended	Atochem Kynar
			125	7	7	<1	>90	no visual change	Atochem Foraflon
			130	120	8			satisfactory resistance	Atochem Foraflon
Zinc Acetate	aqueous solution		121		8			temp. is max. recommended	Atochem Kynar
Zinc Bromide	aq. solution of solid		121		8			temp. is max. recommended	Atochem Kynar
Zinc Chloride	aq. solution of solid		141		8			temp. is max. recommended	Atochem Kynar
Zinc Nitrate	aq. solution of solid		141		8			temp. is max. recommended	Atochem Kynar
Zinc Sulfate	aq. solution of solid		141		8			temp. is max. recommended	Atochem Kynar

REFERENCE

1. PDL staff, *PDL Handbook Series: Chemical Resistance,* Volume 1, William Andrew Publishing, Norwich, NY (1994)

Appendix II: Permeability of Fluoropolymers

II.1 Permeability of Polytetrafluoroethylene (PTFE)

Permeability measurements are rarely made directly on coatings because free standing films are required. Measurements on molded, cast, or extruded films can be used to indicate the performance on coating materials based on the same polymer. One must keep in mind that many coatings contain other materials in addition to the fluoropolymer, which can affect permeation properties in a positive or negative direction. This appendix is an edited version from *Fluoroplastics, Volume 1: Non-Melt Processible Fluoroplastics, The Definitive User's Guide and Databook.*[1]

Table II.1 Hydrogen Permeability vs Temperature and Pressure through DuPont Teflon® Polytetrafluoroethylene

Material Family	POLYTETRAFLUOROETHYLENE								
Material Supplier/ Grade	DUPONT TEFLON®								
MATERIAL CHARACTERISTICS									
Sample Thickness	0.03 mm	0.03 mm	0.03 mm	0.03 mm	0.03 mm	0.03 mm	0.03 mm	0.03 mm	0.03 mm
TEST CONDITIONS									
Penetrant	Hydrogen								
Temperature (°C)	-16	25	68	-17	25	67	-18	25	63
Pressure Gradient (kPa)	1724	1724	1724	3447	3447	3447	6895	6895	6895
Test Method	mass spectrometry and calibrated standard gas leaks; developed by McDonnell Douglas Space Systems Company Chemistry Laboratory								
PERMEABILITY (source document units)									
Gas Permeability ($cm^3 \cdot mm/cm^2 \cdot kPa \cdot sec$)	1.7×10^{-9}	6.34×10^{-9}	1.88×10^{-8}	1.63×10^{-9}	5.9×10^{-9}	1.86×10^{-8}	1.59×10^{-9}	5.94×10^{-9}	1.64×10^{-8}
PERMEABILITY (normalized units)									
Permeability Coefficient ($cm^3 \cdot mm/m^2 \cdot day \cdot atm$)	149	555	1646	143	516	1628	139	520	1436

Table II.2 Nitrogen Permeability vs Temperature and Pressure through DuPont Teflon® Polytetrafluoroethylene

Material Family	POLYTETRAFLUOROETHYLENE								
Material Supplier/Grade	DUPONT TEFLON®								
MATERIAL CHARACTERISTICS									
Sample Thickness	0.03 mm	0.03 mm	0.03 mm	0.03 mm	0.03 mm	0.03 mm	0.03 mm	0.03 mm	0.03 mm
TEST CONDITIONS									
Penetrant	nitrogen								
Temperature (°C)	-23	25	71	-25	25	70	-23	25	68
Pressure Gradient (kPa)	1724	1724	1724	3147	1447	3441	6115	1115	6115
Test Method	mass spectrometry and calibrated standard gas leaks; developed by McDonnell Douglas Space Systems Company Chemistry Laboratory								
PERMEABILITY (source document units)									
Gas Permeability (cm$^3 \cdot$ mm/cm$^2 \cdot$kPa·sec)	9.46×10^{-11}	7.87×10^{-10}	2.9×10^{-9}	8.89×10^{-11}	7.88×10^{-10}	2.89×10^{-9}	9.47×10^{-11}	7.84×10^{-11}	2.87×10^{-9}
PERMEABILITY (normalized units)									
Permeability Coefficient (cm$^3 \cdot$mm/m$^2 \cdot$day·atm)	8.3	68.9	254	7.8	69	253	8.3	68.6	251

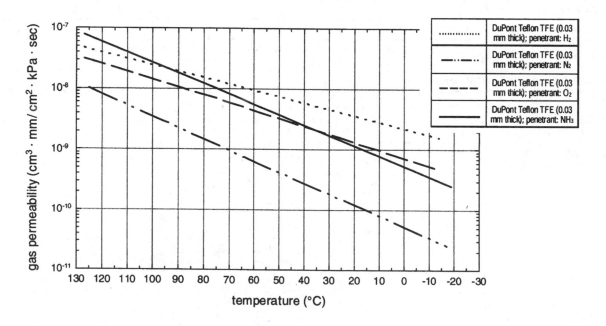

Figure II.1 Gas permeabilility vs temperature through polytetrafluoroethylene.

II.2 Permeability of Ethylene Chlorotrifluoroethylene Copolymer (ECTFE)

Permeability measurements are rarely made directly on coatings because free standing films are required. Measurements on molded, cast, or extruded films can be used to indicate the performance on coating materials based on the same polymer. One must keep in mind that many coatings contain other materials in addition to the fluoropolymer, which can affect permeation properties in a positive or negative direction. This appendix is an edited version from *Fluoroplastics, Volume 2: Melt Processible Fluoroplastics, The Definitive User's Guide and Databook*.[2]

Table II.3 Hydrogen Permeability vs Temperature and Pressure through Ausimont Halar® Ethylene Chlorotrifluoroethylene Copolymer

Material Family	ETHYLENE CHLOROTRIFLUOROETHYLENE COPOLYMER								
Material Supplier/Grade	AUSIMONT HALAR®								
Reference Number	4								
MATERIAL CHARACTERISTICS									
Sample Thickness, mm	0.02								
TEST CONDITIONS									
Penetrant	hydrogen								
Temperature, °C	-22	25	66	-20	25	67	-21	25	68
Pressure Gradient, kPa	1724			3447			6895		
Test Method	Mass Spectrometry and Calibrated Standard Gas Leaks Developed by McDonnell Douglas Space Systems Company Chemistry Laboratories								
PERMEABILITY (source document units)									
Gas Permeability ($cm^3 \cdot mm/cm^2 \cdot kPa \cdot sec$)	1.19×10^{-10}	1.21×10^{-9}	6.58×10^{-9}	1.18×10^{-10}	1.25×10^{-9}	6.65×10^{-9}	1.18×10^{-10}	1.23×10^{-9}	6.74×10^{-9}
PERMEABILITY (normalized units)									
Permeability Coefficient ($cm^3 \cdot mm/m^2 \cdot day \cdot atm$)	10.4	106	576	10.3	109	582	10.3	108	590

Table II.4 Nitrogen Permeability vs Temperature and Pressure through Ausimont Halar® Ethylene Chlorotrifluoroethylene Copolymer

Material Family	ETHYLENE CHLOROTRIFLUOROETHYLENE COPOLYMER								
Material Supplier/Grade	AUSIMONT HALAR®								
Reference Number	4								
MATERIAL CHARACTERISTICS									
Sample Thickness, mm	0.02								
TEST CONDITIONS									
Penetrant	nitrogen								
Temperature, °C	11	25	71	10	25	72	10	25	68
Pressure Gradient, kPa	1724			3447			6895		
Test Method	Mass Spectrometry and Calibrated Standard Gas Leaks Developed by McDonnell Douglas Space Systems Company Chemistry Laboratories								
PERMEABILITY (source document units)									
Gas Permeability ($cm^3 \cdot mm/cm^2 \cdot kPa \cdot sec$)	5.53×10^{-12}	1.29×10^{-11}	2.43×10^{-10}	5.53×10^{-12}	1.49×10^{-11}	4.27×10^{-10}	6.09×10^{-12}	1.43×10^{-11}	2.48×10^{-10}
PERMEABLITY (normalized units)									
Permeability Coefficient ($cm^3 \cdot mm/m^2 \cdot day \cdot atm$)	0.48	1.13	21.3	0.48	1.3	37.4	0.53	1.25	21.7

Table II.5 Oxygen and Ammonia Permeability vs Temperature and Pressure through Ausimont Halar® Ethylene Chlorotrifluoroethylene Copolymer

Material Family	ETHYLENE CHLOROTRIFLUOROETHYLENE COPOLYMER								
Material Supplier/Grade	AUSIMONT HALAR®								
Reference Number	4								
MATERIAL CHARACTERISTICS									
Sample Thickness, mm	0.02								
TEST CONDITIONS									
Penetrant	ammonia			oxygen					
Temperature, °C	-1	25	65	-18	25	55	-15	25	56
Pressure Gradient, kPa	965			1724			3447		
Test Method	Mass Spectrometry and Calibrated Standard Gas Leaks Developed by McDonnell Douglas Space Systems Company Chemistry Laboratories								
PERMEABILITY (source document units)									
Gas Permeability $(cm^3 \cdot mm/cm^2 \cdot kPa \cdot sec)$	3.73×10^{-10}	1.29×10^{-9}	7.5×10^{-9}	5.52×10^{-12}	1.16×10^{-10}	5.16×10^{-10}	5.73×10^{-12}	1.10×10^{-10}	5.26×10^{-10}
PERMEABILITY (normalized units)									
Permeability Coefficient $(cm^3 \cdot mm/m^2 \cdot day \cdot atm)$	32.6	113	617	l0.48	10.2	45.2	0.5	9.6	46.0

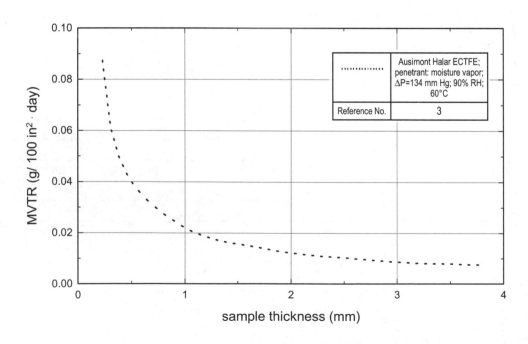

Figure II.2 Moisture vapor permeability rate vs thickness through ethylene chlorotrifluoroethylene copolymer.

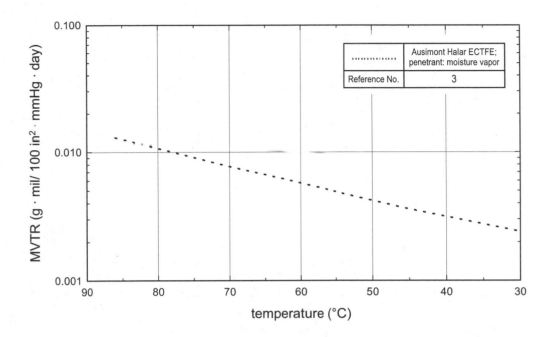

Figure II.3 Moisture vapor permeability rate vs temperature through ethylene chlorotrifluoroethylene copolymer.

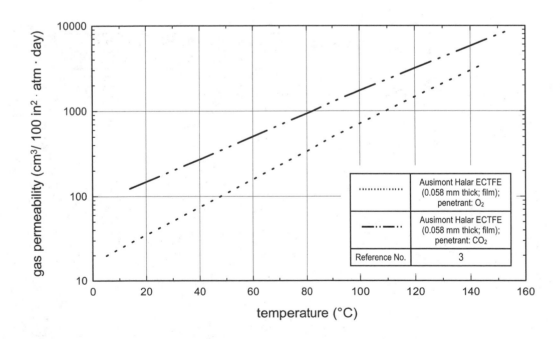

Figure II.4 Carbon dioxide and oxygen permeability vs temperature through ethylene chlorotrifluoroethylene copolymer.

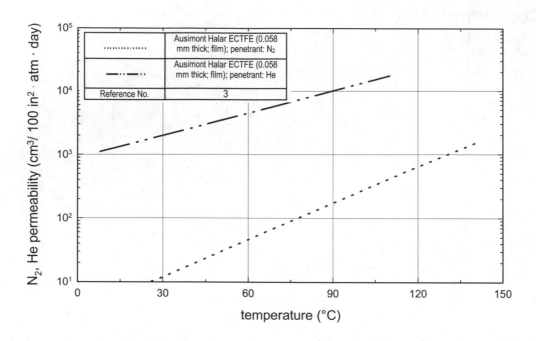

Figure II.5 Nitrogen and helium permeability vs temperature through ethylene chlorotrifluoroethylene copolymer.

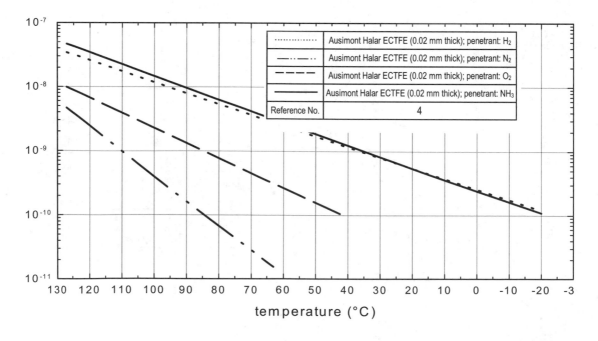

Figure II.6 Gas permeability vs temperature through ethylene chlorotrifluoroethylene copolymer.

II.3 Permeability of Ethylene Tetrafluoroethylene Copolymer (ETFE)

Permeability measurements are rarely made directly on coatings because free standing films are required. Measurements on molded, cast, or extruded films can be used to indicate the performance on coating materials based on the same polymer. One must keep in mind that many coatings contain other materials in addition to the fluoropolymer, which can affect permeation properties in a positive or negative direction. This appendix is an edited version from *Fluoroplastics, Volume 2: Melt Processible Fluoroplastics, The Definitive User's Guide and Databook.*[2]

Table II.6 Carbon Dioxide, Nitrogen, Oxygen, Helium, and Water Vapor Permeability through DuPont Tefzel® Ethylene-Tetrafluoroethylene Copolymer

Material Family	ETHYLENE TETRAFLUOROETHYLENE COPOLYMER				
Material Supplier/Grade	DUPONT TEFZEL®				
Product Form	FILM				
Reference Number	5				
MATERIAL CHARACTERISTICS					
Sample Thickness, mm	0.102				
TEST CONDITIONS					
Penetrant	carbon dioxide	nitrogen	oxygen	helium	water vapor
Temperature, °C	25				
Test Method	ASTM D1434				ASTM E96
PERMEABILITY (source document units)					
Vapor Transmission Rate (g·mil/100 in²·day)					1.65
Gas Permeability (cm³·mil/m²·day)	250	30	100	900	
PERMEABILITY (normalized units)					
Permeability Coefficient (cm³·mm/m²·day·atm)	98.4	11.8	39.4	354	
Vapor Transmission Rate (g·mm/m²·day)					0.65

Table II.7 Oxygen, Nitrogen, Carbon Dioxide, Methane, and Helium Permeability through Ausimont Hyflon®
Ethylene Tetrafluoroethylene Copolymer

Material Family	ETHYLENE TETRAFLUOROETHYLENE COPOLYMER									
Material Supplier/Grade	AUSIMONT HYFLON® 700					AUSIMONT HYFLON® 800				
Features	high molecular weight					low molecular weight				
Reference Number	6									
MATERIAL CHARACTERISTICS										
Melt Flow Index	4 grams / 10 minutes					11 grams /10 minutes				
TEST CONDITIONS										
Penetrant	oxygen	nitrogen	carbon dioxide	methane	helium	oxygen	nitrogen	carbon dioxide	methane	helium
Temperature, °C	23									
Test Method	ASTM D1434									
Test Note	activation energy = 6-8 kcal/mole									
PERMEABILITY (source document units)										
Gas Permeability (cm³·mm/m²·day·atm)	62.646	21.67	232.46	7.88	591	62.646	21.67	232.46	7.88	591
PERMEABILITY (normalized units)										
Permeability Coefficient (cm³·mm/m²·day·atm)	62.6	21.7	1232	7.9	591	62.6	21.7	232	7.9	591

II.4 Permeation of Fluorinated Ethylene Propylene Copolymer (FEP)

Permeability measurements are rarely made directly on coatings because free standing films are required. Measurements on molded, cast, or extruded films can be used to indicate the performance on coating materials based on the same polymer. One must keep in mind that many coatings contain other materials in addition to the fluoropolymer, which can affect permeation properties in a positive or negative direction. This appendix is an edited version from *Fluoroplastics, Volume 2: Melt Processible Fluoroplastics, The Definitive User's Guide and Databook.*[2]

Table II.8 Hydrogen Permeability vs Temperature and Pressure through DuPont Teflon® Fluorinated Ethylene-Propylene Copolymer

Material Family	FLUORINATED ETHYLENE-PROPYLENE COPOLYMER								
Material Supplier/Grade	DUPONT TEFLON®								
Reference Number	4								
MATERIAL CHARACTERISTICS									
Sample Thickness, mm	0.05								
TEST CONDITIONS									
Penetrant	hydrogen								
Temperature, °C	-15	25	68	-13	25	67	-16	25	67
Pressure Gradient, kPa	1724			3447			6895		
Test Method	Mass Spectrometry and Calibrated Standard Gas Leaks Developed by McDonnell Douglas Space Systems Company Chemistry Laboratory								
PERMEABILITY (source document units)									
Gas Permeability ($cm^3 \cdot mm/cm^2 \cdot kPa \cdot sec$)	9.06×10^{-10}	4.11×10^{-9}	1.87×10^{-8}	9.64×10^{-10}	4.35×10^{-9}	1.77×10^{-8}	8.77×10^{-10}	4.4×10^{-9}	1.8×10^{-8}
PERMEABILITY (normalized units)									
Permeability Coefficient ($cm^3 \, mm/m^2 \cdot day \cdot atm$)	79.3	386	1637	84.4	381	1550	76.8	385	1576

Table II.9 Nitrogen Permeability vs Temperature and Pressure through DuPont Teflon® Fluorinated Ethylene-Propylene Copolymer

Material Family	FLUORINATED ETHYLENE-PROPYLENE COPOLYMER								
Material Supplier/Grade	DUPONT TEFLON®								
Reference Number	4								
MATERIAL CHARACTERISTICS									
Sample Thickness, mm	0.05								
TEST CONDITIONS									
Penetrant	nitrogen								
Temperature, °C	-9	2		-7	25	66	-5	25	
Pressure Gradient, kPa	1724			3447			6895		
Test Method	Mass Spectrometry and Calibrated Standard Gas Leaks Developed by McDonnell Douglas Space Systems Company Chemistry Laboratory								
PERMEABILITY (source document units)									
Gas Permeability ($cm^3 \cdot mm/cm^2 \cdot kPa \cdot sec$)	5.06×10^{-11}	3.8×10^{-10}	3.79×10^{-9}	5.64×10^{-11}	3.86×10^{-10}	3.85×10^{-9}	6.39×10^{-11}	3.85×10^{-10}	3.8×10^{-9}
PERMEABILITY (normalized units)									
Permeability Coefficient ($cm^3 \cdot mm/m^2 \cdot day \cdot atm$)	4.4	33.3	332	4.9	33.8	337	5.6	33.7	333

Table II.10 Oxygen and Ammonia Permeability vs Temperature and Pressure through DuPont Teflon® Fluorinated Ethylene-Propylene Copolymer

Material Family	FLUORINATED ETHYLENE-PROPYLENE COPOLYMER								
Material Supplier/Grade	DUPONT TEFLON®								
Reference Number	4								
MATERIAL CHARACTERISTICS									
Sample Thickness, mm	0.05								
TEST CONDITIONS									
Penetrant	ammonia			oxygen					
Temperature, °C	0	25	66	-16	25	52	-16	25	53
Pressure Gradient, kPa	965			1724			3447		
Test Method	Mass Spectrometry and Calibrated Standard Gas Leaks Developed by McDonnell Douglas Space Systems Company Chemistry Laboratory								
PERMEABILITY (source document units)									
Gas Permeability ($crn^3 \cdot mm/cm^2 \cdot kPa \cdot sec$)	3.31×10^{-10}	1.15×10^{-9}	6.3×10^{-9}	1.04×10^{-10}	1.33×10^{-9}	5.16×10^{-9}	1.03×10^{-10}	1.15×10^{-9}	5.31×10^{-9}
PERMEABILITY (normalized units)									
Permeability Coefficient ($cm^3 \cdot mm/m^2 \cdot day \cdot atm$)	29.0	101	552	9.1	116	452	9.0	101	465

Table II.11 Water Vapor, Oxygen, Nitrogen, and Carbon Dioxide Permeability through Fluorinated Ethylene-Propylene Copolymer

Material Family	FLUORINATED ETHYLENE-PROPYLENE COPOLYMER			
Reference Number	4			
TEST CONDITIONS				
Penetrant	water vapor	oxygen	nitrogen	carbon dioxide
Temperature, °C	37.8	25		
Relative Humidity, %	90			
Test Note		STP conditions		
PERMEABILITY (source document units)				
Gas Permeability (cm³·mil/100 in²·day)		750	320	1670
Vapor Transmission Rate (g·mil/100 in²·day)	0.4			
PERMEABILITY (normalized units)				
Permeability Coefficient (cm³·mm/m²·day·atm)		295	126	657
Vapor Transmission Rate (g·mm/m²·day)	0.16			

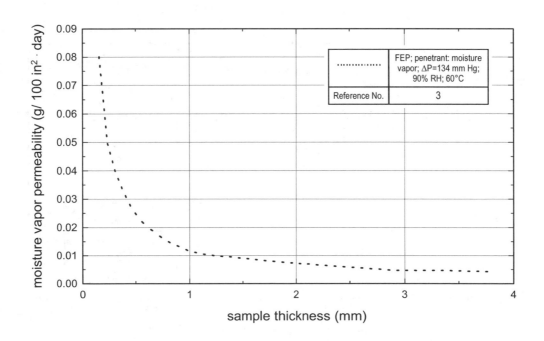

Figure II.7 Moisture vapor permeability rate vs thickness through fluorinated ethylene-propylene copolymer.

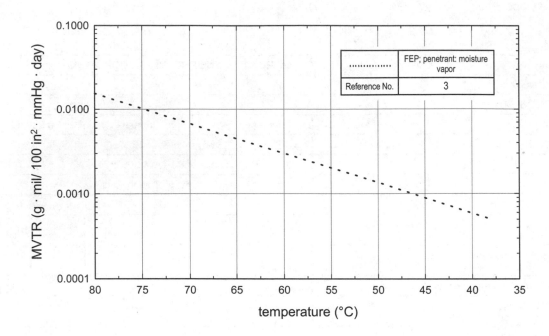

Figure II.8 Moisture vapor permeability rate vs temperature through fluorinated ethylene-propylene copolymer.

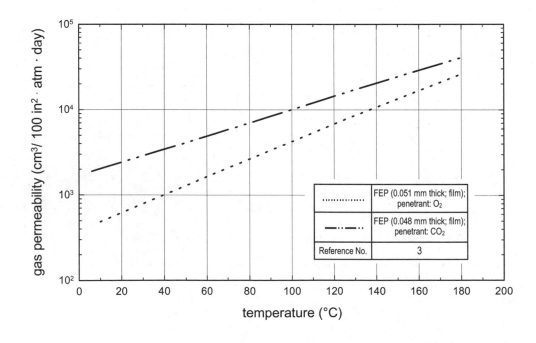

Figure II.9 Carbon dioxide and oxygen permeability rate vs temperature through fluorinated ethylene-propylene copolymer.

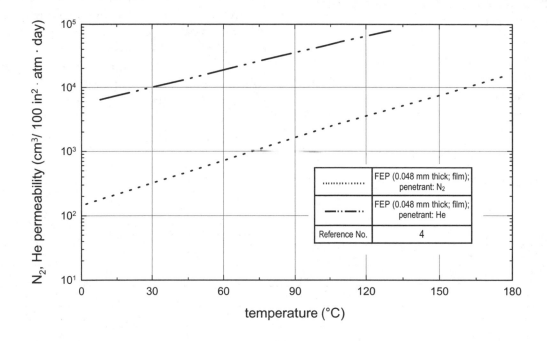

Figure II.10 Nitrogen and helium permeability rate vs temperature after retort through fluorinated ethylene-propylene copolymer.

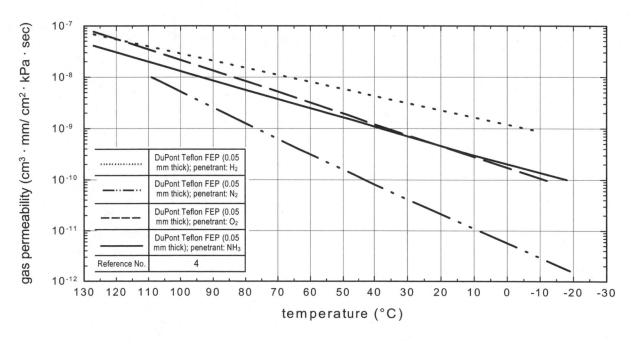

Figure II.11 Gas permeability vs temperature through fluorinated ethylene-propylene copolymer.

II.5 Permeability of Polychlorotrifluoroethylene (PCTFE)

Permeability measurements are rarely made directly on coatings because free standing films are required. Measurements on molded, cast, or extruded films can be used to indicate the performance on coating materials based on the same polymer. One must keep in mind that many coatings contain other materials in addition to the fluoropolymer, which can affect permeation properties in a positive or negative direction. This appendix is an edited version from *Fluoroplastics, Volume 1: Non-Melt Processible Fluoroplastics, The Definitive User's Guide and Databook.*[1]

Table II.12 Carbon Dioxide, Hydrogen, and Hydrogen Sulfide Permeability through 3M Kel-F 81 Polychlorotrifluoroethylene Film

Material Family	POLYCHLOROTRIFLUOROETHYLENE								
Material Supplier/Grade	3M KEL-F 81								
Product Form	FILM								
MATERIAL COMPOSITION									
Note	amorphous form of polymer								
TEST CONDITIONS									
Penetrant	carbon dioxide			hydrogen			hydrogen sulfide		
Temperature (°C)	0	25	50	75	0	25	50	50	75
PERMEABILITY (source document units)									
Gas Permeability ($1\times10^{-10}\cdot cm^3\cdot mm/cm^2\cdot sec\cdot cm$ Hg)	0.35	1.4	2.4	15	3.2	9.8	24	0.35	2.0
PERMEABILITY (normalized units)									
Permeability Coefficient ($cm^3\cdot mm/m^2\cdot day\cdot atm$)	2.3	9.2	15.8	98.5	21.0	64.3	158	2.3	13.1

Table II.13 Hydrogen Permeability vs Temperature and Pressure through 3M Kel-F 81 Polychlorotrifluoroethylene

Material Family	POLYCHLOROTRIFLUOROETHYLENE								
Material Supplier/Grade	3M KEL-F								
MATERIAL CHARACTERISTICS									
Sample Thickness	0.01 mm	0.01 mm	0.01 mm	0.01 mm	0.01 mm	0.01 mm	0.01 mm	0.01 mm	0.01 mm
TEST CONDITIONS									
Penetrant	hydrogen								
Temperature (°C)	-15	25		-12	25	67	-16	25	70
Pressure Gradient (kPa)	1721	1724	1721	1417	1447	3447	6895	6895	6895
Test Method	mass spectrometry and calibrated standard gas leaks; developed by McDonnell Douglas Space Systems Company Chemistry Laboratory								
PERMEABILITY (source document units)									
Gas Permeability ($cm^3 \cdot mm/cm^2 \cdot kPa \cdot sec$)	6.39×10^{-11}	4.07×10^{-10}	2.33×10^{-9}	6.69×10^{-11}	4.13×10^{-10}	2.25×10^{-9}	5.77×10^{-11}	4.14×10^{-10}	2.49×10^{-9}
PERMEABILITY (normalized units)									
Permeability Coefficient ($cm^3 \cdot mm/m^2 \cdot day \cdot atm$)	5.6	35.6	204	5.9	36.2	197	5.1	36.2	218

Table II.14 Nitrogen Permeability vs Temperature and Pressure through 3M Kel-F 81 Polychlorotrifluoroethylene

Material Family	POLYCHLOROTRIFLUOROETHYLENE					
Material Supplier/Grade	3M KEL-F					
MATERIAL CHARACTERISTICS						
Sample Thickness	0.01 mm	0.01 mm	0.01 mm	0.01 mm	0.01 mm	0.01 mm
TEST CONDITIONS						
Penetrant	nitrogen					
Temperature (°C)	25	68	25	69	25	70
Pressure Gradient (kPa)	1724	1724	3447	3447	6895	6895
Test Method	mass spectrometry and calibrated standard gas leaks; developed by McDonnell Douglas Space Systems Company Chemistry Laboratory					
PERMEABILITY (source document units)						
Gas Permeability ($cm^3 \cdot mm/cm^2 \cdot kPa \cdot sec$)	1.77×10^{-13}	4.15×10^{-11}	1.77×10^{-13}	4.36×10^{-11}	1.77×10^{-13}	4.45×10^{-11}
PERMEABILITY (normalized units)						
Permeability Coefficient ($cm^3 \cdot mm/m^2 \cdot day \cdot atm$)	0.02	3.63	0.02	3.82	0.02	3.9

Table II.15 Oxygen and Ammonia Permeability vs Temperature and Pressure through 3M Kel-F 81 Polychlorotrifluoroethylene

Material Family	POLYCHLOROTRIFLUOROETHYLENE					
Material Suppler/Grade	3M KEL-F					
MATERIAL CHARACTERISTICS						
Sample Thickness	0.01 mm	0.01 mm	0.01 mm	0.01 mm	0.01 mm	0.01 mm
TEST CONDITIONS						
Penetrant	ammonia			oxygen		
Temperature (°C)	25	59	25	52	25	52
Pressure Gradient (kPa)	965	965	1724	1724	3447	3447
Test Method	mass spectrometry and calibrated standard gas leaks; developed by McDonnell Douglas Space Systems Company Chemistry Laboratory					
PERMEABILITY (source document units)						
Gas Permeability (cm³· mm/cm²·kPa·sec)	1.2×10^{-11}	2.76×10^{-10}	2.95×10^{-12}	9.42×10^{-11}	2.84×10^{-12}	9.42×10^{-11}
PERMEABILITY (normalized units)						
Permeability Coefficient (cm³·mm/m²·day·atm)	1.05	24.2	0.26	8.2	0.25	8.25

Table II.16 Oxygen, Nitrogen, and Carbon Dioxide Permeability through Allied Signal ACLAR Polychlorotrifluoroethylene

Material Family	POLYCHLOROTRIFLUOROETHYLENE							
Material Supplier/Grade	ALLIED SIGNAL ACLAR							
Material Supplier/Grade	33C			22C		22A		
Product Form	FI LM							
Features	transparent	transparent	transparent	transparent	transparent	transparent	transparent	transparent
TEST CONDITIONS								
Penetrant	oxygen	carbon dioxide	oxygen	nitrogen	carbon dioxide	oxygen	nitrogen	carbon dioxide
Temperature (°C)	25	25	25	25	25	25	25	25
Test Note	STP conditions							
PERMEABILITY (source document units)								
Gas Permeability (cm³·mil/100in²·day)	17	16	15	2.5	40	12	2.5	30
PERMEABILITY (normalized units)								
Permeability Coefficient (cm³·mm/m²·day·atm)	2.8	6.3	5.9	1.0	15.7	4.7	1.0	11.8

Table II.17 Oxygen, Nitrogen, and Helium Permeability through 3M Kel-F 81 Polychlorotrifluoroethylene

Material Family	POLYCHLOROTRIFLUOROETHYLENE							
Material Supplied Grade	3M KEL-F 81							
Product Form	FILM							
MATERIAL COMPOSITION								
Note	amorphous form of polymer							
TEST CONDITIONS								
Penetrant	nitrogen			helium	oxygen			
Temperature (°C)	25	50	75	25	0	25	50	75
PERMEABILITY (source document units)								
Gas Permeability (1×10^{-10} cm³·mm/cm²·sec·cm Hg)	0.05	0.30	0.91	21.7	0.07	0.40	1.40	5.70
PERMEABILITY (normalized units)								
Permeability Coefficient (cm³·mm/m²·day·atm)	0.33	1.97	5.98	142.5	0.46	2.63	9.19	37.43

Table II.18 Water Vapor Permeability through Allied Signal ACLAR Polychlorotrifluoroethylene

Material Family	POLYCHLOROTRIFLUOROETHYLENE						
Material Supplier / Trade Name	ALLIED SIGNAL ACLAR						
Grade	33C		22C			22A	88A
Product Form	TRANSPARENT FILM						
MATERIAL CHARACTERISTICS							
Sample Thickness	0.019mm	0.051 mm	0.0254 mm	0.051 mm	0.19 mm	0.038mm	0.019mm
TEST CONDITIONS							
Penetrant	water vapor						
Temperature (°C)	37.8	37.8	37.8	37.8	37.8	37.8	37.8
Relative Humidity (%)	90	90	90	90	90	90	90
Test Method	ASTM E96, method E; measured on sealed pouches						
PERMEABILITY (source document units)							
Vapor Transmission Rate (g/m²·day)	0.43-0.59	0.15-0.31	0.47-0.93	0.24-0.62	0.09-0.13	0.32-0.62	0.70-0.86
Vapor Transmission Rate (g/day·100 in²)	0.028-0.038	0.010-0.020	0.030-0.060	0.016-0.040	0.006-0.007	0.020-0.040	0.045-0.055
PERMEABILITY (normalized units)							
Vapor Transmission Rate (g·mm/m²·day)	0.008-0.011	0.0077-0.0158	0.0119-0.0236	0.0122-0.0316	0.0171-0.0247	0.0122-0.0236	0.0133-0.0163

Table II.19 Water Vapor Transmission through 3M Kel-F 81 Polychlorotrifluoroethylene

Material Family	POLYCHLOROTRIFLUOROETHYLENE			
Material Supplier/ Grade	3M KEL-F 81			
Product Form	FI LM			
MATERIAL COMPOSITION				
Note	amorphous form of polymer			
TEST CONDITIONS				
Penetrant	water vapor			
Temperature (°C)	25	50	75	100
PERMEABILITY (source document units)				
Gas Permeability ($1\times10^{-10} \cdot cm^3 \cdot mm/cm^2 \cdot sec \cdot cm\ Hg$)	1	10	28	100
Vapor Transmission Rate ($g \cdot mil/m^2 \cdot atm \cdot day$)	0.19	1.76	4.56	15.20
PERMEABILITY (normalized units)				
Permeability Coefficient ($cm^3 \cdot mm/m^2 \cdot day \cdot atm$)	6.57	65.7	184	657
Vapor Transmission Rate ($g \cdot mm/m^2 \cdot day$)	0.005	0.043	0.116	0.386

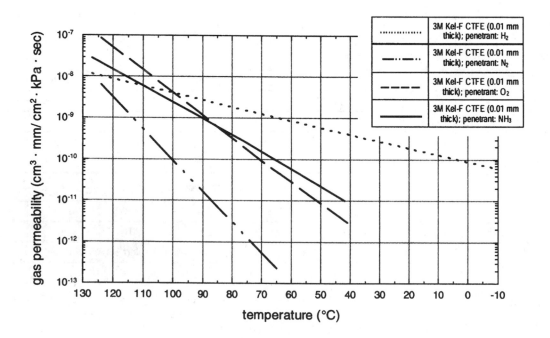

Figure II.12 Gas permeabilility vs temperature through polychlorotrifluoroethylene.

II.6 Permeability of Perfluoroalkoxy Copolymer (PFA)

Permeability measurements are rarely made directly on coatings because free standing films are required. Measurements on molded, cast, or extruded films can be used to indicate the performance on coating materials based on the same polymer. One must keep in mind that many coatings contain other materials in addition to the fluoropolymer, which can affect permeation properties in a positive or negative direction. This appendix is an edited version from *Fluoroplastics, Volume 2: Melt Processible Fluoroplastics, The Definitive User's Guide and Databook.*[2]

Table II.20 Gas Permeability of Oxygen, Carbon Dioxide, and Nitrogen through DuPont Company Teflon® PFA Perfluoroalkoxy Film

Material Family	PERFLUOROALKOXY RESIN		
Material Supplier/ Grade	DUPONT TEFLON® PFA		
Product Form	FILM		
Reference Number	8		
TEST CONDITIONS			
Penetrant	carbon dioxide	nitrogen	oxygen
Temperature (°C)	25	25	25
Test Method	ASTM D1434	ASTM D1434	ASTM D1434
PERMEABILITY (source document units)			
Gas Permeability (cm³·mil/100 in²·day)	2260	291	881
Gas Permeability (cm³· mm/m²·day·Pa)	0.00878	0.00113	0.00342
PERMEABILITY (normalized units)			
Permeability Coefficient (cm³· mm/m²·day·atm)	890	115	347

II.7 Permeability of Polyvinylidene Fluoride (PVDF)

Permeability measurements are rarely made directly on coatings because free standing films are required. Measurements on molded, cast, or extruded films can be used to indicate the performance on coating materials based on the same polymer. One must keep in mind that many coatings contain other materials in addition to the fluoropolymer, which can affect permeation properties in a positive or negative direction. This appendix is an edited version from *Fluoroplastics, Volume 2: Melt Processible Fluoroplastics, The Definitive User's Guide and Databook*.[2]

Table II.21 Ammonia, Helium, Chlorine, and Hydrogen Permeability through Solvay Solef® Polyvinylidene Fluoride Film

Material Family	POLYVINYLIDENE FLUORIDE			
Material Supplier/Grade	SOLVAY SOLEF®			
Product Form	FILM			
Manufacturing Method	cast film			
MATERIAL CHARACTERISTICS				
Sample Thickness (mm)	0.1			
TEST CONDITIONS				
Penetrant	ammonia	helium	chlorine	hydrogen
Temperature (°C)	23			
Test Method	ASTM D1434			
PERMEABILITY (source document units)				
Gas Permeability ($cm^3 \cdot N/m^2 \cdot bar \cdot day$)	65	850	12	210
PERMEABILITY (normalized units)				
Permeability Coefficient ($cm^3 \cdot mm/m^2 \cdot day \cdot atm$)	6.6	86	1.2	21.3

Table II.22 Carbon Dioxide, Nitrogen, Oxygen, and Water Vapor Permeability through Solvay Solef® 1008 Polyvinylidene Fluoride Film

Material Family	POLYVINYLIDENE FLUORIDE			
Material Supplier/ Grade	SOLVAY SOLEF® 1008			
Product Form	FILM			
Features	Translucent			
MATERIAL CHARACTERISTICS				
Sample Thickness, mm	0.1			
TEST CONDITIONS				
Penetrant	carbon dioxide	nitrogen	oxygen	water vapor
Temperature, °C	23			38
Test Method	ASTM D1434			ASTM E96, proc. E
PERMEABILITY (source document units)				
Vapor Transmission Rate (g/m²·day)				7.5
Gas Permeability (cm³·N/m²·bar·day)	70	30	21	
PERMEABILITY (normalized units)				
Permeability Coefficient (cm³·mm/m²·day·atm)	7.09	3.04	2.13	
Vapor Transmission Rate (g·mm/m²·day)				0.75

Table II.23 Freon®, Nitrous Oxide, Hydrogen Sulfide, and Sulfur Dioxide Permeability through Solvay Solef® Polyvinylidene Fluoride Film

Material Family	POLYVINYLIDENE FLUORIDE						
Material Supplier/ Grade	SOLVAY SOLEF®						
Product Form	FILM						
Features	cast film						
MATERIAL CHARACTERISTICS							
Sample Thickness, mm	0.025						
TEST CONDITIONS							
Penetrant	Freon® 12	Freon® 114	Freon® 115	Freon® 318	nitrous oxide	hydrogen sulfide	sulfur dioxide
Temperature, °C	23						
Test Method	ASTM D1434						
PERMEABILITY (source document units)							
Gas Permeability (cm³·N/m²·bar·day)	6.3	10	4	17	900	60	60
PERMEABILITY (normalized units)							
Permeability Coefficient (cm³·mm/m²·day·atm)	0.16	0.25	0.1	0.18	22.8	1.52	1.52

Table II.24 Water Vapor, Oxygen, and Carbon Dioxide Permeability through Atochem Foraflon® Polyvinylidene Fluoride Film

Material Family	POLYVINYLIDENE FLUORIDE				
Material Supplier/Grade	ATOCHEM FORAFLON®				
Product Form	EXTRUDED FILM				
Reference Number	10				
MATERIAL CHARACTERISTICS					
Sample Thickness, mm	0.02	0.028	0.04	0.037	0.034
TEST CONDITIONS					
Penetrant	Water vapor			oxygen	carbon dioxide
Temperature, °C	38			30	
Test Method	NFH 00044			IS0 2556	
PERMEABILITY (source document units)					
Vapor Transmission Rate (g/m²·day)	34	22	16		
Gas Permeability (cm³/m²·day)				140	890
PERMEABILITY (normalized units)					
Permeability Coefficient (cm³·mm/m²·day·atm)				5.18	30.26
Vapor Transmission Rate (g·mm/m²·day)	0.68	0.62	0.64		

Table II.25 Water Vapor, Oxygen, Nitrogen, and Carbon Dioxide Permeability through Polyvinylidene Fluoride

Material Family	POLYVINYLIDENE FLUORIDE			
Reference Number	7			
TEST CONDITIONS				
Penetrant	water vapor	oxygen	nitrogen	carbon dioxide
Temperature, °C	23	25		
Relative Humidity, %	90			
Test Method		STP conditions		
PERMEABILITY (source document units)				
Gas Permeability (cm³·mm/100 in²·day)		1.4	9	5.5
Gas Permeability (cm³·mm/m²·day·atm)		0.55	3.5	2.2
Vapor Transmission Rate (g·mil/100 in²·day)	2.6			
Vapor Transmission Rate (g/day·100 in²)	1.0			
PERMEABILITY (normalized units)				
Permeability Coefficient (cm³·mm/m²·day·atm)		0.55	3.5	2.2
Vapor Transmission Rate (g·mm/m²·day)	1.0			

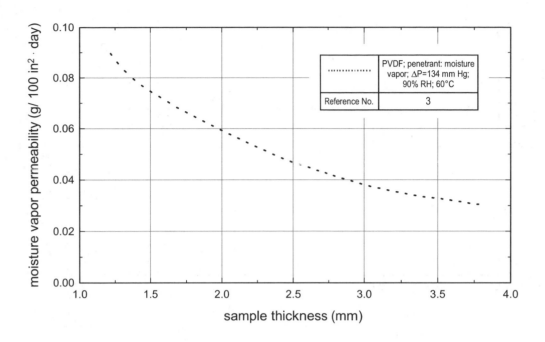

Figure II.13 Moisture vapor permeability vs thickness through polyvinylidene fluoride.

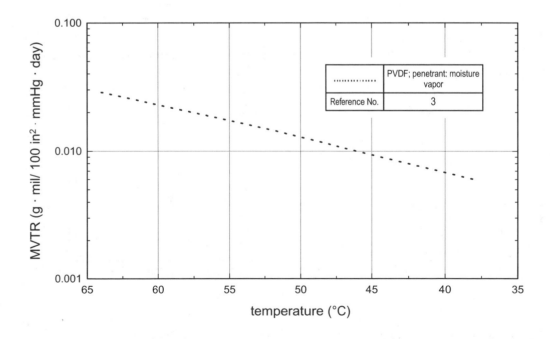

Figure II.14 Moisture vapor permeability vs temperature through polyvinylidene fluoride.

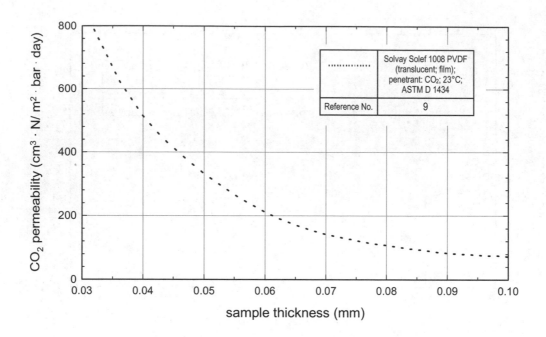

Figure II.15 Carbon dioxide permeability vs thickness through polyvinylidene fluoride.

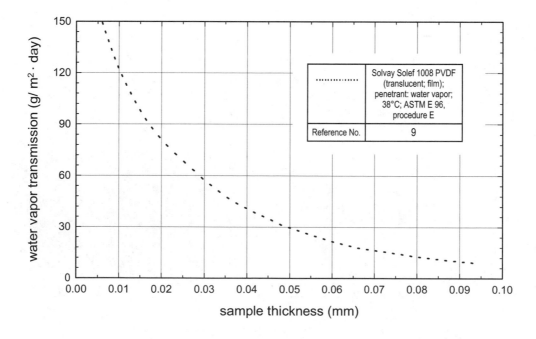

Figure II.16 Water vapor permeability vs thickness through polyvinylidene fluoride.

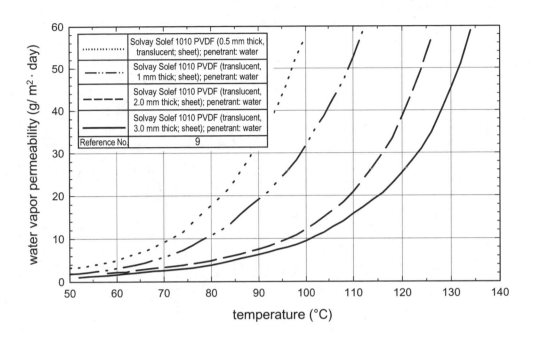

Figure II.17 Water vapor permeability vs temperature through polyvinylidene fluoride.

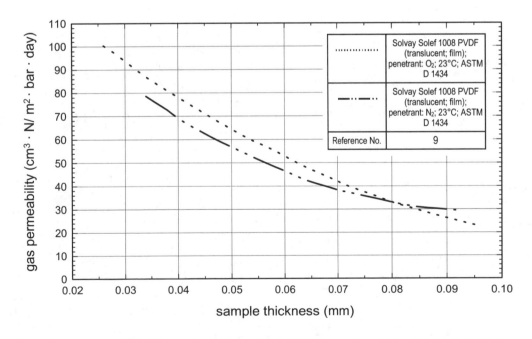

Figure II.18 Nitrogen and oxygen permeability vs thickness through polyvinylidene fluoride.

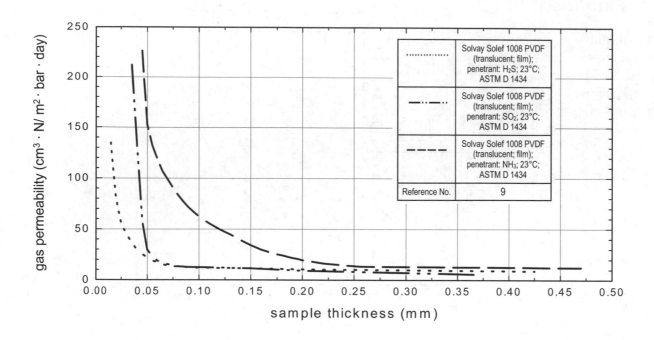

Figure II.19 Gas permeability vs thickness through polyvinylidene fluoride.

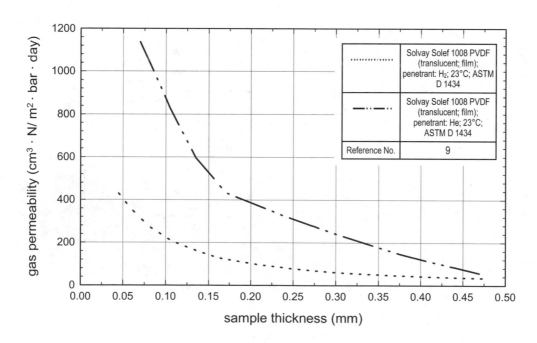

Figure II.20 Helium and hydrogen permeability vs thickness through polyvinylidene fluoride.

II.8 Permeability of Polyvinyl Fluoride (PVF)

Permeability measurements are rarely made directly on coatings because free standing films are required. Measurements on molded, cast, or extruded films can be used to indicate the performance on coating materials based on the same polymer. One must keep in mind that many coatings contain other materials in addition to the fluoropolymer, which can affect permeation properties in a positive or negative direction. This appendix is an edited version from *Fluoroplastics, Volume 2: Melt Processible Fluoroplastics, The Definitive User's Guide and Databook*.[2]

Table II.26 Water Vapor, Oxygen, Nitrogen, and Carbon Dioxide Permeability through Polyvinyl Fluoride

Material Family	POLYVINYL FLUORIDE			
Reference Number	7			
TEST CONDITIONS				
Penetrant	water vapor	oxygen	nitrogen	carbon dioxide
Temperature, °C	37.8	25		
Relative Humidity, %	90			
Test Method		STP conditions		
PERMEABILITY (source document units)				
Gas Permeability (cm³·mil/100 in²·day)		3.0	0.25	11
Gas Permeability (cm³·mm/m²·day·atm)		1.2	0.10	4.3
Vapor Transmission Rate (g·mil/100 in²·day)	3.24			
Vapor Transmission Rate (g/day·100 in²)	1.3			
PERMEABILITY (normalized units)				
Permeability Coefficient (cm³·mm/m²·day·atm)		1.2	0.1	4.3
Vapor Transmission Rate (g·mm/m²·day)	1.3			

REFERENCES

1. Ebnesajjad, S., *Fluoroplastics, Vol. 1: Non-Melt Processible Fluoroplastics, The Definitive User's Guide and Databook*, William Andrew, Inc., Norwich, NY (2000)

2. Ebnesajjad, S., *Fluoroplastics, Vol. 2: Melt Processible Fluoroplastics, The Definitive User's Guide and Databook*, William Andrew, Inc., Norwich, NY (2003)

3. Chemical Resistance of Halar® Fluoropolymer, supplier technical report (AHH), Ausimont

4. Adam, S. J., and David, C. E., Permeation Measurement of Fluoropolymers Using Mass Spectroscopy and Calibrated Standard Gas Leaks, *23rd International SAMPE Tech. Conf, Conf. Proceedings - SAMPE* (1991)

5. Tefzel® Fluoropolymers Design Handbook, supplier design guide No. E-31301-1, DuPont Co. (1973)

6. Hyflon® ETFE 700/800 Properties and Applications, supplier design guide, Ausimont USA, Inc.

7. Aclar® Performance Films, Supplier Technical Report No. SFI-14, Revised 9-89, Allied Signal Engineered Plastics (1989)

8. Handbook of Properties for Teflon® PFA, supplier design guide (E-96679) - DuPont Company (1987)

9. Solvay Polyvinylidene Fluoride, supplier design guide (B-1292c-B-2.5-0390) Solvay (1992)

10. Foraflon® PVDF, supplier design guide (694.E/07.87/20), Atochem S. A. (1987)

Appendix III: Permeation of Automotive Fuels Through Fluoroplastics

III.1 Introduction

This appendix is an edited version from *Fluoroplastics, Volume 2: Melt Processible Fluoroplastics, The Definitive User's Guide and Databook.*[1]

III.2 IVA Test Method

Modified SAE J-30 (Sec. 6.12) Method, 2000.

III.3 Fuel Types

- Fuel C - Reference fuel, 50150 blend, by volume, of iso-octane and toluene.
- M20: 20 vol% methanol in Fuel C.
- Sour Gas: 0.08 molar t-butyl hydroperoxide in Fuel C.

The majority of the test was conducted using M20. The resins studied were:

- Teflon® 62: Polytetrafluoroethylene (PTFE); fine powder, paste extrusion resin.
- Teflon® FEP 100: Fluorinated ethylene propylene copolymer (FEP); general purpose, medium viscosity, melt extrusion resin.
- Teflon® PFA 340: Copolymer of tetrafluoroethylene and a perfluoroalkoxy monomer (PFA); general purpose, medium viscosity, melt extrusion resin.
- Tefzel® 200: Ethylene tetrafluoroethylene copolymer (ETFE); general purpose, melt extrusion resin.
- Vestamid® L2121: Nylon 12 polyamide; 7% plasticizer, melt extrusion resin.
- Zyth® CFE 3011: Nylon 12, 12 polyamide; 13% plasticizer, melt extrusion resin.

Table III.1 Permeation Rates (g·mm)/(day·m²)

Hose	M20	Fuel C	Sour Gas
PTFE	0.23	0.15	0.11
PTFE w/overbraid	0.22		
Conductive PTFE	0.18		
PFA (0.76 mm)	0.14		0.16
PFA (0.53 mm)	0.12		
PFA (0.28 mm)	0.13		
Conductive PFA	0.08		
ETFE	0.13	0.09	
FEP	0.18		
Nylon 12	1.35		
Nylon 12,12	1.32		
Note: Tube diameters 6-10 mm, wall thickness 0.75-0.85 mm except those noted.			

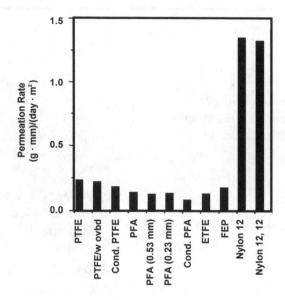

Figure III.1 Permeation of M20 fuel through fluoropolymers and nylons. Note: ovbd = overbraided.

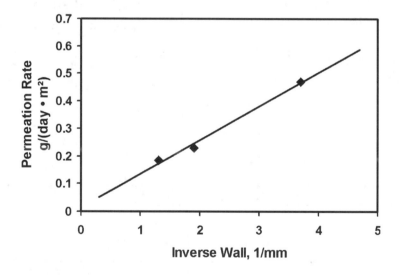

Figure III.2 Permeation of M20 fuel through PFA vs inverse wall thickness.

REFERENCES

1. Ebnesajjad, S., *Fluoroplastics, Vol. 2: Melt Processible Fluoroplastics, The Definitive User's Guide and Databook*, William Andrew, Inc., Norwich, NY (2003)

2. Teflon®/Tefzel® Technical Information published by DuPont, No. H34374 (Sep 1992)

3. Society of Automotive Paper Series 910104, Part 11. Fluoropolymer Resins: Permeation of Automotive Fuels, D. R. Goldberry

4. Chillous, S. E., and Will, R. R., presented at Int. Congress & Exposition, Detroit, MI (2/25/91-3/1/91)

Appendix IV: Permeation of Chemicals Through Fluoroplastics

IV.1 Introduction

This appendix is an edited version from *Fluoroplastics, Volume 2: Melt Processible Fluoroplastics, The Definitive User's Guide and Databook.*[1]

Table IV.1 Permeation Rate of Chlorine Gas through Fluoroplastic Films

Polymer	Thickness, microns	Chlorine
		Permeation Rate, g/m²/24 hr @25°C
Granular PTFE	250	1.974
	2,250	0.358
	4,450	0.255
Fine Powder PTFE	250	5.55
	2,250	0.369
	4,450	0.289
FEP	4,450	0.190
PFA	250	1.605
	2,250	0.569
	4,450	0.265
ETFE	250	1.164
	2,250	0.254
	4,450	0.250
ECTFE	4,450	0.199
PVDF	250	1.018
	5,250	0.167

Table IV.2 Permeation Rate of Nitric Acid through Fluoroplastic Films

Polymer	Thickness, microns	20% Nitric Acid	
		Permeation Rate, g/m²/24 hr	
		@25°C	@45°C
Granular PTFE	250		0.395
PFA	250	0.397	0.610
ETFE	250	0.469	-
	2,250	0.035	0.107
ECTFE	250	0.072	1.453
	2,250	0.061	0.037
PVDF	250	0.344	3.703
	2,250	-	0.265

Table IV.3 Permeation Rate of Methylene Chloride through Fluoroplastic Films

Polymer	Thickness, microns	Methylene Chloride Permeation Rate, g/m²/24 hr	
		@ 25°C	@ 45°C
Granular PTFE	250	3.85	9.08
Fine Powder PTFE	250	20.6	60.8
PFA	250	2.34	10.6
ETFE	250	33.1	113.6
ECTFE	250	59.5	634.6
PVDF	250	8.55	36.06
Polypropylene	250	504.2	2,250

Table IV.4 Permeation Rate of Phenol through Fluoroplastic Films

Polymer	Thickness, microns	Phenol Permeation Rate, g/m²/24 hr		
		@ 25°C	@ 45°C	@ 75°C
Granular PTFE	250	0.050	0.247	5.854
Fine Powder PTFE	250	0.084	0.991	-
PFA	250	0.013	0.237	-
ETFE	250	0.158	1.562	15.3
ECTFE	250	0.067	-	-
PVDF	250	0.218	3.394	-
Polypropylene	250	0.027	0.734	-

Table IV.5 Permeation Rate of Benzene through Fluoroplastic Films

Polymer	Thickness, microns	Benzene Permeation Rate, g/m²/24 hr		
		25°C	45°C	75°C
Granular PTFE	250	2.591	29.4	89.26
	2,250	0.777	-	3.286
	4,450	0.0335	-	-
ETFE	250	5.326	16.4	94.9
	2,250	0.118	0.011	5.655
	4,450	0.068	-	-

Table IV.6 Permeation Rate of Methyl Ethyl Ketone through Fluoroplastic Films

Polymer	Thickness, microns	Methyl Ethyl Ketone		
		Permeation Rate, g/m²/24 hr		
		25°C	45°C	75°C
Granular PTFE	250	7.726	20.4	29.1
	2,250	0.316	-	
	4,450	0.028	-	-
ETFE	250	6.882	42.3	426.8
	2,250	0.034	-	-
	4,450	0.023	-	-
ECTFE	250	27.6	66.1	519.6
	2,250	0.033	-	-
PVDF	250	482.1	1866.8	8247.0
	2,250	0.168	-	-

Table IV.7 Permeation Rate of Water through Fluoroplastic Films

Polymer	Thickness, microns	Water		
		Permeation Rate, g/m'²/24 hr		
		@ 25°C	@ 45°C	@ 75°C
Granular PTFE	250	0.295	0.845	5.568
	2,250	0.050	-	-
ETFE	250	0.672	2.513	17.18
	-	-	-	-
	4,450	0.053	-	-

REFERENCES

1. Ebnesajjad, S., *Fluoroplastics, Vol. 2: Melt Processible Fluoroplastics, The Definitive User's Guide and Databook*, William Andrew, Inc., Norwich, NY (2003)

2. Teflon® Industrial Coatings, The Facts: Permeation, Its Effects on Teflon® ETFE Coatings on Corrosive Fume Exhaust Ducts, DuPont Co. (1999)

Trade Names

Trademark	Property of	Trademark	Property of
Acrysol®	Rohm & Haas Corp.	Gore-Tex®	W.L. Gore & Assoc.
Aerosil®	Degussa	Granodine®	Henkel Technologies
Afflair®	EM Industries	Greblon®	Weilburger Coatings (Grebe Group)
Aflon®	Asahi Glass Co.	Greblon Alpha®	Weilburger Coatings (Grebe Group)
AgION®	Ag ION Technologies		
Algoflon®	Ausimont Corp.	Greblon Beta®	Weilburger Coatings (Grebe Group)
Aquadag®	Acheson		
Autograph®	DuPont Co.	Greblon Gamma®	Weilburger Coatings (Grebe Group)
Bakelite®	Bakelite AG		
Bonderite®	Henkel Technologies	Gylon®	B.F. Goodrich Corp.
Cab-O-Sil®	Cabot Corporation	Halar®	Ausimont Corp.
Coraflon®	PPG	Hiflon®	Hindustan Fluoropolymer Co.
Cryscoat®	Chemetall Oakite	Hostaflon®	Dyneon Corp.
Delrin®	DuPont Co.	Hyflon®	Ausimont Corp.
Dobanol®	Shell Chemicals	Hylar®	Ausimont Corp.
Dykor®	Whitford Worldwide	Iriodin®	EMD Chemicals
Dyneon®	3M	Isopar®	Exxon-Mobile Corp.
Eclipse®	Whitford Worldwide	Kapton®	DuPont Co.
EMAWELD®	Ashland Specialty Chemical Co.	Kel F®	3M
		Kevlar®	DuPont Co.
Emralon®	Acheson	KYNAR®	Arkema
Excalibur®	Whitford Worldwide	Lucite®	Lucite International
Fluon®	Asahi Glass Co.	Lumiflon®	Asahi Glass
Fluorad®	3M	Mearl®	Mearl Company
Fluorinert®	3M	Natrosol®	Hercules Corp.
Fluorocomp®	Teleflex Corp.	Neoflon®	Daikin Corp.
Fluoromelt®	Asahi Glass Co.	Nomex®	DuPont Co.
Fluoroplast®	KCCE Corp.	Plexiglass®	AtoHaas
Fluorotech®	Acton Corp.	Polyflon®	Daikin Corp.
ForaFlon®	Atofina Corp.	Polymist®	Ausimont Corp.
Frekote®	Frekote Co.	QuanTanium®	Whitford Worldwide
Freon®	DuPont Co.	Quantum2®	Whitford Worldwide
Galvalume®	BIEC International Inc	Rheocin®	Sud-Chemie
Gardobond®	Chemetall Oakite	Ryton®	Phillips 66
Genetron®	Allied Signal Corp.	Scotch®	3M
Gore-Tex®	W.L. Gore & Assoc.		

Trademark	Property of	Trademark	Property of
Scotch-Brite®	3M	Teflon-S®	DuPont Co.
Serdox®	Elementis Specialities	Tefzel®	DuPont Co.
Siltem®	General Electric Company	Tergitol®	Dow Chemical Company
SilverStone®	DuPont Co.	Torlon®	Solvay Advanced Polymers.
Skandia®	Akzo Nobel	Triton®	Union Carbide Corp.
Skandia® Marrlite Plus	Akzo Nobel	Tyvek®	DuPont Co.
Skandia® Marrlite	Akzo Nobel	Tyzor®	DuPont Co.
Skandia® Stratos	Akzo Nobel	Ultem®	General Electric Company
Skandia® Tech	Akzo Nobel	Ultralon®	Whitford Worldwide
Spectragraphics®	DuPont Co.	Vespel®	DuPont Co.
SilverStone Supra®	DuPont Co.	Viton®	DuPont Co.
Surfynol®	Air Products	Whitford®	Whitford Worldwide
Tedlar®	DuPont Co.	Xylac®	Whitford Worldwide
Tefal®	T-fal	Xylan®	Whitford Worldwide
Teflon®	DuPont Co.	Xylar®	Whitford Worldwide
Teflon II®	DuPont Co.	Zonyl®	DuPont Co.
Teflon® AF	DuPont Co.		

A

Abrasion - Wearing, grinding, or rubbing away by friction.

Abrasion Resistance - Wear rate or abrasion rate is an important property of materials during motion in contact with other materials. Abrasion or wear resistance is measured by a number of methods such as ASTM Method D3389, also known as *Taber Test*.

Adhesive Failure - Failure of an adhesive bond at the adhesive-substrate interface.

Agglomerates - Groups of pigment or fluoropolymer particles that are loosely bound together that generally can be separated by dispersion equipment.

Amorphous Polymer - Polymers having non-crystalline or low-order molecular structure or morphology. Amorphous polymers may have some molecular order but usually are substantially less ordered than crystalline polymers.

ASTM - American Society of Testing Materials.

Average Particle Size - The average diameter of particles as determined by various test methods.

B

Binder - A polymer that acts as an adhesive to join elements of matrix coatings.

Bulk Density -The mass per unit volume in powder form, including the air trapped between particles.

Burn-off - A method of removing a coating or degrading a coating to make removal easier.

C

C8 - An alternative name for perfluoroammonium octanoate.

Carrier - The liquid portion of a coating (solvent or water) in which solids are dissolved, dispersed, or suspended.

Chemical Resistance - Degradation of a material caused by chemical reaction.

Chain Transfer Agent - A substance that is able to cause the transfer of a radical to a different molecule in a chain polymerization. It provides an atom to the radical at the growing end of a polymer chain, and in doing so is made into a radical which can start the growth of a new polymer chain. Examples include solvents, impurities, or a modifier deliberately added for this purpose.

Coagulation - This is a process for separation of PTFE (polytetrafluoroethylene) solids from its dispersion. The emulsion or dispersion containing this polymer (dispersion polymerization) has to be broken (destabilized) in order to cause precipitation of PTFE particles. Dilution to reduce solids concentration below 20%, addition of water-soluble organic compounds, and addition of soluble inorganic salts are the common techniques to break PTFE emulsions.

Coalescence - Refers to the mechanism for melting and consolidation of a fluoropolymer coating. After the fluoropolymer melts, adjacent particles begin to combine (i.e., coalesce) under the driving force of surface tension.

Contact Angle - The angle that the droplet or edge of the liquid forms with the solid plane is called the *contact angle*. It is a means of estimating the nonstick properties of a coating by measuring the ability of a liquid such as water or hexadecane to wet its surface. As surface energy of a surface decreases (as in a nonstick coating), the contact angle increases.

Convection - The mass movement of particles arising from the movement of a streaming fluid due to difference in a physical property such as density, temperature, etc. Mass movement due to a temperature difference results in heat transfer, as in the upward movement of a warm air current.

Copolymerization - A polymerization where more than one monomer takes part in the reaction to form the polymer chain.

Corona Gun - A powder gun that uses corona charging.

Corona Charge - An electrostatic charge induced on powder particles by passing them through an electrostatic field generated by high voltage.

Corrosion - The process of metal oxidation (decomposition) in which metal ions react with oxygen to form metal oxides. Fluoropolymer coatings provide excellent barriers against most corrosives.

Cracking - Appearance of external and/or internal cracks in the material as a result of stress that exceeds the strength of the material. The stress can be external and/or internal and can be caused by a variety of adverse conditions: structural defects, impact, aging, corrosion, etc., or a combination thereof. Also called *resistance to cracking, grazing, cracking resistance.*

Critical Cracking Thickness (CCT) - The maximum thickness which can be coated in a single layer (pass) of polytetrafluoroethylene dispersion without crack formation. This thickness is measured after sintering has been completed.

Crosslinking - Reaction or formation of covalent bonds between chain-like polymer molecules or between polymer molecules and low-molecular compounds such as carbon black fillers. As a result of crosslinking polymers, such as thermosetting resins, may become hard and infusible. Crosslinking is induced by heat, UV or electron-beam radiation, oxidation, etc. Crosslinking can be achieved either between polymer molecules as in unsaturated polyesters, or with the help of multifunctional crosslinking agents such as diamines that react with functional side groups of the polymers. Crosslinking can be catalyzed by the presence of transition metal complexes, thiols, and other compounds.

Crystalline Phase - This is an organized structural arrangement for polymer molecules. In this arrangement, polymer chains are aligned into a closely packed, ordered state called *crystalline phase.*

Crystallinity - Crystalline content of a polymer expressed in weight percent.

Cure Schedule - The time/temperature relationship required to cure a coating. Cure schedules can include multiple ramps and holds at different temperatures.

Curing - The process of bonding or fusing a coating to a substrate with heat and developing specified properties in the coating.

D

Deflagration - A violent reaction whereby tetrafluoroethylene is degraded into carbon and tetrafluoromethane.

Degradation - Loss or undesirable change in plastic properties as a result of aging, chemical reactions, wear, use, exposure, etc. The properties include color, size, strength, etc.

Dielectric Constant - The dielectric constant of an insulating material is the ratio of the capacitance of a capacitor insulated with that material to the capacitance of the same capacitor insulated with a vacuum.

Dielectric Strength - Ability of a coating to resist the passage of electric current through it.

Differential Scanning Calorimetry - DSC is a technique in which the energy absorbed or produced is measured by monitoring the difference in energy input into the substance and a reference material as a function of temperature. Absorption of energy produces an endotherm; production of energy results in an exotherm. DSC may be applied to processes involving an energy change, such as melting, crystallization, resin curing, and loss of solvents, or to processes involving a change in heat capacity, such as the glass transition.

Dispersion - A dispersion is often defined as a uniform mixture of solids particles and a liquid. It may contain other agents such as a surfactant and a resin soluble in the liquid (solvent). An example of a dispersion is a house paint. A feature of most dispersions is stability, which means little or no settling of the solid particles.

Dispersion Polymerization - This technique is a heterogenous regime where a significant amount of surfactant is added to the polymerization medium. Characteristics of the process include small, uniform polymer particles, which may be unstable and coagulate if they are not stabilized. Hydrocarbon oil is added to the dispersion polymerization reactor to stabilize the polytetrafluoroethylene emulsion. Temperature and agitation control are easier in this mode than suspension polymerization. Polytetrafluoroethylene fine powder and dispersion are produced by this technique.

Dry Blending - A process for manufacturing powder coatings in which materials are blended physically together without melting.

DSC - See *Differential Scanning Calorimetry.*

E

Elasticity - Property whereby a solid material changes its shape and size under action of opposing forces, but recovers its original configuration when the forces are removed.

Electron Beam Radiation - Ionizing radiation propagated by electrons that move forward in a narrow stream with approximately equal velocity. Also called *electron beam.*

Electrostatic Spray - A coating or painting method of spraying and charging a coating so that it is deposited on a grounded substrate. (See *Corona Charge* and *Tribocharging*).

Elongation - The increase in gauge length of a specimen in tension, measured at or after the fracture, depending on the viscoelastic properties of the material. Note: Elongation is usually expressed as a percentage of the original gauge length. Also called *ultimate elongation, tensile negation, breaking elongation.*

Elongation at Break - The increase in distance between two gauge marks, resulting from stressing the specimen in tension, at the exact point of break. The measurement is taken at the exact point of break according to ASTM D638.

Emulsion - See *Dispersion.*

Encapsulation - This term means to enclose as in a capsule. Polytetrafluoroethylene (PTFE) can be used to encapsulate metal articles to impart chemical resistance to them. Examples include encapsulated metal gaskets and butterfly valve gates. The metal provides mechanical strength and resistance to creep.

End Groups - The functional groups appear at the ends of polymer chains and in effect "end" the chain growth.

Epoxides - Organic compounds containing three-membered cyclic group(s) in which two carbon atoms are linked with an oxygen atom as in an ether. This group is called an epoxy group and is quite reactive, allowing the use of epoxides as intermediates in preparation of certain fluorocarbons and cellulose derivatives and as monomers in preparation of epoxy resins.

F

Faraday Cage Effect - Repulsion of charged particles because of the part's concave shape. Charges build at the entry area, preventing penetration into the cavity.

Fibrillation - This phenomenon occurs when polytetrafluoroethylene fine powder particles are subjected to shear usually at a temperature above its transition point (19°C). For example, when fine powder particles rub against each other, groups of polymer chains are pulled out of crystallites. These fibrils can connect polymer particles together. They have a width of less than 50 nm.

Fillers - Pigments and other solids used to alter properties of coatings.

Film - A product (e.g., plastic) that is extremely thin compared to its width and length. There are supported and unsupported films such as coatings and packagings, respectively.

Film Formation - A continuous film formed due to heated polymer particles melting and coalescing or crosslinking.

Fine Powder PTFE - Polytetrafluoroethylene (PTFE) polymerized by dispersion polymerization method.

Finishes - Highly formulated dispersions of polytetrafluoroethylene containing a variety of fillers such as pigments, resins, extenders, and others. Finishes are used to coat different surfaces such as cookware, housewares, and industrial equipment.

Flame Treatment - In adhesive bonding, a surface preparation technique in which the plastic is briefly exposed to a flame. Flame treatment oxidizes the surface through a free radical mechanism, introducing hydroxyl, carbonyl, carboxyl, and amide functional groups to a depth of ~ 4–6 nm, and produces chain scissions and some crosslinking. Commonly used for polyolefins, polyacetals, and polyethylene terephthalate, flame treatment increases wettability and interfacial diffusivity.

Flash Point - The lowest temperature at which a solvent will generate sufficient vapors to ignite in the presence of flame.

Flashing - A brief post coating application step to drive off solvents or carriers prior to full cure. This helps prevent bubbling.

Fluidized Bed Coating - A method of applying a coating to an article in which the article is immersed in a fluidized bed (a fixed container in which powder is aerated) of powdered coating. Dipping directly into the fluidized powder can coat preheated objects or wet primed parts. In electrostatic fluidized bed coating, the part is usually not heated but is charged and passed over a fluidized bed of power which has the opposite charge.

Fluid Energy Mill - A mill that utilizes high-speed air to reduce the size of solid particles.

Free Radical - An atom or group of atoms with an odd or unpaired electron. Free radicals are highly reactive and participate in free radical chain reactions such as combustion and polymer oxidation reactions. Scission of a covalent bond by thermal degradation or radiation in air can produce a molecular fragment named a free radical. Most free radicals are highly reactive because of their unpaired electrons, and have short half lives.

$$R - R' \rightarrow R\cdot + R'$$

FTIR - Fourier transform infrared spectroscopy is a spectroscopic technique in which a sample is irradiated with electromagnetic energy from the infrared region of the electromagnetic spectrum (wavelength ~0.7 to 500 mm). The sample is irradiated with all infrared wavelengths simultaneously, and mathematical manipulation of the Fourier transform is used to produce the absorption spectrum or "fingerprint" of the material. Molecular absorptions in the infrared region are due to rotational and vibrational motion in molecular bonds, such as stretching and bending. FTIR is commonly used for the identification of plastics, additives, and coatings.

G

Gamma Ray Irradiation - A technique for reduction in the molecular weight of polytetrafluoro-ethylene by exposing this polymer to gamma rays from a source such as ^{60}Co.

Granular Polytetrafluoroethylene - The name used to refer to the products of suspension polymerization of tetrafluoroethylene.

H

Halogenated Solvents - Organic liquids containing at least one atom of a halogen (Cl, F, I, Br) are called halogenated solvents.

Hammer Mill - A mill often used in producing polytetrafluoroethylene-filled compounds. It consists of a rotor equipped with a set of small hammers, rotating inside a basket made from mesh screen. The resin and filler blend are placed or fed continuously into the basket and subjected to the hammer action. After sufficient grinding, the mixture passes through the screen and is discharged.

Heat Deflection Temperature - The temperature at which a material specimen (standard bar) is deflected by a certain degree under specified load. Also called *tensile heat distortion temperature, heat distortion temperature, HDT, deflection temperature under load.*

Homopolymer - A polymer that contains only a single type of monomer (i.e., propylene).

Hot Flocking - Flocking deposition. An application method of applying powder by spray to a substrate heated above the melt point of the powder.

HVLP (High Volume, Low Pressure) - A spray technique utilizing high volume air in combination with low air pressure to increase transfer efficiency and reduce air pollution.

Hydrocarbon - A chemical compound that contains only hydrogen and carbon atoms.

Hydrofluoric Acid - HF is a highly corrosive acid.

Hydrophilic Surface - Surface of a hydrophilic substance that has a strong ability to bind to, adsorb, or absorb water; a surface that is readily wettable with water. Hydrophilic substances include carbohydrates such as starch.

Hysteresis Loop - Hysteresis means a retardation of the effect when the forces acting upon a body are changed. For example, the viscosity of shear thinning liquids tends to decrease as shear increases, but it will not increase to the similar value at a given shear if, at the end of the shear rate increase period, the shear rate is decreased back to the initial value. The plot obtained by plotting the viscosity against shear rate is called a *hysteresis loop*.

I

Infrared Oven - An oven equipped with infrared lamps where heat is generated by infrared rays.

Initiator - An chemical that causes a chemical reaction to start and which enters into the reaction to become part of the resultant polymer.

Intercoat Adhesion - A coating's ability to adhere to previously applied films, including primers.

K

Kesternich – An acid corrosion test, used for acid rain simulation (DIN 50018) named after the German scientist who developed the Kesternich Cabinet and test method.

L

Leveling - The elimination of surface irregularities such as brush marks, roller pattern, or spray pattern to achieve a smooth coating surface.

Limiting Oxygen Index (LOI) - LOI is defined as the required minimum percentage of oxygen in a mixture with nitrogen, which would allow a flame to be sustained by an organic material such as a plastic.

Linings - Inserts, usually made from plastics, to protect metallic or nonmetallic substrates. Linings or liners are either inserted or formed in place and are usually thicker than coatings fabricated from a dispersion.

Liquid Crystal Polymer (LCP) - Among the stiffest and highest strength plastics are those with substantial aromatic groups along their backbones. These polymers have high melting points, high glass transition temperatures and, usually, good chemical resistance. Examples of these polymers are polyaramids and liquid crystal polymers. The polymer backbones are so stiff that the crystal structures are partially retained even in the liquid phase. The mechanical properties of solid LCPs are directional and can be quite high.

Lower Explosive Limit (LEL) - The lowest percentage at which organic vapors or particles suspended in air will ignite if a source of ignition is introduced. It is also referred to as minimum explosive concentration (MEC).

M

Mrad - Means megarad. It is a radiation dose unit, in energy per unit mass, that simply measures absorbed energy. It is one million rads or 2.30 calories per gram.

Melting Point - The temperature at which the solid crystalline and liquid phases of a substance are in thermodynamic equilibrium. The melting point is usually referred to normal pressure of 1 atm.

Mica - Mica is a crystalline platey filler made by wet or dry grinding of muscovite or phlogopite, minerals consisting mainly of aluminum and potassium orthosilicates, or by chemical reaction between potassium fluorosilicate and alumina. Used as a filler in thermosetting resins to impart good dielectric properties and heat resistance, and in thermoplastics such as polyolefins to improve dimensional stability, heat resistance, and mechanical strength. Mica fillers also reduce vapor permeability and increase wear resistance. Mica fillers having increased flake size or platiness increase flexural modulus, strength, heat deflection temperature, and moisture resistance. Surface modified grades of mica are available for specialty applications.

Micron (Micrometer) - A unit of length equal to 1×10^{-6} meter. Its symbol is Greek small letter mu (μ).

Microporosity - Defects such as small voids or inclusions in fluoropolymer parts which can be detected by a microscope or the use of a fluorescent dye.

Mil - One thousandth (0.001) of an inch (25.4 microns). It is the most common non-metric measurement of coating thickness.

Moisture Vapor Permeation - Refers to permeation of water vapor through films and membranes, which can be measured by a number of standard methods (e.g., ASTM).

Molecular Weight - The molecular weight (formula weight) is the sum of the atomic weights of all the atoms in a molecule (molecular formula). Also called *MW, formula weight, average molecular weight.*

Molecular Weight Distribution - The relative amounts of polymers of different molecular weights that comprise a given specimen of a polymer. It is often expressed in terms of the ratio between weight- and number-average molecular weights, Mw/Mn.

Monomer - The individual molecules from which a polymer is formed (i.e., ethylene, propylene, tetrafluoroethylene).

Multilayer Coating - A coating that is produced by multiple passes of the substrate through the coating process. After each pass the thickness of the coating increases. Multilayer coating is a means of overcoming critical cracking thickness when relatively thick coatings are required.

N

Nanometer - A unit of length equal to 1×10^{-9} meter. Often used to denote the wavelength of radiation, especially in the UV and visible spectral region. Unit is abbrevieated *nm.*

Newtonian Fluid - A term to describe an ideal fluid in which shear stress and shear rate is proportional (e.g., water).

Nonpolar - In molecular structure, a molecule in which positive and negative electrical charges coincide. Most hydrocarbons, such as polyolefins, are nonpolar.

Nucleophile - Nucleophiles or nucleophilic reagents are basic, electron-rich reagents. Negative ions and chemical groups can be nucleophiles, in addition to neutral compounds such as ammonia and water. Both ammonia and water molecules contain a pair of unshared electrons.

O

Optical Properties - The effects of a material or medium on light or other electromagnetic radiation passing through it, such as absorption, reflection, etc.

Organic Compound - A chemical compound that contains one or more carbon atoms in its molecular structure.

OSHA - Occupational Safety and Health Administration.

Ozone - O_3.

P

Peel Strength - The bond strength of a film adhered by an adhesive to a substrate is measured by different techniques and is called *peel strength.* An extensiometer can be used to measure peel strength.

Perfluorinated Fluoropolymers - Polymer consisting of only carbon and fluorine (and an occasional oxygen atom) atoms are called *perfluorinated fluoropolymers.*

Perfluoroammonium Octanoate - (C8)

Permeability - The capacity of material to allow another substance to pass through it; or the quantity of a specified gas or other substance which passes through under specified conditions.

pH - An expression of the degree of acidity or alkalinity of a substance. Neutrality is pH 7, acid solutions are less than 7, and alkaline solutions are greater than 7.

Phenolic - A thermosetting resin or plastic made by condensation of a phenol with an aldehyde.

Polar - In molecular structure, a molecule in which the positive and negative electrical charges are permanently separated. Polar molecules ionize in solution and impart electrical conductivity to the solution. Water, alcohol, and sulfuric acid are polar molecules; carboxyl and hydroxyl are polar functional groups.

Polyarylene Polymers - Examples of these plastics include polyetherketone, polyetheretherketone, polyetherketoneketone, polyphenylenesulfide, and others.

Polyarylsulfone - Polyarylsulfone is a thermoplastic containing repeating sulfone and ether groups in its wholly aromatic backbone. It has excellent resistance to high and low temperatures, good impact strength, improved resistance to environmental stress cracking, good dielectric properties, rigidity, and resistance to acids and alkalis. Polyarylsulfone is nonflammable, but is attacked by some organic solvents. Processed by injection molding, compression molding, and extrusion. Used in high-temperature electrical and electronic applications such as circuit boards and lamp housings, piping, and auto parts.

Polymer - Polymers are high molecular weight substances with molecules resembling linear, branched, crosslinked, or otherwise shaped chains consisting of repeating molecular groups. Synthetic polymers are prepared by polymerization of one or more monomers. The monomers comprise low-molecular-weight reactive substances, often containing more than one reactive molecular bond or chemical bond. Natural polymers have molecular structures similar to synthetic polymers but are not man-made, occur in nature, and have various degrees of purity. Also called *synthetic resin, synthetic polymer, resin, plastic.*

Polymer Fume Fever - A condition that occurs in humans as a result of exposure to degradation products of polytetrafluoroethylene and other fluoropolymers. The symptoms of exposure resemble those of flu and are temporary. After about twenty-four hours, the flu-like symptoms disappear.

Porosity - Porosity is defined as the volume of voids per unit volume of a material or as the volume of voids per unit weight of material.

Primer - Also called *primer coating,* the paint layer directly on the substrate which has the primary function of helping the coating system stick to the substrate.

R

Radiation Dose - Amount of ionizing radiation energy received or absorbed by a material during exposure. Also called *radiation dosage, ionizing radiation dose.*

Radical - An atom or group of atoms that has at least one unpaired electron and is therefore unstable and highly reactive.

Repulsive Intermolecular Forces - Forces generated when atoms or molecules approach each other closely.

Rheology - A science that studies and characterizes flow of polymers, resins, gums, and other materials.

Rotational Lining - See also *Rotational Molding.* Rotational lining is a process by which a hollow object is lined with a plastic. The surface of the part, contrary to rotational molding process, is prepared to adhere the liner to the mold wall.

Rotational Molding - Also known as rotocasting or rotomolding, is a process for manufacturing hollow plastic parts. A typical procedure for rotational molding is as follows: Very fine plastic powder is placed in a mold and the closed mold is heated above the melting point of the powder while the mold is rotated in two planes at right angle to each other. The heating continues until the polymer powder fuses and melts to form a homogeneous layer of uniform thickness. The mold is rotated while it is cooled down to the removal temperature. At the end mold rotation is stopped and the part is removed.

S

Salt Fog - The ASTM B117 test procedure that simulates the corrosive environment caused by road salt and marine spray.

Sand Blasting (also **Grit Blasting**) - The process of surface cleaning and roughening substrates by propelling harder materials onto the substrate. It provides a mechanical aid to coating adhesion.

Semicrystalline Plastic - A plastic material characterized by localized regions of crystallinity.

Shear - Displacement of a plane of a solid body parallel to itself, relative to other parallel planes within the body; deformation resulting from this displacement.

Shelf Life - Time during which a physical system such as material retains its storage stability under specified conditions. Also called *storage life.*

Silicone - Silicones are polymers, the backbone of which consists of alternating silicon and oxygen atoms. These are more correctly called *polysiloxanes.* Also called *siloxane, silicone rubber, silicone plastic, silicone fluid, SI, polysiloxane.*

Sintering - Consolidation and densification of polytetrafluoroethylene particles above its melting temperature is called *sintering*. Also see *Coalescence*.

Softening Point - Temperature at which the material changes from rigid to soft or exhibits a sudden and substantial decrease in hardness.

Solubility - The solubility of a substance is the maximum concentration of a compound in a binary mixture at a given temperature forming a homogeneous solution. Also called *dissolving capacity*.

Solubility Parameter - Solubility parameter characterizes the capacity of a substance to be dissolved in another substance (e.g., of a polymer in a solvent). It represents the cohesive energy of molecules in a substance and determines the magnitude and the sign of the heat of mixing two substances in given concentrations. The magnitude and the sign of the heat of mixing determine the sign of the free energy of mixing. The solution occurs when the sign of the free energy of mixing is negative.

Steric Hindrance - A spatial arrangement of the atoms of a molecule that blocks reaction of the molecule with another molecule.

Stick Slip - This is a jerking action that occurs in a moving part such as a bearing in overcoming a higher static coefficient of friction than a dynamic coefficient of friction before movement begins.

Strain - The per unit change, due to force, in the size or shape of a body referred to its original size or shape. Note: Strain is nondimensional but is often expressed in unit of length per unit of length or percent. Also called *mechanical strain*.

Stress Cracking - Appearance of external and/or internal cracks in the material as a result of stress that is lower than its short-term strength.

Stress Relaxation - Time-dependent decrease in stress in a solid material as a result of changes in internal or external conditions. Also called *stress decrease*.

Substrate - Any surface to be coated.

Surface Appearance - The smoothness, gloss, and presence or lack of surface defects in a coating.

Supercritical Carbon Dioxide - Refers to carbon dioxide that has been heated to above its critical temperature and pressure. Supercritical CO_2 is a potent solvent for great many organic substances. It is also a suitable medium for polymerization of fluorinated monomers.

Surface Energy - See *Surface Tension*.

Surface Roughness - The closely spaced unevenness of a solid surface (pits and projections); can be quantified by various methods (e.g., by using a profilometer in coatings).

Surface Tension - The surface tension is the cohesive force at a liquid surface measured as a force per unit length along the surface, or the work which must be done to extend the area of a surface by a unit area (e.g., by a square centimeter). Also called *free surface energy*.

Surfactant - Derived from *surface active agent*. Defined as substances that aggregate or absorb at the surfaces and interfaces of materials and change their properties. These agents are used to compatibilize two or more immiscible phases such as water and oil. In general, one end of a surfactant is water-soluble and the other end is soluble in an organic liquid.

Suspension Polymerization - Refers to a heterogeneous polymerization regime in which the product of the reaction is a solid forming a suspension in the liquid medium of reaction. Little or no surfactant is added to the reaction medium. Characteristics of the process include high agitation rate and poor particle size control. An advantage of this reaction is high purity of the polymer product as compared to that of the dispersion method.

T

Tensile Properties - Properties describing the reaction of physical systems to tensile stress and strain.

Tensile Strength - The maximum tensile stress that a specimen can sustain in a test carried to failure. Note: The maximum stress can be measured at or after the failure or reached before the fracture, depending on the viscoelastic behavior of the material. Also called *ultimate tensile strength, tensile ultimate strength, and tensile strength at break.*

Thermal Conductivity - The time rate of heat transfer by conduction across a unit area of substance at unit thickness and unit temperature gradient.

Thermal Expansion Coefficient - The change in volume per unit volume resulting from a change in temperature of the material. The mean coefficient of thermal expansion is commonly referenced to room temperature.

Thermal Properties - Properties related to the effects of heat on physical systems such as materials and heat transport. The effects of heat include the effects on structure, geometry, performance, aging, stress-strain behavior, etc.

Thermal Stability - The resistance of a physical system, such as a material, to decomposition, deterioration of properties, or any type of degradation in storage under specified conditions.

Thermoplastic - Thermoplastics are resin or plastic compounds which, after final processing, are capable of being repeatedly softened by heating and hardened by cooling by means of physical changes. There are a large number of thermoplastic polymers belonging to various classes such as polyolefins and polyamides. Also called *theremoplastic resin.*

Thermoset - Thermosets are resin and plastic compounds that, after final processing, are substantially infusible and insoluble. Thermosets are often liquids at some stage in their manufacture or processing and are cured by heat, oxidation, radiation, or other means, often in the presence of curing agents and catalysts. Curing proceeds via polymerization and/or crosslinking. Cured thermosets cannot be resoftened by heat. There are a large number of thermosetting polymers belonging to various classes such as alkyd and phenolic resins. Also called *thermosetting resin, thermoset resin.*

Thixotropic Liquids - These liquids exhibit lower viscosity as shear rate increases. A practical example is house paint, which appears thinner when stirred. See also *Hysteresis Loop.*

Tribo Gun - A powder coating gun that uses tribocharging to charge the powder.

Tribocharging - The process of creating a static electric charge on powder particles by action against a nonconductive material.

Tribological Characteristics - These characteristics deal with friction or contact related phenomena in materials. Coefficient of friction and wear rate are the most important tribological characteristics of a material.

U

Ultraviolet Radiation - Electromagnetic radiation in the 40–400 nm wavelength region. Sun is the main natural source of UV radiation on the earth. Artificial sources are many, including fluorescent UV lamps. Ultraviolet radiation causes polymer photodegradation and other chemical reactions. Note: UV light comprises a significant portion of the natural sunlight. Also called *UV radiation, UV light, ultraviolet light.*

Upper Explosive Limit (UEL) - The highest point at which organic vapors or particles suspended in air will ignite if a source of ignition is introduced.

V

Van der Waals Forces - Weak attractive forces between molecules, weaker than hydrogen bonds and much weaker than covalent bonds.

Viscosity - The internal resistance to flow exhibited by a fluid, the ratio of shearing stress to rate of shear. A viscosity of one poise is equal to a force of one dyne per square centimeter that causes two parallel liquid surfaces one square centimeter in area and one centimeter apart to move past one another at a velocity of one centimeter per second.

Voids - See *Porosity.*

W

Wear - Deterioration of a surface due to material removal caused by any of various physical processes, mainly friction against another body.

Weight Solids - It is the amount of a substance, relative to the total weight, which remains after all volatile components of the substance have been evaporated. It is usually expressed as a percentage.

Wettability - The rate at which a substance (particle, fiber) can be made wet under specified conditions.

Wetting - The spreading out (and sometimes absorption) of a fluid onto (or into) a surface. In adhesive bonding, wetting occurs when the surface tension of the liquid adhesive is lower than the critical surface tension of the substrates being bonded. Good surface wetting is essential for high-strength adhesive bonds. Poor wetting is evident when the liquid beads up on the part surface. Wetting can be increased by preparation of the part surface prior to adhesive bonding.

Wrap - A characteristic of liquid and powder coatings in electrostatic application where the coating adheres to areas of the substrate not in direct line-of-sight of the delivery of the coating material from the spray gun.

Y

Yield Deformation - The strain at which the elastic behavior begins, while the plastic is being strained. Deformation beyond the yield deformation is not reversible.

Index